QP 601 .E5153 1992
Enzyme assays

2/9/93

# Enzyme Assays

## The Practical Approach Series

SERIES EDITORS

**D. RICKWOOD**
*Department of Biology, University of Essex*
*Wivenhoe Park, Colchester, Essex CO4 3SQ, UK*

**B. D. HAMES**
*Department of Biochemistry and Molecular Biology,*
*University of Leeds, Leeds LS2 9JT, UK*

Affinity Chromatography
Anaerobic Microbiology
Animal Cell Culture (2nd Edition)
Animal Virus Pathogenesis
Antibodies I and II
Biochemical Toxicology
Biological Membranes
Biomechanics—Structures and Systems
Biosensors
Carbohydrate Analysis
Cell Growth and Division
Cellular Calcium
Cellular Neurobiology
Centrifugation (2nd Edition)
Clinical Immunology
Computers in Microbiology
Crystallization of Proteins and Nucleic Acids
Cytokines
The Cytoskeleton
Directed Mutagenesis
DNA Cloning I, II, and III
Drosophila
Electron Microscopy in Biology
Electron Microscopy in Molecular Biology
Enzyme Assays
Essential Molecular Biology I and II
Fermentation
Flow Cytometry
Gel Electrophoresis of Nucleic Acids (2nd Edition)
Gel Electrophoresis of Proteins (2nd Edition)
Genome Analysis
HPLC of Macromolecules
HPLC of Small Molecules
Human Cytogenetics I and II (2nd Edition)
Human Genetic Diseases
Immobilised Cells and Enzymes
Iodinated Density Gradient Media

Light Microscopy in Biology
Lipid Modification of Proteins
Liposomes
Lymphocytes
Lymphokines and Interferons
Mammalian Cell Biotechnology
Mammalian Development
Medical Bacteriology
Medical Mycology
Microcomputers in Biochemistry
Microcomputers in Biology
Microcomputers in Physiology
Mitochondria
Molecular Genetic Analysis of Populations
Molecular Neurobiology
Molecular Plant Pathology I and II
Monitoring Neuronal Activity
Mutagenicity Testing
Neurochemistry
Nucleic Acid and Protein Sequence Analysis
Nucleic Acid Hybridisation
Nucleic Acids Sequencing
Oligonucleotide Synthesis
PCR
Peptide Hormone Action
Peptide Hormone Secretion
Photosynthesis: Energy Transduction
Plant Cell Culture
Plant Molecular Biology
Plasmids
Postimplantation Mammalian Embryos
Prostaglandins and Related Substances
Protein Architecture
Protein Function
Protein Purification Applications
Protein Purification Methods
Protein Sequencing
Protein Structure
Protein Targeting
Proteolytic Enzymes
Radioisotopes in Biology
Receptor Biochemistry
Receptor–Effector Coupling
Receptor–Ligand Interations
Ribosomes and Protein Synthesis
Solid Phase Peptide Synthesis
Spectrophotometry and Spectrofluorimetry
Steroid Hormones
Teratocarcinomas and Embryonic Stem Cells
Transcription and Translation
Virology
Yeast

To

Janet and Janet

# Enzyme Assays

## A Practical Approach

---

Edited by

ROBERT EISENTHAL

and

MICHAEL J. DANSON

*Department of Biochemistry,*
*University of Bath*
*Bath BA2 7AY*
*UK*

Oxford New York Tokyo

D. HIDEN RAMSEY LIBRARY
U.N.C. AT ASHEVILLE
ASHEVILLE, N. C. 28814

*Oxford University Press,
Walton Street, Oxford OX2 6DP*

*Oxford New York Toronto
Delhi Bombay Calcutta Madras Karachi
Petaling Jaya Singapore Hong Kong Tokyo
Nairobi Dar es Salaam Cape Town
Melbourne Auckland*

*and associated companies in
Berlin Ibadan*

*Oxford is a trade mark of Oxford University Press*

*Published in the United States
by Oxford University Press, New York*

*© Oxford University Press 1992*

*Users of books in the Practical Approach series are advised that prudent laboratory safety procedures should be followed at all times. Oxford University Press make no representation, express or implied, in respect of the accuracy of the material set forth in books in this series and cannot accept any legal responsibility or liability for any errors or omissions that may be made.*

*All rights reserved. No part of this publication may be reproduced, stored in a retrieval system, or transmitted, in any form or by any means, electronic, mechanical, photocopying, recording, or otherwise, without the prior permission of Oxford University Press.*

*This book is sold subject to the condition that it shall not, by way of trade or otherwise, be lent, re-sold, hired out, or otherwise circulated without the publisher's prior consent in any form of binding or cover other than that in which it is published and without a similar condition including this condition being imposed on the subsequent purchaser.*

*British Library Cataloguing in Publication Data
A catalogue record for this book is
available from the British Library*

*Library of Congress Cataloging-in-Publication Data
Enzyme assays : a practical approach / edited by Robert Eisenthal and
Michael J. Danson.
(The Practical approach series)
1. Enzymes—Analysis. 2. Enzymes—Purification. I. Eisenthal,
Robert. II. Danson, Michael J. III. Series.
QP601.E5153 1992 574.19'25—dc20 91–29790*

*ISBN 0–19–963142–5
ISBN 0–19–963143–3 (pbk.)*

*Typeset by Footnote Graphics, Warminster, Wilts
Printed in Great Britain by
Information Press, Eynsham, Oxon*

# Preface

Virtually all chemical reactions in living systems are catalysed by enzymes, and the assay of enzyme activity is probably one of the most frequently encountered procedures in biochemistry. Most enzyme assays are carried out for the purpose of estimating the amount of active enzyme present in a cell or tissue, or as an essential part of an investigation involving the purification of an enzyme (see also *Protein Purification: A Practical Approach*). They are also a manifestly integral component of the determination of kinetic parameters, or the investigation of catalytic mechanism.

All too frequently, however, the investigator may choose an assay that is inappropriate to the purpose. It is hoped that this book will help the experimentalist select, and if necessary modify, existing assays, and interpret the data obtained correctly and to the maximum advantage. There is no ideal assay for any enzyme and, in general, the appropriateness of an assay will depend on the nature of the enzyme, its purity, and the purpose of the assay. For following the progress of a purification, convenience and speed may be the prime considerations for which a sacrifice in accuracy or precision may be tolerated. For kinetic and mechanistic work, accuracy and reproducibility are obviously essential. This book will also aid in the design of new assay methods that may be more suitable to the purpose of the investigation than those appearing in the literature, or in the improvement of existing assays.

The assay of enzyme activity is essentially a kinetic measurement and as such there are many pitfalls for the unwary. The first chapter of the book deals with the general principles of enzyme assay and is a comprehensive account of how to avoid these pitfalls, whilst also alerting the worker at the bench to intrinsic properties of the enzyme that may manifest themselves through kinetic assays.

The range of techniques used to measure the rate of an enzyme-catalysed reaction is vast and will depend on the nature of the chemical change and the ingenuity of the investigator. Within these limits a wide scope of methodology is available, and the six subsequent chapters discuss the instrumental techniques most frequently used. The techniques described in Chapters 2–7 are, admittedly, discussed in detail in many excellent texts, review articles, and monographs, and reference to these is made in the individual chapters. However, although theory and applications are discussed in those articles, they do not in general address the unique problems arising from the use of these techniques in enzyme assays.

As several thousand enzyme-catalysed reactions are known, it would have been impossible in a book of this size to deal with all the possible applications of the techniques described here to every known enzyme. However, the techniques chapters in this book contain experimental protocols that have been carefully chosen to represent the various types of enzyme-catalysed reactions amenable to assay using that particular technique. These then can be adapted to assay enzymes other than those specifically described. The theory underlying each method is introduced together with a description of the instrumentation, sensitivity, and sources of error. The methods discussed cover those most used for enzyme assay and include photometric, electrochemical, radiochemical, and HPLC techniques. The assay of enzymes after gel electrophoresis is an important application, and a separate chapter is devoted to special methods for detecting enzyme activity under these conditions.

Most enzymes are intracellular, and their measured activities may well depend on the method used to disrupt the cells. An associated problem is maintaining enzyme activity in cell extracts or in purified, or partially purified, fractions. The catalytic activity of an expiring enzyme is of little use. Accordingly, a chapter is included on the techniques involved in enzyme extraction, and in stabilizing enzyme activity.

Determination of kinetic parameters is usually undertaken to characterize an enzyme, to provide a quantitative evaluation of substrate specificity, and to study kinetic mechanisms. The increasing availability of desk-top computers and associated software for analysing kinetic data has sometimes led, in our experience, to uncritical application of statistical methods. Such an approach may well mask features of the data that might reveal interesting properties of an enzyme. The penultimate chapter describes how one should apply statistical methods in a rational manner to the analysis of kinetic data, an important topic missing from many enzymology texts.

The final chapter is a critical discussion of buffers and methods of protein estimation and will provide a realistic basis for choosing a system appropriate to the enzyme under investigation.

In summary, this book is a guide to the principles and practice of enzyme assays. It is intended for all those in the life sciences who are concerned with practical enzymology.

*Bath*
October 1991

R.E.
M.J.D.

# Contents

# 4. High performance liquid chromatographic assays 123

*Shabih E. H. Syed*

# Contributors

KEITH BROCKLEHURST
Department of Biochemistry, University of London, Queen Mary and Westfield College, Mile End Road, London E1 4NS.

J. B. CLARK
Department of Neurochemistry, Institute of Neurology, University of London, Queen Square, London WC1N 3BG.

OTHMAR GABRIEL
Department of Biochemistry and Molecular Biology, Georgetown University Medical Centre, 3900 Reservoir Road NW, Washington DC 20007–2197, USA.

DOUGLAS M. GERSTEN
Department of Pathology, Georgetown University Medical Centre, 3900 Reservoir Road NW, Washington DC 20007–2197, USA.

PETER J. F. HENDERSON
Department of Biochemistry, University of Cambridge, Tennis Court Road, Cambridge, CB2 1QW.

ROBERT A. JOHN
Department of Biochemistry, University College, Cardiff, PO Box 78, Cardiff, CF1 1XL.

K. G. OLDHAM
The Dianthus Group, Tamarind House, Crossways, Cowbridge, S. Glamorgan, CF7 7LJ. [formerly: Biomedical Division, Amersham International plc, Cardiff Laboratories, Cardiff, Wales, CF4 7YT, UK.]

N. C. PRICE
School of Natural Sciences, University of Stirling, Stirling, Scotland, FK9 4LA.

LEWIS STEVENS
School of Natural Sciences, University of Stirling, Stirling, Scotland, FK9 4LA.

SHABIH E. H. SYED
Department of Biochemistry, University of Leicester, University Road, Leicester, LE1 7RH.

KEITH F. TIPTON
Biochemistry Department, Trinity College, Dublin 2, Ireland.

*Contributors*

P. J. WATKINS
Cardiff Institute of Higher Education, Western Avenue, Cardiff, CF5 2SG.

P. D. J. WEITZMAN
Cardiff Institute of Higher Education, Western Avenue, Cardiff, CF5 2SG.

# Abbreviations

| | |
|---|---|
| $A_{340}$ | Absorbance at 340 nm |
| Ace | 2-[(2-Amino-2-oxoethyl)-amino]ethanesulphonic acid |
| ACV | δ-(L-α-Aminoadipyl)-L-cystinyl-D-valine |
| Ada | *N*-(2-Acetamido)-2-aminodiacetic acid |
| ADP | Adenosine diphosphate |
| AMP | Adenosine monophosphate, Adenylate |
| amp | Ampere |
| APAD | Acetylpyridine adenine dinucleotide (oxidized) |
| APADH | Acetylpyridine adenine dinucleotide (reduced) |
| ATEE | Acetyltyrosine ethyl ester |
| ATP | Adenosine triphosphate |
| ATPase | Adenosine triphosphatase |
| BAEE | Benzoylarginine ethyl ester |
| Bes | *N*,*N*-Bis(2-Hydroxyethyl)-2-aminoethanesulphonic acid |
| Bicine | *N*,*N*-Bis (2-hydroxyethyl)glycine |
| Bis-Tris | Bis(2-hydroxyethyl)amino-tris(hydroxymethyl)methane |
| BSA | Bovine serum albumin |
| $BV_{ox}$, $BV_{red}$ | Benzylviologen (oxidized, reduced) |
| Caps | 3-(Cyclohexylamino)-2-hydroxy-1-propanesulphonic acid |
| CDP | Cytidine diphosphate |
| Ches | 2-(*N*-Cyclohexylamino)ethanesulphonic acid |
| CI | covalently immobilized |
| Ci | Curie (2.2 $\times$ $10^6$ decompositions per second) |
| CM | Carboxymethyl |
| CoA,CoASH | Coenzyme A |
| DAD | Diode array detector |
| DAP | Diaminopimelic acid |
| dATP | Deoxyadenosine triphosphate |
| DBM | Diazabenzyloxymethyl |
| DCI | 3,4-Dichloroisocoumarin |
| dCTP | Deoxycytidine triphosphate |
| DEAE | Diethylaminoethyl |
| DEHPA | Bis(diethylhexyl)phosphoric acid |
| dGTP | Deoxyguanosine triphosphate |
| DHF | Dihydrofolate |
| DHFR | Dihydrofolate reductase |
| DHQ | Dihydroquinozolinium |
| Dipso | 3-[*N*,*N*-Bis(2-hydroxyethyl)amino]-2-hydroxypropanesulphonic acid |

| | |
|---|---|
| DME | Dropping mercury electrode |
| DMSO | Dimethylsulphoxide |
| DNase | Deoxyribonuclease |
| DOPA | Dihydroxyphenylalanine |
| d.p.m. | Disintegrations per minute |
| dTDP | Deoxythymidine diphosphate |
| DTNB | 5,5′-Dithiobis(2-nitrobenzoate) |
| dTTP | Deoxythymidine triphosphate |
| $E$ | Absorbance (extinction) coefficient |
| $E_{1/2}$ | Half-wave potential |
| EC | Enzyme commission |
| EDTA | Ethylenediamine tetra-acetic acid |
| EGTA | Ethyleneglyco-bis(β-aminoethyl ether)N,N,N′,N′-tetraacetic acid |
| ELISA | Enzyme-linked immunosorbent assay |
| $f$ | Activity coefficient |
| FCCP | Carbonyl cyanide *p*-trifluoromethoxyphenylhydrazone |
| FMN | Flavin mononucleotide (oxidized) |
| $FMNH_2$ | Flavin mononucleotide (reduced) |
| $g$ | Relative centrifugal force |
| GDP | Guanosine diphosphate |
| GOT | Glutamatc oxalacetate transaminase |
| GTP | Guanosine triphosphate |
| Hepes | *N*-(2-Hydroxyethyl)piperazine-*N*′-(2-ethanesulphoric acid) |
| Hepps | *N*-(2-Hydroxyethyl)piperazine-*N*′-(2-propanesulphonic acid) |
| Heppso | *N*-(2-Hydroxyethyl)piperazine-*N*′-(2-hydroxypropanesul-phonic acid) |
| Hip | Hippuric acid |
| HPLC | High performance liquid chromatography |
| I | Intensity of light |
| $I$ | Ionic strength |
| Kat | Katal |
| $k_{cat}$ | Catalytic rate constant |
| $K_i$ | Inhibition constant |
| $K_m$ | Michaelis constant |
| LLD | Lower limit of detection |
| Mes | 2-(*N*-Morpholino)ethanesulphonic acid |
| Mops | 3-(*N*-Morpholino)propanesulphonic acid |
| Mopso | 3-(*N*-Morpholino)-2-hydroxypropanesulphonic acid |
| MPDP | 1-Methyl-4-phenyl-2,3-dihydropyridine |
| $M_r$ | Relative molecular mass (molecular weight) |
| MTT | 3-(4,5-Dimethylthiazol-2-yl)-2,5 diphenyltetrazolium bromide |

| | |
|---|---|
| $MV_{ox}$,$MV_{red}$ | Methylviologen (oxidized, reduced) |
| NAD | Nicotinamide adenine dinucleotide (oxidized) |
| NADH | Nicotinamide adenine dinucleotide (reduced) |
| NADP | Nicotinamide adenine dinucleotide phosphate (oxidized) |
| NAT | *N*-acetyl transferase |
| NBT | Nitroblue tetrazolium |
| σ | Standard error or standard deviation |
| OAB | *O*-Aminobenzaldehyde |
| OAT | Ornithine aminotransferase |
| OPA | *O*-Phthaldehyde |
| PABA | *p*-Aminobenzoate |
| PAGE | Polyacrylamide gel electrophoresis |
| PEP | Phosphenolypyruvate |
| PFK | Phosphofructokinase |
| $P_i$ | Inorganic orthophosphate |
| Pipes | Piperazine-*N*,*N′*-bis(2-ethanesulphonic acid) |
| PK | Pyruvate kinase |
| PMS | Phenazine methosulphate |
| PMSF | Phenylmethanesulfonylfluoride |
| Popso | Piperazine-*N*,*N′*-bis(2-hydroxypropanesulphonic acid) |
| $PP_i$ | Inorganic pyrophosphate |
| RI | Refractive index |
| RIA | Radioimmuno assay |
| RNase | Ribonuclease |
| RPC | Reverse phase chromatography |
| SCE | Standard calomel electrode |
| SDS | Sodium dodecyl sulphate (Sodium lauryl sulphate) |
| SEC | Size exclusion chromatography |
| SPA | Scintillation proximity assays |
| Taps | *N*-Tris(hydroxymethyl)methyl-3-aminopropane sulphonic acid |
| Tapso | 3-[*N*-Tris(hydroxymethyl)methylamino]-2-hydroxypropane-sulphonic acid |
| TBA | tert-Butylammoniuim hydroxide |
| TCA | Trichloroacetic acid |
| TCC | 2,3,5-Triphenyltetrazoluim chloride |
| TEMED | *N*,*N*,*N′N′*-Tetramethylethylene diamine |
| Tes | 2-(Tris[hydroxymethyl]methylamino)ethanesulphonic acid |
| THF | Tetrahydrofolate |
| THF | Tetrahydrofuran (Chapter 3 only) |
| TLC | Thin layer chromatography |
| Torr | mmHg |
| Tricine | *N*-Tris(hydroxymethyl) methylglycine |
| Tris | Tris (hydroxymethyl) aminomethane |

| | |
|---|---|
| UDP | Uridine diphosphate |
| UV | Ultraviolet |
| $v$ | Velocity |
| V | Volts |
| $v_o$ | Initial velocity |
| $V_{max}$ | Maximum velocity |
| ε | Absorption (extinction) coefficient |
| λ | Wavelength |
| $\lambda_{max}$ | Wavelength of maximum light absorption |

# 1

# Principles of enzyme assay and kinetic studies

KEITH F. TIPTON

## 1. Introduction

The activity of an enzyme may be measured by determining the rate of product formation or substrate utilization during the enzyme-catalysed reaction. For many enzymes there are several alternative assay procedures available and the choice between them may be made on the grounds of convenience, cost, the availability of appropriate equipment and reagents, and the level of sensitivity required. It would not be possible in this account to give detailed descriptions of the assay mixtures and procedures that have been devised for individual enzymes. The examples that will be presented here are intended to illustrate general features of specific types of assay. Convenient recipes for the assay of individual enzymes can be found in a variety of sources, such as *Methods in Enzymology* (1), *Methods in Enzymatic Analysis* (2), and *The Enzyme Handbook* (3), as well as in the original literature.

This chapter will discuss the general principles of enzyme assay procedures and the problems that may arise in their application and interpretation. It may seem to be a catalogue of potential disasters, but it is essential to be sure that any assay procedure used gives a true measure of the activity of an enzyme. Far too many otherwise carefully-conducted experimental studies have been rendered meaningless because of failure to ensure that the enzyme assay is giving valid results. All the potential problems may be avoided by careful experimental design and adequate controls. Furthermore, some aberrant behaviour seen in enzyme assays can give valuable information on the properties of the enzyme being studied.

## 2. Behaviour of assays

### 2.1 Reaction progress curves

When the time-course of product formation, or substrate utilization, is determined a curve such as that shown in *Figure 1* is usually obtained. The

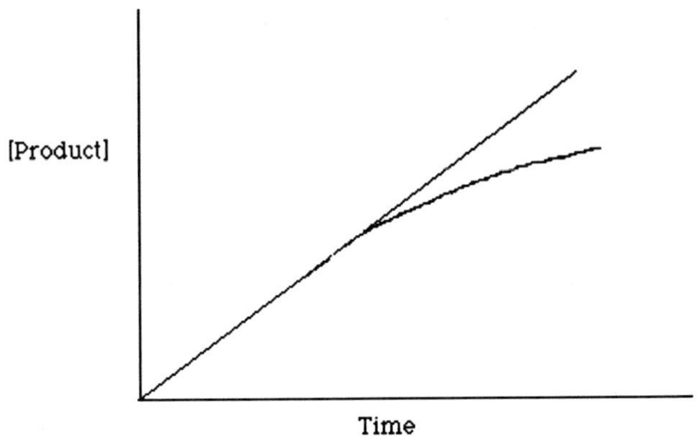


**Figure 1.** A typical progress curve of an enzyme-catalysed reaction.

time-course is initially linear but the rate of product formation starts to decline at longer times. There are several possible reasons for this departure from linearity and these will be discussed in turn.

### 2.1.1 Substrate depletion

The reaction may be slowing down because of substrate depletion. As the substrate concentration falls the enzyme will become less and less saturated and the velocity will fall, ultimately becoming zero when all the substrate has been used. If the reaction is slowing down simply because of substrate exhaustion, the addition of more substrate should delay thc fall-off. In some assays the substrate is continuously regenerated (Section 6.1.2 *iii*). Clearly, at any given enzyme concentration the period of linearity would be expected to be longer at higher substrate concentrations. If initial substrate concentrations much below the $K_m$ value are used it may be difficult to obtain a prolonged period of linearity unless highly-sensitive assays are used to allow product formation to be detected under conditions where there is a negligible change in substrate concentration. It is a useful practice to calculate whether the total change observed corresponds to that expected from the amount of substrate initially present.

### 2.1.2 Equilibrium

A reversible reaction may be slowing down because it is approaching equilibrium, where the rate of the backward reaction (converting product to substrate) will increase until, at equilibrium, it is equal to the rate of the forward (substrate to product) reaction. A decline in rate due to this cause can be prevented by the presence of any system that removes the product. This might be achieved by the use of a second enzyme-catalysed reaction such as in a coupled enzyme assay (Section 6.1.2) or by the presence of a reagent which reacts with the products. In the case of the oxidation of ethanol by alcohol dehydrogenase (EC 1.1.1.1):

$$CH_3CH_2OH + NAD^+ \rightleftharpoons CH_3CHO + NADH + H^+$$

the addition of semicarbazide to trap the acetaldehyde formed, as a semicarbazone, can reduce the curvature. The reaction produces hydrogen ions and rapidly approaches equilibrium at neutral pH values. Assay at higher pH values will prolong the linear phase.

### 2.1.3 Product inhibition

Products of enzyme-catalysed reactions are frequently reversible inhibitors of the reaction and a great deal of valuable information on the kinetic mechanism obeyed by an enzyme can be obtained from studying the nature of such inhibition (for example, 4–7). As in the previous case, the use of a system that removes the product should prevent curvature due to this cause.

### 2.1.4 Instability

One of the components of the assay system may be unstable and be steadily breaking down. This could be one of the substrates or the enzyme itself. The simplest way to check for this is to incubate the assay mixture for a series of times under conditions identical to those used in the assay itself but without one of the components (enzyme or substrate), before starting the reaction by the addition of the missing component. If the rates of the reaction are the same whichever component is missing during the pre-incubation period, a loss of linearity due to this cause can probably be excluded. If they are not, this approach should indicate which of the components is unstable. It is important to ensure that the conditions of the pre-incubation are identical to those of the assay itself; for example, many compounds are light-sensitive and this can be a particular problem where relatively high intensities of light are used, such as is possible in fluorimetry. Thus the pre-incubation should be carried out at the same level of illumination as in the assay. A further problem that can be encountered with optical assays is that the use of narrow slit widths can result in a localized destruction of only a small proportion of the material in the assay curvette; for example, the fluorescence of tryptophan solutions may decline with time but removal of the cuvette and shaking it can result in an apparent return to the original level of fluorescence if the photo-destruction is limited to only a very small proportion of the total solution. A convenient method for determining whether an enzyme is stable during assay is described in Section 2.3.

In some cases a component of the assay mixture may appear to be less stable under the pre-incubation conditions than it is in the complete assay mixture. This could result from the binding of substrate stabilizing the enzyme. In the case of light-sensitive compounds the absorbance of light by some other components of the assay may protect the photo-labile compound by decreasing the amount of light to which it is exposed. We have observed such behaviour in our studies on the oxidation of 1-methyl-4-phenyl-2,3-dihydropyridine (MPDP) by the enzyme monoamine oxidase (EC 1.4.3.4).

MPDP absorbs at 340 nm and its oxidation may be followed by recording the decrease in absorbance at that wavelength. However, it is an extremely photo-labile compound and is rapidly oxidized when illuminated at 340 nm. This can lead to a situation where the high rate of decline in absorbance at 340 nm observed in the absence of the enzyme actually decreases when crude preparations of the enzyme are added because absorbance of the incident light by the enzyme preparation decreases that reaching the MPDP. When assays are carried out by alternative methods that do not involve irradiation of the substrate, the enzyme can indeed be shown to catalyse the oxidation of MPDP.

### 2.1.5 Time-dependent inhibition

An enzyme might be less stable when catalysing the reaction than it is under the pre-incubation conditions described in Section 2.1.4. Such an effect would result in a decline in the rate of the reaction with time, whereas the individual components of the assay mixture might appear quite stable during the pre-incubation experiments. In such cases the addition of more enzyme to the assay after the reaction had ceased would be expected to cause the reaction to re-start. If the amount of enzyme added is the same as that originally used it would be expected that the resulting rate would be the same as that obtained when the assay was originally started, unless there had been a significant depletion of substrate(s) or accumulation of inhibitory products during the reaction.

Several amino acid decarboxylases have been shown to give rise to progress curves such as that shown in *Figure 1* because of the conversion of the pyridoxal phosphate coenzyme to the pyridoxamine form during the progress of the assay. In this case the departure from linearity may be delayed by adding an excess of pyridoxal phosphate to the assay mixture (8).

Enzyme-activated irreversible inhibitors, which are also known as mechanism-based inhibitors, $k_{cat}$ inhibitors, or suicide inhibitors, are substrate analogues that are not intrinsically reactive but which are converted by the action of a specific enzyme to a highly-reactive species that combines irreversibly (or very tightly) with it (see 8–12 for reviews). Some inhibitors of this type react stoichiometrically with the enzyme to cause inhibition. However, since these compounds are substrate analogues which must be involved in part of the enzyme-catalysed reaction in order to generate the reactive inhibitor, it is not surprising that others function as both substrate and inhibitor for the enzyme according to the overall reaction:

$$\mathrm{E + I \rightleftharpoons E.I \rightarrow (E.I)^{*}} \begin{cases} \nearrow \ \mathrm{E{-}I} \\ \searrow \ \mathrm{E + Products} \end{cases} \qquad (1)$$

where I is the enzyme-activated inhibitor, E.I is the initial non-covalent complex (analogous to the enzyme–substrate complex), and (E.I)* represents an activated complex which can either react to give the irreversibly inhibited species (E–I) or breakdown to form products and the free enzyme (E). If the formation of products is followed, a curve such as that shown in *Figure 1* will result. Addition of more of the substrate/inhibitor would not re-start the reaction, but addition of more enzyme would do so. Analysis of the behaviour of such systems can give the kinetic parameters describing the inhibitory process together with the *partition ratio*, which corresponds to the number of mol of product formed by one mol enzyme before it is inhibited (for accounts of the kinetic analysis of such behaviour see 12–14). Several inhibitors of the enzyme monoamine oxidase, for example, have been shown to act in this way (15, 16). The substrate for this enzyme, 2-phenylethylamine, has been shown to act as a time-dependent inhibitor at higher concentrations whereas at lower concentrations of this substrate, where these time-dependent inhibitory effects are less important, the progress curves are non-linear because of the substrate depletion (17). This type of behaviour emphasizes the necessity of checking the linearity of progress curves over a range of substrate concentrations, not just at the lowest substrate concentration that is to be used.

#### 2.1.6 Assay method artefact

If the specific detection procedure used ceases to respond linearly to increasing product concentrations this can lead to a decline in the measured rate of the reaction with time. In spectrophotometric or fluorimetric assays the absorbance of the product may reach such high levels that the apparatus no longer responds linearly to increasing concentrations (18, 19). Many convenient enzyme assays involve the use of one or more auxiliary enzymes to allow the reaction to be followed (Section 6.1.2 *iii*). In such cases departure from linearity may result from failure of the auxiliary system to respond linearly to increasing product formation or time. This could be due to many of the causes described in here or simply due to it approaching its maximum velocity. Clearly, if such coupled-assay procedures are to be used, it is essential to carry out careful control experiments to prove that the system is capable of providing a true measure of the activity of the enzyme under study under all conditions to be used. This important aspect is discussed in more detail in Section 6.1.2 *iii*.

#### 2.1.7 Change in assay conditions

If the assay conditions are not constant the rate of product formation might be expected to change. If, for example, the reaction under study involves the formation or consumption of hydrogen ions, the pH of the reaction mixture may change during the course of the reaction unless it is adequately buffered. If this resulted in a change of pH away from the optimum pH of the reaction this would lead to a decrease in the rate of the reaction. Clearly, such a problem may be avoided by the use of adequate buffers, but it is important to check the pH of a reaction mixture before and at the end of a reaction time-course to ensure that such effects are not occurring. The practice of measuring the pH at the beginning and the end of a progress curve and assuming that the operating pH value is the mean between these two values is not valid since the pH does not respond linearly to changes in hydrogen ion concentration. Furthermore, if the initial rate of the reaction is to be measured the operative pH should be that at the start of the reaction, not some arbitrary intermediate value occurring at a later stage.

## 2.2 Initial-rate measurements

As can be seen from the above discussion the decrease in the rate of product formation with time can be the result of one or more of a number of effects. At very short times, however, these effects should not be significant and thus if one measures the initial, linear, rate of the reaction by drawing a tangent to the early part of the progress curve (see *Figure 1*), these complexities should be avoided. Frequently the linear portion of an assay is sufficiently prolonged to allow the initial rate to be estimated accurately simply by drawing a tangent to, or taking the first-derivative of, the early part of the progress curve. Where loss of linearity occurs relatively rapidly because of depletion of substrate or approach to equilibrium (Sections 2.1.1 and 2.1.2), the period of linearity may be prolonged by decreasing the enzyme concentration to slow down the rate of product formation, and increasing the sensitivity of the assay method if necessary.

It has been commonly assumed that restricting measurements of reaction rates to a period in which less than 10–20% of the total substrate consumption has occurred will provide a true measure of the initial rate. Consideration of the possible causes for non-linearity discussed in the previous section will show that such an approach may not be valid. Even if the only reason for departure from linearity were depletion of substrate, consideration of the Michaelis–Menten relationship will indicate that such an approach will only give a valid approximation if the initial substrate concentration is in excess of the $K_m$ value. Methods for determining initial rates from such non-linear progress curves have been reviewed (20, 21).

In cases where curvature makes it difficult to estimate the initial rate with accuracy, it may be possible to do so by fitting the observed time-dependence of product formation to a polynomial equation and deriving the initial slope at $t = 0$ (22). Alternative less-sophisticated approaches involve laying a glass rod or a small mirror approximately at right-angles to the early part of the progress curve. If the rod, or mirror, is moved until the reflection of the line is continuous with the line itself it will be exactly perpendicular to the progress curve. Thus, if a line is drawn along the surface of the rod or mirror, the initial-rate tangent should intersect with this line at 90°. Alternatively, the negative reciprocal of the line drawn to the surface of the rod, or mirror, will correspond to the initial rate. In either case it is important to check that the initial rate line passes through the origin (Product = 0 at $t = 0$).

Such approaches may be of value in several cases but in practice it may not be easy to estimate the zero time of the assay precisely. Starting an assay by adding one of the components and ensuring adequate mixing can lead to significant uncertainty about the exact time that the reaction was started. Furthermore, initial parts of a progress curve may be difficult to determine; for example, with spectrophotometric or fluorimetric assays of crude enzyme preparations there may be an appreciable period, during which particles are settling, before a rate can be measured accurately. Such problems can be further compounded in cases where there is either a burst or a lag before the true rate of the reaction is established (Section 2.4).

Because of these potential problems it is desirable, if at all possible, to adjust the conditions such that a linear response is maintained for a sufficient time to allow the direct measurement of initial rate. In cases where this cannot be achieved it is necessary to consider the possible causes of such non-linearity and to analyse the progress curves appropriately; for example, in the case of a compound acting as both a substrate and an enzyme-activated irreversible inhibitor a full analysis of the entire progress curve can be used provided that the decline in velocity is solely due to such inhibition (12).

## 2.3 Integrated rate equations

If the decrease in the rate of an enzyme-catalysed reaction with time were solely due to the depletion of substrate it would be possible to correct for this fall-off by use of the Michaelis–Menten relationship. Several attempts have been made to do this, but they will of course only be valid if substrate depletion is the sole cause of curvature in the time-course of product formation. It has sometimes been assumed that a more pronounced curvature of time-courses at lower substrate concentrations indicates that the decline in velocity is due to substrate depletion. However, such observations do not show whether substrate depletion is the only cause of the departure from linearity. Furthermore, a similar effect would be expected if the enzyme were unstable under the conditions of the assay, but was stabilized by its interaction with substrate. In such cases the stabilization would be greatest at higher substrate concentrations where the enzyme was more saturated. If depletion of substrate is the only cause of curvature in the time-course of the reaction it may be described by an integrated form of the Michaelis–Menten equation:

$$v = \frac{\mathrm{d}p}{\mathrm{d}t} = \frac{V_{\mathrm{max}}}{1 + (K_{\mathrm{m}}/s)} = \frac{V_{\mathrm{max}}}{1 + [K_{\mathrm{m}}/(s_0 - p)]} \tag{2}$$

where $V$ and $K_m$ are the maximum velocity and Michaelis constant, respectively, $v$ is the initial velocity, $p$ is the product concentration, $s_0$ is the initial substrate concentration, and $s$ is the substrate concentration remaining at any time $t$.

Integration of this equation gives:

$$V_{max}t = p + \{K_m \ln [s_0/(s_0 - p)]\} \tag{3}$$

and

$$\frac{2.303}{t} \times \log [s_0/(s_0 - p)] = \frac{V_{max}}{K_m} - \left(\frac{1}{K_m} \times \frac{p}{t}\right) \tag{4}$$

Thus, if the amount of product formed is measured at a series of times and $(2.303/t) \log [s_0/(s_0 - p)]$ is plotted against $p/t$, a straight line will be obtained with a slope of $-1/K_m$ and an intercept on the base line of $V$. At low substrate concentrations, where $s_0 \ll K_m$, the equation simplifies to:

$$\frac{2.303}{t} \times \log [s_0/(s_0 - p)] = V_{max}/K_m \tag{5}$$

and thus a graph of $2.303 \log[s_0/(s_0 - p)]$ against $t$ will be a straight line of slope $V_{max}/K_m$ which passes through the origin. Under these conditions it will not be possible to determine $V_{max}$ and $K_m$ separately.

The integrated rate equation has been extended to include cases where a reversible reaction approaches equilibrium and also to take account of inhibition by the products of the reaction (23 gives a detailed account). Its use is attractive in that it should allow full use to be made of all the data comprising the reaction progress curve, rather than just the small portion of it representing the initial rate of the reaction. Furthermore, it should allow detailed kinetic analysis to be undertaken with much less work than is required when initial rate measurements are used. Consideration of the form of the equation describing the reaction progress curve also indicates that the second-derivative of such a curve will show a minimum which corresponds, on the substrate concentration axis, to half the $K_m$ value (24).

The problem of applying the integrated rate equation is that it is only valid if the departure from linearity is due only to substrate depletion, or in cases where more elaborate forms of the equation are applied (23), due to approaching equlibrium or product inhibition. However, as discussed in Section 2.1, the reasons for non-linear progress curves can be considerably more complex.

Selwyn (25) has presented a valuable method for determining whether an enzyme is stable during assay. He pointed out that for any enzyme-catalysed reaction the rate of product formation will depend on the enzyme concentration ($e$) and some function (f) of the concentrations of substrate ($s$), product ($p$), and any inhibitor ($i$) or activator ($a$) present.

$$\frac{dp}{dt} = e \times f(s,a,i,p) \tag{6}$$

Thus under conditions where $a$ and $i$ are constant and the concentration of substrate does not significantly change, this equation can be integrated to give:

$$e \times t = f(p) \tag{7}$$

which indicates that the amount of product formed should depend only on the enzyme concentration and the time. Thus a graph of product concentration against $e \times t$ should describe the same curve whatever initial concentration of enzyme is used. However, if the enzyme is unstable during the course of the assay, Equation 7 will no longer hold since the active enzyme concentration will also be time-dependent. In such cases the graphs of $p$ versus $e \times t$ will give different curves for each starting enzyme concentration. This latter behaviour would be expected if non-linearity of an assay were due to instability or time-dependent inhibition of the enzyme (Sections 2.1.4 and 2.1.5), whereas a single curve would be expected for cases outlined in Sections 2.1.1, 2.1.2, and 2.1.3.

## 2.4 Bursts and lags in progress curves

With some enzymes there may be either a burst of product formation or a lag before the linear phase of the reactions is obtained, as illustrated in *Figure 2*. This may be an artefact of the assay system being used, but in other cases such behaviour can give interesting information about the enzyme-catalysed reaction itself. The possible causes of such behaviour are listed below.

### 2.4.1 Inadequate temperature control

Frequently it is necessary to keep the enzyme solution and perhaps some other components of the assay mixture cold in order to ensure their stability. Addition of an ice-cold component to a reaction mixture that has been equilibrated to the assay temperature can lead to a drop in temperature and thus to a slower rate of reaction which increases as the temperature of the mixture rises to the equilibration value. It should also be remembered that reaction mixtures placed in temperature-controlled vessels do not immediately adjust to the new temperature and apparent lags may be seen if the mixture is not given an adequate time to adjust to the chosen temperature before initiating the reaction. The converse behaviour can sometimes be seen if a reaction mixture is equilibrated in a water bath which is also used for circulating a water jacket around the reaction vessel. In such cases there may be a significant drop in temperature between the water bath and reaction vessel, leading to an apparent initial burst of activity before the temperature of the pre-equilibrated mixture falls to that in the reaction vessel.

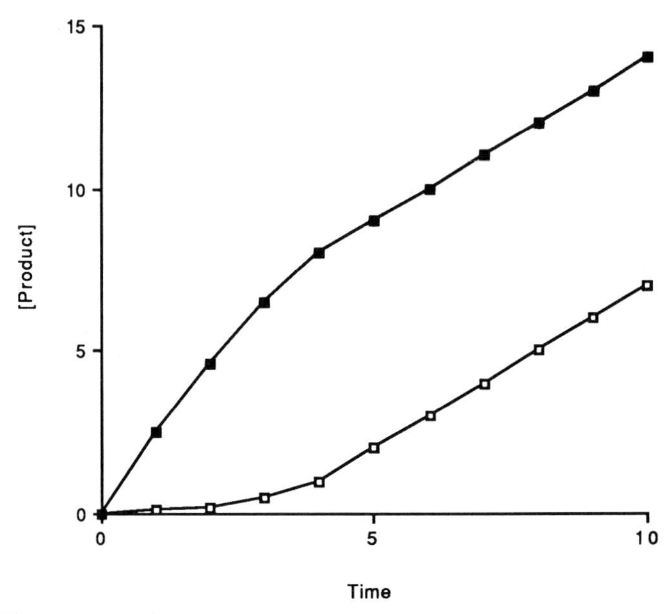


**Figure 2.** Time-courses of enzyme-catalysed reactions showing burst (■) or lag (□) phases before the steady-state rate is obtained.

If it is essential that a component of the assay be kept at a temperature different from that required for the assay, correction may be made for the effect on the temperature of the final assay mixture. Making the reasonable assumption that the heat capacities of all solutions making up the assay mixture are the same:

$$\mathbf{a}T_1 + \mathbf{b}T_2 = (\mathbf{a} + \mathbf{b})T_3$$

where $T_1$ and $T_2$ are the temperatures of the component solutions, $T_3$ is the assay temperature, and **a** and **b** are the volumes of the solution components. Using this relationship, one can calculate the initial temperatures of the assay components that will give, on mixing, the required assay temperature.

### 2.4.2 Settling of particles

When crude tissue preparations are assayed by spectrophotometry or fluorimetry the measurements during the first few minutes of the assay may be erratic due to the settling of particles from the solution. This may sometimes be misinterpreted as a burst or lag phase in the reaction. In such cases remixing the contents of the assay cuvette after the reaction has become linear should result in a second phase of aberrant behaviour.

### 2.4.3 Slow detector response

A lag phase may result if the initial response of the detection system is too slow. This type of behaviour will be discussed in terms of coupled enzyme assays in Section 6.1.2.

### 2.4.4 Slow dissociation of a reversible inhibitor (or activator)

Although most reversible enzyme activators and inhibitors will dissociate extremely rapidly from the enzyme when the enzyme-inhibitor mixture is diluted, those that show extremely high affinity for the enzyme may show significant time-dependence in their rates of dissociation from it, and association with it (26, 27). In such cases dilution of the enzyme-inhibitor mixture into the assay may show a lag as the inhibitor slowly dissociates to its equilibrium value. Conversely, if enzyme is added to a reaction mixture containing inhibitor the rate may slowly decrease until the binding equilibrium has been established.

### 2.4.5 Pre-steady-state transients

A burst or lag phase in the time-course of product formation can be due to the time taken for the concentrations of the intermediate enzyme–substrate and enzyme–product complexes to rise to their steady-state levels. Usually such transients are rapid and only detectable by use of specialized equipment, such as a stopped-flow apparatus, which allows measurements of reaction to be made within milliseconds, or less, of mixing (28, 29). Occasionally, however, such processes can occur sufficiently slowly to be observed on the time-scale associated with normal enzyme assays. One of the best-known examples of such behaviour is the hydrolysis of *p*-nitrophenylacetate by chymotrypsin (EC 3.4.21.1) (30).

A rather more complicated transient phenomenon has been observed with the enzyme arylsulphatase A (EC 3.1.6.1). In this case the reaction between enzyme and substrate results in the formation of an inactive covalently-modified form of the enzyme which is slowly hydrolysed to regenerate the active enzyme. Thus the initial rate slowly decays, in a first-order process, to give a final steady-state rate that depends on the partition between the active and inactive covalently-modified enzyme forms (31).

### 2.4.6 Relief of substrate inhibition or activation

Many enzymes are inhibited by high concentrations of one or more of their substrates (6, p. 120). If the initial substrate concentration added to an assay mixture is sufficient to cause some degree of inhibition, the rate of the reaction will tend to increase with time as substrate utilization decreases the inhibition. Alternatively, an initial burst phase in the progress curve can occur if the substrate also behaves as an activator at higher concentrations. Bursts or lags due to such causes should be eliminated by reducing the initial substrate concentration to a level where inhibition, or activation, does not occur. High-substrate inhibition or activation should, of course, be readily detected by their characteristic effects on the dependence of initial velocity on substrate concentration (for example, 6, p. 126). A commonly-used coupled assay for phosphofructokinase (EC 2.7.1.11) involves the use of phosphoenolpyruvate, pyruvate kinase, NAD, and lactate dehydrogenase (Section 6.1.2). However, the enzyme from several sources is allosterically inhibited by phosphoenolpyruvate which can lead to a lag in the progress curve (32, 33).

A similar lag phase in the reaction progress curve can occur if the substrate solution is contaminated by a small amount of another substrate for the enzyme which has a high affinity for it but which is broken down slowly (6, p. 72).

### 2.4.7 Activation by product

A progress curve that curves upward may be observed if one of the products of the reaction is an activator. This type of behaviour can, for example, occur in the assay of phosphofructokinase which is activated by the product fructose-1, 6-bisphosphate (32). In this case, however, there is a further complication because the substrate ATP is an allosteric inhibitor of the enzyme which can lead to lag phases such as those discussed above (Section 2.4.6).

### 2.4.8 Substrate interconversions

If a compound exists in more than one form, only one of which is an effective substrate for the enzyme, a slow interconversion between these forms can lead to burst or lag phases in the progress curves; for example, a lag in the progress curve of the reaction catalysed by fructokinase (EC 2.7.1.3) is observed when freshly-prepared solutions of fructose are used as substrate. This is because in such solutions the sugar is essentially all in the pyranose form, which is not a substrate, and only mutarotates rather slowly to give the active furanose form. If fructose solutions are allowed sufficient time for the mutarotation equilibrium between the two forms to be established the lag phase in the progress curve is no longer apparent (34). Such effects can also give rise to burst phases if a substrate exists in a slow equilibrium between active and inactive forms where an initial rapid phase, corresponding to the utilization of the active form of the substrate, would be followed by a slower phase determined by the rate of isomerization from inactive to active forms. Clearly, in cases of substrates that can exist in different isomeric forms, and hydration states or polymeric forms, such effects should be taken into account if bursts or lags are observed.

### 2.4.9 Hysteretic effects

Frieden (35) used the term hysteresis to refer to burst or lag phases in progress curves resulting from slow isomerization of the enzyme. He argued that such behaviour may have important regulatory significance. Detailed treatments of the behaviour of such systems have been presented (36–38).

Hysteretic effects can yield a number of differently-shaped reaction progress curves. However, it is important to exclude the other possible causes of bursts or lags before concluding that the effect is due to hysteresis. If it is possible to monitor changes in the conformation of the enzyme in solution, correlation of the time-courses of such effects with those seen in the progress curves may provide evidence for hysteretic behaviour. If the effect is due to a slow conformational change induced by one of the substrates, the behaviour may depend on the way in which the reaction is started. Thus hysteretic effects might be observed if the reaction is started by the addition of enzyme to a mixture containing all the other substrates, whereas if the enzyme were pre-incubated with the substrate responsible for inducing the conformational change, before starting the reaction with another substrate, no such effect might be seen.

Hysteresis may occur if an enzyme exists in a slow association–dissociation equilibrium in which the two polymerization states differ in their activities. In this case the magnitude of the burst or lag may depend on the enzyme concentration since this will affect the degree of association. Furthermore, the hysteresis may be dependent on how the reaction is initiated. If the reaction is started by addition of a sample of enzyme from a concentrated stock solution then the effects might be different from when the enzyme was diluted into an incomplete assay mixture and allowed to equilibrate before starting the reaction with another component. Such behaviour has, for example, been shown to account for the hysteresis observed with hexokinase (EC 2.7.1.1) (39) and glutaminase (EC 3.5.1.2) (40). The mitochondrial form of aldehyde dehydrogenase (EC 1.2.1.3) can show extremely long lag phases before the reaction becomes detectable (41). With preparations from some species it appears that an enzyme association–dissociation phenomenon may contribute to the lag (42) whereas the polymerization state does not appear to be a factor with the enzyme from some other species (43).

### 2.4.10 Summary

With assays that show burst or lag phases, it is important to determine the cause in order to know which phase of the reaction corresponds to the true 'initial rate' of the reaction. The term *initial rate* is normally used to refer to the steady-state rate of the reaction that is established after any pre-steady-state events have occurred. In the cases described in Sections 2.4.1, 2.4.2, and 2.4.3 the initial rate corresponds to the linear phase of the reaction that is established after any apparent burst or lag. The same would apply for the case in Section 2.4.4 but analysis of the behaviour could give valuable information on the rates of ligand association and dissociation. Where a lag or burst is due to pre-steady-state transients (Section 2.4.5), the initial rate (steady-state) is that obtained after the transient phase, although more complete analysis of the curve can give valuable information about the values of individual rate constants (28, 29). In contrast, the true initial rate is that obtained at the start of the reaction for the case in Section 2.4.6 where the rate at the substrate concentration initially present is required. Similarly, in the case in Section 2.4.7 the initial rate corresponds to that at the start of the reaction since, by definition, no significant product formation should occur during this phase. Where slow substrate interconversions occur (Section 2.4.8) the problem becomes one of determining the true substrate concentration at which the initial rate has been measured.

Genuine hysteretic effects (Section 2.4.9) are much more difficult to analyse since the various phases of the reaction progress curve may be controlled by different conformational or aggregation studies of the enzyme. If, for example, there are two forms of the enzyme that have different activities, it might be possible to obtain rate data for both species from the different phases of the progress curve, perhaps by the use of computer-aided curve-fitting procedures (44). In practice, however, the results of such an analysis might be difficult to interpret since true initial-rate conditions may not apply at the later stages of the progress curve. Furthermore, a more detailed knowledge of the mechanisms underlying the observed transients would be necessary before any such analysis could yield meaningful results.

## 2.5 Blank rates

### 2.5.1 Possible causes

It is not uncommon to observe an apparent rate of reaction in the absence of one of the components of the complete assay mixture. It is important to understand the causes of such blank rates in order to make appropriate corrections, since for any accurate studies it is essential to ensure that the determined rates are due only to the specific enzyme-catalysed reaction under investigation. It is possible that a blank rate will only occur with certain components of an incomplete assay mixture and thus it is necessary to test for such rates using different combinations of the system; for example, by omitting the enzyme and each of the substrates in turn. Some of the more common causes of blank rates are listed below:

*i. Settling of particles*

Spectrophotometric and fluorimetric assays of enzyme activities in crude tissue preparations, such as homogenates or subcellular organelle preparations, will be affected by the settling of particles causing changes in absorbance and light-scattering. After a sufficient time for the particles to settle these changes should cease, but they will be re-started on mixing the assay system again, as will occur when the full reaction is started by the addition of the mixing component. It may be possible to use detergents to reduce this problem by rendering the particles soluble, but it will of course be necessary to check whether the detergent used has any effect on the activity of the enzyme under study.

*ii. Precipitation*

Gradual precipitation of material in the assay mixture can lead to similar problems in optical assays as those caused by settling of particles. In some cases such effects may be confused with genuine reaction rates. It is thus important to be aware of possible artefacts of this type and to inspect the assay cuvette at the end of the 'reaction' for signs of turbidity or visible precipitate formation. Changes in absorbance at wavelengths distant from those where any reaction-dependent changes should occur can be used to monitor turbidity changes directly. Such effects may result from the enzyme or another component of the mixture not being fully soluble under the assay conditions or from interactions between different components leading to precipitation. Magnesium or calcium ions are added to many assay mixtures because they are essential for the activity of a number of enzymes. However, if such mixtures contain strong phosphate buffer then precipitation will occur when the solubility product of calcium or magnesium phosphate is exceeded. A more confusing situation can occur if one of the products of the reaction is not very soluble and precipitates during the later stages of the reaction, giving rise to accelerating progress curves. Although the blank rates arising from precipitation directly affect optical assays, such behaviour could also invalidate the results obtained with other assay procedures.

*iii. Contamination of one of the components of the assay mixture*

The presence of one of the substrates in the enzyme solution can give a blank rate with an incomplete reaction mixture. Crude tissue preparations are likely to contain endogenous substrates and this will lead to a reaction in the absence of added substrate. If the degree of contamination is quite small, the blank rate from this source would be expected to be non-linear and to cease when the endogenous substrate is exhausted. If the enzyme is stable under the assay conditions it may be possible to wait until the blank rate dies away before starting the assay. Alternatively, if the contaminating substrate is a small molecule, it should be possible to remove it by dialysis or gel filtration. Problems from this source would be expected to decrease on purification of the enzyme. Some commercially-available enzyme preparations contain substrate, which has been added for stability; it is thus necessary to remove such material by dialysis or gel filtration before assay.

The possibility of contamination of reagents with substrate cannot be excluded. For enzymes which use $CO_2$ or bicarbonate as a substrate, great care must be taken to remove all such material from each component of the mixture. Volatile substrates such as ammonia or aldehydes can be particularly difficult sources of contamination in laboratories where such compounds are in frequent use. Cross-contamination can also occur unless care is taken to ensure that a different pipette is always used for each component of the assay and reaction vessels are thoroughly cleaned.

*iv. Adsorption to assay vessels*

Many proteins adhere to glass and in cases where a vessel has already been used for one assay this can result in the presence of sufficient adsorbed enzyme to give a rate in a subsequent assay in the absence of added enzyme. Adsorption can be so strong that rinsing with distilled water is insufficient to remove the bound enzyme and more vigorous procedures such as acid washing are required. The use of silicone-treated glass or plastic vessels may minimize this problem but we have found that not all plastics are inert in this respect. If they are suitable for the assay, disposable plastic cuvettes are recommended.

Contamination of the assay vessels with one of the substrates can also lead to blank values; for example, in radiochemical assays it is necessary to ensure that apparently clean reaction vessels or scintillation vials do not contain any significant amounts of adsorbed radioactive material.

*v. Non-enzymic reactions*

Solutions of NAD(P)H are unstable at pH values below neutrality, leading to a spontaneous fall in absorbance at 340 nm. Similarly, many *p*-nitrophenyl esters are relatively unstable in aqueous solution and steadily hydrolyse to liberate *p*-nitrophenol. In these cases the blank rates due to the non-enzymic reactions should be subtracted from the rates given in the presence of enzyme. In spectrophotometric assays correction can most conveniently be done by using a double-beam (or ratio-recording) spectrophotometer that automatically records the difference between the absorbance of the sample and that of the blank. A steady drift in the response of the recording apparatus can also give rise to an apparent blank rate and it is important to check the stability from time to time in the absence of reactants. The reaction of exogenous factors can also lead to blank rates; for example, in poorly-buffered solutions the absorption of $CO_2$ will lead to a drift to lower pH values, which would be reflected as blank rates when enzymes are assayed by determining changes in pH or by use of a pH-stat. Reaction between different components of an assay mixture can also give rise to blank rates; for example, aldehydes can react non-enzymically with $NAD^+$ to give a product that has a similar absorbance to NADH. This reaction can cause significant problems in determining the activity of aldehyde dehydrogenase at alkaline pH values (45) but it is not significant at neutral or acid pH values.

*vi. Contaminating enzymes*

The presence of another enzyme in the preparation which catalyses an interfering reaction can give rise to a blank rate. If the substrate for the contaminating enzyme is also a contaminant of the preparation it may be possible to remove it by dialysis or gel filtration. However, this is not always possible; for example, the assay of dehydrogenases in crude tissue preparations may be difficult because of the presence of NADH–cytochrome-*c* reductase (EC 1.6.99.3), and in this case cytochrome-*c* is not readily removed by dialysis. In such cases it may be necessary to use an inhibitor of the contaminating enzyme, e.g. rotenone, taking care to ensure that it has no effect on the enzyme under study, or to purify the enzyme in order to remove the contaminating material. It may not be satisfactory simply to use an alternative assay procedure that does not detect the activity of the contaminating enzyme because the latter reaction may result in significant depletion of the substrate.

In some cases the contaminating enzyme may require no substrates other than those present for the assay of the enzyme under study. As an example, an assay for the enzyme pyruvate carboxylase (EC 6.4.1.1) involves the use of malate dehydrogenase to couple the oxaloacetate produced to the oxidation of NADH, which may be followed spectrophotometrically (*Figure 3*). If the enzyme preparation is contaminated with lactate dehydrogenase this will also catalyse the oxidation of NADH in converting pyruvate to lactate. Clearly, in

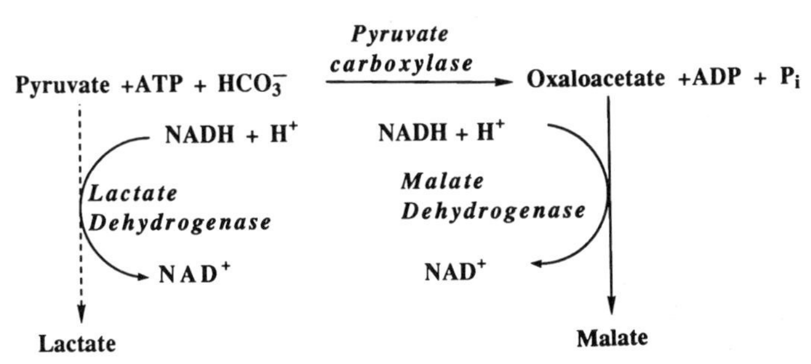


**Figure 3.** A coupled assay for pyruvate carboxylase. The broken line shows the interfering reaction that will take place in the presence of contaminating lactate dehydrogenase.

this case it is not possible to exclude pyruvate from the assay mixture since it is a substrate for the enzyme being assayed. It would, however, be possible to use an alternative assay, such as the incorporation of radioactively-labelled bicarbonate into oxaloacetate, because the interference from lactate dehydrogenase might not be expected to be important in the absence of added NADH. The coupled assay can only be used satisfactorily if the pyruvate carboxylase preparation is purified to a state where it is free from contaminating lactate dehydrogenase.

### 2.5.2 Correction for blank rates

As will be clear from the above discussion it is important to understand the cause of a blank rate before one may make the appropriate corrections for it. In many cases it is possible to obtain the true rate of the enzyme-catalysed reaction by simply subtracting the blank rate given in a suitable incomplete mixture from that obtained with the full assay. This approach assumes that the blank rate is an artefact that is unconnected with the activity of the enzyme under study, that it continues linearly for the total period of the assay, and that it will be unchanged in the full assay. In cases where these assumptions are valid, failure to subtract the blank rate will yield apparent anomalies in kinetic behaviour. If the blank rate occurs in the absence of the enzyme, failure to subtract it will give a plot of initial velocity against enzyme concentration that does not pass through the origin but which shows a finite activity at zero enzyme concentration (*Figure 4*). Failure to subtract a blank rate that occurs in the absence of one of the substrates can give behaviour that does not conform to the Michaelis–Menten equation (Section 5.2).

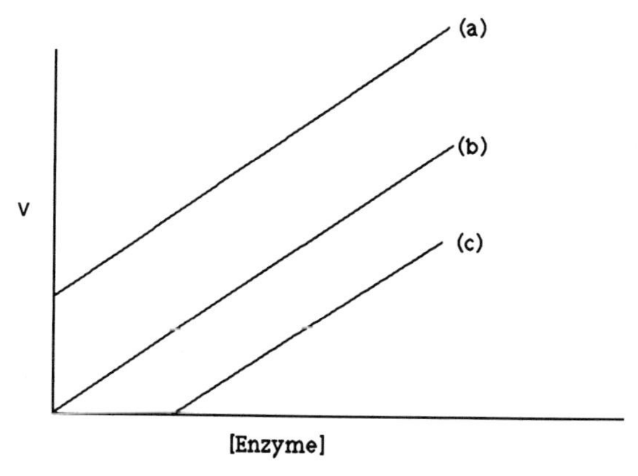


**Figure 4.** The dependence of initial velocity on the enzyme concentration. Line (b) shows the expected dependence; lines (a) and (c) show possible results from incorrect treatment of blank rates. In line (a) a blank rate occurring in the absence of the enzyme has not been subtracted and in line (c) a blank rate that occurs in the absence of one of the substrates, but is suppressed in the full assay, has been subtracted.

If an apparent blank rate is due to the settling of particles (Section 2.5.1 *i*) subtraction of the initial blank rate from the initial rate obtained after starting the reaction may be adequate, but such rates are normally irregular and it would be better to await the decline of the blank rate and the stabilization of the assay before measuring the rate. It would be inappropriate to subtract the blank rate if it were due to contamination of the enzyme (Section 2.5.1 *iii*) or assay vessel (Section 2.5.1 *iv*) with one of the substrates and the complete assay mixture contained saturating concentrations of that substrate. In these cases the blank rate is due to the enzyme itself and its subtraction would therefore result in an underestimation of the activity.

A more complicated system where it is inappropriate to subtract an apparent blank rate can occur if an enzyme can catalyse the decomposition of one of its substrates alone but that reaction is suppressed by the presence of the second substrate. This can, for example, occur in the assay of pyruvate carboxylase (see *Figure 3*). The enzyme has a relatively weak ATPase activity and will catalyse the hydrolysis of ATP in the absence of the other substrates. However, there is competition between this and the full reaction and it is effectively suppressed in the complete assay mixture (46). Thus, if the enzyme were assayed by measuring the formation of ADP or inorganic phosphate, there would be a blank rate which should not be subtracted from the rate seen in the full mixture. Subtraction of the blank rate would be expected to give a dependence of initial velocity on enzyme concentration which did not pass through the origin, giving an activity of zero at a finite enzyme concentration (see *Figure 4*).

### 2.5.3 Masking of an assay

In some cases the activity of a contaminating enzyme may interfere with the assay of an enzyme. The blank rates that can occur in the assay of enzymes utilizing NADH in the presence of contaminating NADH–cytochrome *c* reductase were discussed in Section 2.5.1 *vi*. Such contamination would, of course, affect attempts to assay dehydrogenases in the direction of NADH formation by catalysing the re-oxidation of the NADH formed. This could lead to an underestimation of the true reaction rate or a complete masking of the reaction. In such cases it would be necessary to work in the presence of an inhibitor of the contaminating enzyme. As discussed in Section 2.5.1 *vi*, use of an alternative assay that does not rely on measurement of NADH formation may not be satisfactory.

# 3. The effects of enzyme concentration

## 3.1 Direct proportionality

As enzymes are catalysts the initial velocity of the reaction would be expected to be proportional to the concentration of the enzyme. This is indeed the case for most enzyme-catalysed reactions where a graph of initial velocity against total enzyme concentration will be a straight line passing through the origin (zero activity at zero enzyme concentration; see *Figure 4*). There are, however, some cases where the simple relationship does not appear to hold and it is thus important to check for linearity in any studies. In some cases departure from linearity may be artefactual resulting from changes of the pH or ionic strength of the assay mixture as increasing amounts of the enzyme solution are added, and it is important to check that such effects are not occurring. In other cases the behaviour can be more interesting. A graph of initial velocity against enzyme concentration can show either upward or downward curvature, as illustrated in *Figures 5* and *6*. Some common causes that can result in such behaviour are considered below.

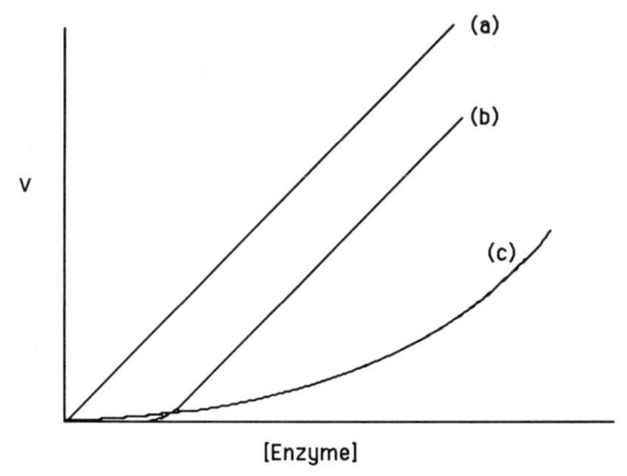


**Figure 5.** Upwardly-curving dependence of initial velocity on enzyme concentration. Curve (a) shows the normally-expected relationship; curve (b) represents the case where there is an irreversible inhibitor contaminating the assay mixture; curve (c) shows the possible behaviour if there were a reversible activator present in the enzyme preparation.

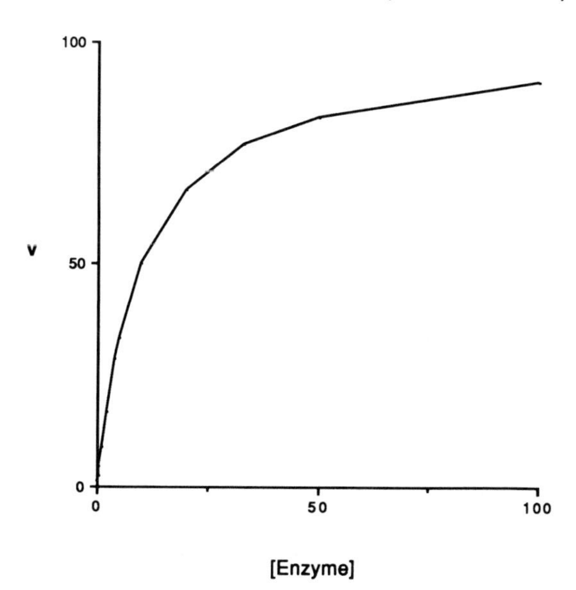


**Figure 6.** Downwardly-curving dependence of initial velocity on enzyme concentration.

## 3.2 Upward curvature

There are two common causes for this type of behaviour.

### 3.2.1 The presence of a small amount of an irreversible inhibitor of the enzyme in the assay mixture

In this case small amounts of enzyme added will be completely inhibited and activity will only be detected after sufficient enzyme has been added to react with all the inhibitor present. This will give rise to a curve of the type shown in *Figure 5*, curve (b). A number of enzymes are irreversibly inhibited by heavy metal ions and contamination of buffer or substrate solutions by these ions is a common cause of such behaviour. Since such irreversible inhibition is time-dependent, the order of addition of components to the assay mixture may affect the observed results. If the enzyme is pre-incubated in an incomplete assay mixture before starting the reaction by addition of a substrate, a curve such as that shown in curve (b) of *Figure 5* might result. However, if the reaction were started by the addition of enzyme, a non-linear time course of the reaction would be expected (Section 2.1), and if it were possible to estimate the initial velocity before the irreversible inhibition became significant a linear dependence on enzyme concentration would result.

### 3.2.2 The presence of dissociable activator in the enzyme solution

This can be represented by the equilibrium between activator (A) and enzyme (E):

$$\mathrm{E} + \mathrm{A} \overset{K_a}{\rightleftharpoons} \mathrm{EA} \tag{8}$$

where the dissociation constant $K_a$ will be given by:

$$K_a = \frac{[\mathrm{E}][\mathrm{A}]}{[\mathrm{EA}]} \text{ and thus: } [\mathrm{EA}] = \frac{[\mathrm{E}][\mathrm{A}]}{K_a} \tag{9}$$

In these equations [E] is the concentration of free enzyme. The total enzyme concentration $[\mathrm{E_T}]$ will be:

$$[\mathrm{E_T}] = [\mathrm{E}] + [\mathrm{EA}] \tag{10}$$

Since the activator is present in the enzyme solution it will be present in a constant proportion to the enzyme concentration and thus its concentration can be expressed as $x[\mathrm{E_T}]$. Thus substituting into equation (9) gives:

$$[\mathrm{EA}] = \frac{[\mathrm{E_T}]^2}{[\mathrm{E_T}] + (K_a/x)} \tag{11}$$

and thus the concentration of the EA complex will not increase linearly with the enzyme concentration, giving rise to a curve such as that shown in *Figure 5*, curve (c). The precise shape of the curve obtained will depend on the kinetic mechanism of activation and whether the free enzyme has any significant activity. If an excess of the activator were added to the assay mixture, this should displace the dissociation equilibrium so that essentially all the enzyme would exist as the EA complex and lead to a linear velocity–enzyme concentration relationship. There are several examples of this type of behaviour (6, p. 48).

A particular case of this behaviour occurs when an enzyme exists in an associating equilibrium with the aggregated form being the more active. Thus in this case the endogenous activator may be regarded as being the enzyme itself. Ox heart phosphofructokinase has been shown to behave in this way (47), giving a curve similar to that shown in curve (c) of *Figure 5*. In this case the allosteric activator AMP promotes association, and a linear dependence of initial velocity on enzyme concentration is observed in the presence of high concentrations of AMP. In contrast, the allosteric inhibitor citrate promotes disaggregation and results in a more pronounced curvature.

## 3.3 Downward curvature

There are three common cases that can give rise to curves such as that shown in *Figure 6*, where the reaction rate reaches an apparent maximum at higher enzyme concentrations.

### 3.3.1 The detection method may become rate-limiting at higher enzyme concentrations

If a coupled assay (Section 6.1.2) is used, for example, the activity of the coupling enzyme could become limiting so that the addition of further amounts of the enzyme under study could not result in further increases in the measured velocity. Similar effects may be obtained with other assay methods if there is a failure to respond linearly to increasing reaction rates (Section 2.1.6). An over-damped recorder or over-long time constant may lead to a maximum rate of response. In optical assays high enzyme concentrations can increase the initial absorbance of the solution to such a high level that the instrument is no longer responding linearly (Chapter 2).

### 3.3.2 Failure to measure the true initial rate of the reaction

This can lead to an apparent maximum value in the velocity–enzyme concentration curve. If a discontinuous assay is used which measured the amount of product formed after a fixed time, it may be that at higher enzyme concentrations the reaction has gone to completion or reached equilibrium within the assay time used. In that case the addition of more enzyme would not result in any increase in the measured product formation. This further emphasizes the necessity of ensuring that the assay procedure measures the initial rate of the reaction under all conditions used.

### 3.3.3 The presence of a dissociable inhibitor in the enzyme solution

This is the converse of the dissociable activator case discussed in Section 3.2.2. Increasing the enzyme concentration will lead to a proportional increase in that of the inhibitor and hence the amount of enzyme that is in the inactive enzyme–inhibitor complex will increase. If the complex is completely inactive, a graph of initial velocity against enzyme concentration will tend to an apparent maximum, as shown in *Figure 6*, whereas if it has a finite activity the curve will tend to a constant slope that is less than the initial slope. Several examples of such behaviour are well-documented (e.g. 6, p. 51). If the inhibitor is a small molecule it should be possible to remove it from the enzyme solution by dialysis or gel filtration, but this would not be possible if the concentration-dependent inhibitor were the result of polymerization of the enzyme to give an inactive, or less active, form.

# 4. Expression of enzyme activity

In order to express the activity of an enzyme in absolute terms it is necessary to ensure that the assay procedure used is measuring the true initial velocity and that this is proportional to the enzyme concentration. Under these conditions the ratio of velocity/enzyme concentration will be a constant that can be used to express the activity of an enzyme quantitatively. This can be valuable for comparing the data obtained for the same enzyme from different laboratories, assessing the effects of physiological or pharmacological challenges on cells or tissues, monitoring the extent of purification of enzymes, and comparing the activities of different enzymes, or of the same enzyme with different substrates.

## 4.1 Units and specific activity

The activity of an enzyme may be expressed in any convenient units, such as absorbance change per unit time per mg enzyme protein, but it is preferable to have a more standardized unit in order to facilitate comparisons. The most commonly used quantity is the *Unit*, sometimes referred to as the International Unit or Enzyme Unit. One Unit of enzyme activity is defined as that catalysing the conversion of 1 μmol substrate (or the formation of 1 μmol product) in 1 min. The specific activity of an enzyme preparation is the number of Units per mg protein. Since some workers use the term unit to refer to more arbitrary measurements of enzyme activity, it is essential that it is defined in any publication.

If the relative molecular mass of an enzyme is known it is possible to express the activity as the *molecular activity*, defined as the number of Units per μmol enzyme; in other words the number of mol of product formed, or substrate used, per mol enzyme per min. This may not correspond to the number of mol substrate converted per enzyme active-site per min since an enzyme molecule may contain more than one active site. If the number of active sites per mol is known the activity may be expressed as the *catalytic centre activity*, which corresponds to mol substrate used, or product formed, per min per catalytic centre (active site). The term *turnover number* has also been used quite frequently but there appears to be no clear agreement in the literature as to whether this refers to the molecular or the catalytic centre activity.

## 4.2 The Katal

Although the Unit of enzyme activity, and the quantities derived from it, have proven to be most useful, the Nomenclature Commission of the International Union of Biochemistry has recommended the use of the Katal (abbreviated to Kat). This differs from the units described above in that the second, rather than the minute, is used as the unit of time in conformity with the International System of units (SI Units).

1 Kat corresponds to the conversion of 1 mol of substrate per second. Thus it is an inconveniently large quantity compared to the Unit. The relationships between Kats and Units are

$$1 \text{ Kat} = 60 \text{ mol} \times \text{min}^{-1} = 6 \times 10^7 \text{ Units}$$
$$1 \text{ Unit} = 1 \ \mu\text{mol} \times \text{min}^{-1} = 16.67 \text{ nKat}$$

In terms of molecular or catalytic centre activities the Kat is, however, not such an inconveniently large quantity and it is consistent with the general expression of rate constants in $s^{-1}$.

## 4.3 Stoichiometry

When expressing the activity of an enzyme it is important to bear in mind the stoichiometry of the reaction. Some enzyme-catalysed reactions involve two molecules of the same substrate; e.g. adenylate kinase (EC 2.7.3.4) catalyses the reaction:

$$2\text{ADP} \rightleftharpoons \text{AMP} + \text{ATP}$$

and carbamoylphosphate synthetase (ammonia) (EC 6.4.3.16) catalyses:

$$\text{HCO}_3^- + 2\text{ATP} + \text{NH}_4^+ \rightarrow \text{Carbamoylphosphate} + 2\text{ADP} + \text{P}_i.$$

In the former case the activity will be twice as large if it is expressed in terms of ADP utilization than if expressed in terms of the formation of either of the products. In the latter case the value expressed in terms of disappearance of ATP or formation of ADP would be twice that obtained if any of the other substrates or products were measured. Thus it is important to specify the substrate or product measured and the stoichiometry when expressing the specific activity of an enzyme.

## 4.4 Conditions for activity measurements

Although the quantity velocity/enzyme concentration is a useful constant for comparative purposes, it will only be constant under defined conditions of pH, temperature, and substrate concentration. A temperature of 30°C has become widely used as the standard for comparative purposes, but in some cases it may be desirable to use a more physiological temperature. There is no clear recommendation as to pH and substrate concentration except that these should be stated and, where practical, should be optimal. It may, however, be appropriate to use physiological pH values, which may differ from the optimum pH, if the results are to be related to the behaviour of the enzyme *in vivo*. Since the activities of some enzymes are profoundly affected by the buffer used and the ionic strength of the assay mixture, the full composition of the assay should be specified. In cases where there is a non-linear dependence of initial velocity upon enzyme concentration, and the artefacts referred to in Section 3 are excluded, it may be possible to work in the presence of an excess of the appropriate activator to obtain linearity (Section 3.2.2). However, when enzyme association–dissociation is involved it may not be possible to find an appropriate effector and an alternative approach would be to investigate conditions of pH, buffer composition, or ionic strength under which the strength of the inter-subunit interactions are such that the polymerization/depolymerization becomes unimportant over the concentration range used. In the case of phosphofructokinase mentioned in Section 3.2.2, for example, linearity can be achieved either by inclusion of an excess of the allosteric AMP or by working at a higher pH value.

# 5. The effects of substrate concentration

## 5.1 The Michaelis–Menten relationship

The Michaelis–Menten equation predicts a hyperbolic relationship between initial velocity and substrate concentration and the kinetic behaviour of many enzymes is described by this relationship. However, it would not be possible to provide a detailed treatment of enzyme kinetics within this account and this section will concentrate on the practical problems that can arise when apparently complex behaviour is observed. A discussion on the analysis of kinetic data is included in Chapter 10 of this book and the reader is referred to several comprehensive accounts for detailed treatments of the steady-state kinetics of enzyme-catalysed reactions (4–7, 48–51). Although the double-reciprocal plot is recognized to be a poor procedure for determining enzyme kinetic parameters (Chapter 10), it is useful for illustrative purposes and will be used for such in this account. The direct-linear plot, which is a superior procedure for the calculation of data, is less clear for their presentation and therefore will not be used here.

In studying the variation of the initial velocity over a range of substrate concentrations many of the pitfalls to be avoided are similar to those discussed in terms of studies of the effects of variation of enzyme concentration. Thus it is important to ensure that other factors, such as pH and ionic strength, which may affect the activity of the enzyme remain constant. Although it is generally a simple matter to ensure that changing the substrate concentration does not affect the pH of the reaction mixture, it may be less easy to control the ionic strength if the substrate is a charged, or multi-charged species. It may be possible to work at such high ionic strengths that the changes due to substrate addition are insignificant. Where such an approach is not possible it will be necessary to carry out separate control experiments on the effects of ionic strength on the activity of an enzyme. Changes in the dielectric of the reaction mixture should also be controlled in cases where one of the substrates is non-polar or if it is added in solution in an organic solvent.

## 5.2 Failure to obey the Michaelis–Menten equation

Departure from the simple hyperbolic behaviour predicted by the Michaelis–Menten equation can result from a number of different causes. In each case it is necessary to ensure that the behaviour seen is a genuine phenomenon rather than an artefact. In this section some of the common cases of such complex behaviour will be discussed in turn.

### 5.2.1 High-substrate inhibition

It is not uncommon for enzymes to be inhibited by high concentrations of one or more of their substrates, leading to kinetic plots such as those shown in *Figure 7*. Such behaviour can be useful in helping to deduce the kinetic mechanism involved (6, 7, 52) but it can restrict the range of substrate concentrations that can be used for determining $K_m$ and $V$ values. The treatment of high-substrate inhibition data has been discussed in detail (6, p. 126). If such inhibition is observed, it is necessary to carry out appropriate controls to ensure that it is a property of the enzyme and its substrate rather than an artefact arising from failure to control the pH, ionic strength, or dielectric of the assay medium correctly. It is also necessary to show that the inhibition is due to the substrate itself rather than to an inhibitory contaminant since, as discussed in Section 6.4.3, contamination of the substrate with a compound that is either a non-competitive (mixed) or uncompetitive inhibitor of the enzyme will lead to behaviour resembling high-substrate inhibition. A particular case to guard against concerns substrates that chelate metal ions. If the enzyme requires free metal ions for activity, an excess of a chelating substrate, such as ATP or citrate, may reduce their concentrations to levels where the enzyme is unable to function. This may be remedied by ensuring that the metal ions are always present in excess. If that is not possible, it may be necessary to calculate the concentrations of free and complexed species (53) to ensure that the free metal ion concentration remains sufficient for activity. The complexities that can arise when the metal–substrate complex is the true substrate for the enzyme will be discussed in more detail in Section 5.2.2 *vi*.

### 5.2.2 Sigmoid kinetics

A sigmoid dependence of initial velocity upon substrate concentration (*Figure 8*) may indicate that the enzyme obeys co-operative kinetics. However, there are several other possible causes of such behaviour and it is necessary to carry

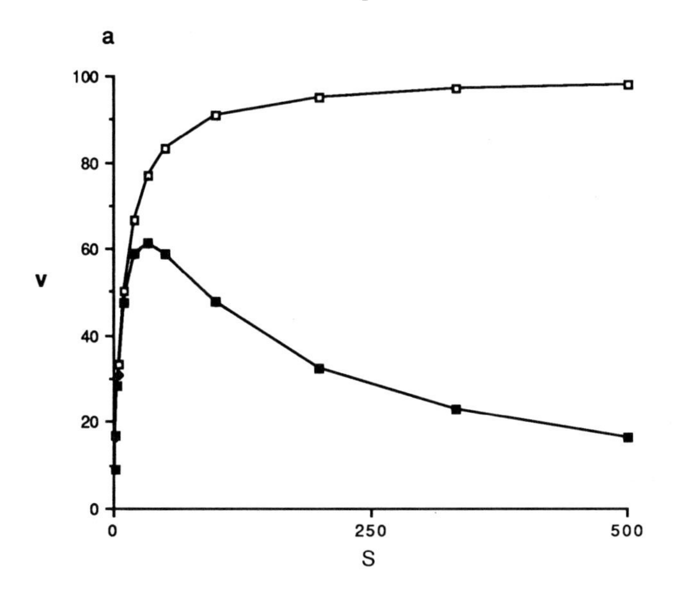


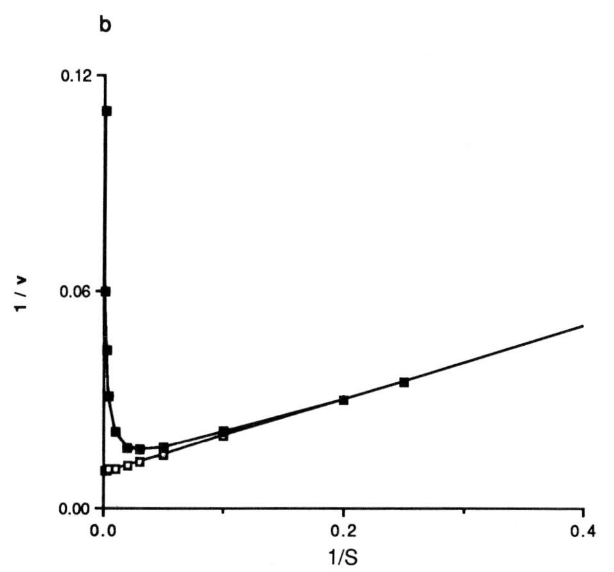


**Figure 7.** High-substrate inhibition. The curves obey the equation:

$$v = V_{max}/[1 + (K_m/S) + (S/K_i)]$$

where for the open symbols $K_i = \infty$ and for the closed symbols it is set to 100. In both cases $V_{max}$ and $K_m$ are 100 and 10, respectively.

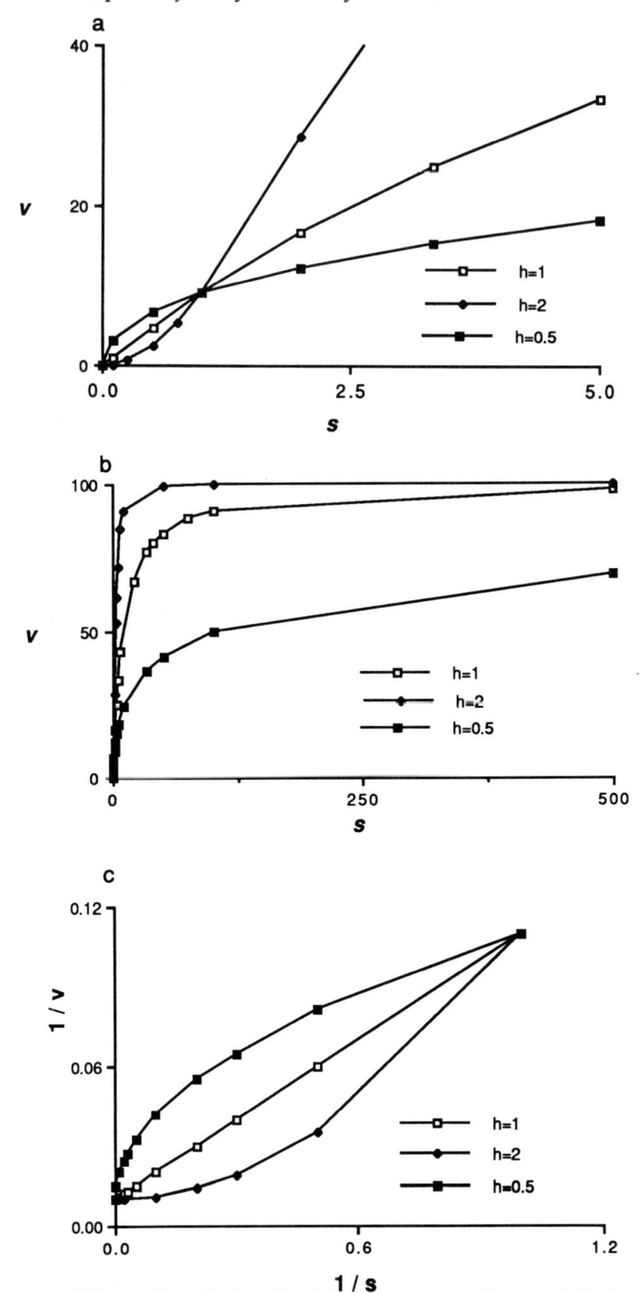


**Figure 8.** Co-operativity in the effects of substrate concentration on initial velocity. The curves were fitted to the Hill equation (Equation 13) with $V$ and $K$ being equal to 100 and 10, respectively. The behaviour at $h = 1$ (no co-operativity), $h = 2$ (positive co-operativity), and $h = 0.5$ (negative co-operativity) is shown. Panels (a) and (b) show the same data over different ranges of substrate concentration, panel (c) shows the data in double-reciprocal form.

out careful control experiments before ascribing such behaviour to this cause. The effects that can result in such kinetic behaviour will be considered below.

*i. True co-operativity*

Co-operativity, as strictly defined, is a phenomenon reflecting the equilibrium binding of substrates (54) where the binding of one molecule of a substrate to an enzyme can either facilitate (positive co-operativity) or hinder (negative co-operativity) the binding of subsequent molecules of the same substrate. Positive co-operativity will give rise to kinetic behaviour such as that shown in *Figure 8*. In order to ensure that such behaviour is due to positive co-operativity, it will be necessary to carry out substrate binding studies under equilibrium conditions since the steady-state initial velocities may not bear any simple relationship to the equilibrium saturation curve for substrate binding.

Most co-operative enzymes also exhibit allosteric behaviour, i.e. their activities are affected by the binding of molecules (allosteric effectors) to sites distinct from the active site. Allosteric effects are, however, distinct from co-operativity and may occur in enzymes that show no co-operativity. Full discussions of the methods available for analysing co-operative behaviour and distinguishing between the possible models that may account for such effects are available elsewhere (e.g. 6, p. 399; 49, p. 151; 55, 56).

It is common to present data for co-operative enzymes in terms of the simple model advanced by Hill (57) in an attempt to explain the sigmoid saturation curve for oxygen binding to haemoglobin:

$$Y_s = \frac{[S]^h}{K + [S]^h} \quad (12)$$

where $Y_s$ represents the fractional saturation of the enzyme with substrate and $h$ is the Hill constant, which does not correspond to the number of substrate-binding sites present in the molecule (6, p. 399; 55, 56). In the case of a co-operative enzyme where initial velocities are measured the corresponding Hill equation can be written as:

$$v = \frac{V_{max}[S]^h}{K + [S]^h} \quad (13)$$

which can be transformed to a linear relationship as:

$$\log \frac{v}{V_{max} - v} = h \log [S] - \log K \quad (14)$$

Such plots are known as Hill plots. Although the Hill equation has been shown to be based on an inadequate model, because it envisages the simultaneous binding of substrate molecules to the enzyme (e.g. 6, p. 399; 55, 56) the plot is still widely used to express co-operativity. A Hill constant of greater than unity indicates positive co-operativity whereas one of less than unity is given in cases of negative co-operativity. If there is no co-operativity the value of $h$ will be unity and equation 13 will reduce to the simple Michaelis–Menten equation.

Despite its widespread use, the Hill equation is an invalid model for co-operative systems and it has been shown that for any system that involves the sequential binding of substrates, the plot will be linear only over a restricted range of substrate concentrations with the slopes tending to unity at very high and very low substrate concentrations (6, p. 399; 55, 56, for further discussion). Thus departure from the linearity predicted by the Hill equation may be expected.

*ii. Alternative pathways in a steady-state system*

Any enzyme reaction mechanism in which there are alternative pathways by which the substrates can interact to give the complex that breaks down to give products will, under steady-state conditions, give rise to a complex initial-rate equation. The simplest example concerns an enzyme with two substrates, Ax and B, which are converted to the products A and Bx, respectively. A reaction mechanism in which the two substrates can bind to the enzyme in a random order:

$$\begin{array}{ccccc} & \nearrow\!\!\!\swarrow & EAx & \searrow\!\!\!\nwarrow & \\ E & & & & EAxB \longrightarrow E + A + Bx \\ & \searrow\!\!\!\nwarrow & EB & \nearrow\!\!\!\swarrow & \end{array} \quad (15)$$

will give a steady-state initial-rate equation of the form

$$v = \frac{p[Ax][B] + q[Ax]^2[B] + r[Ax][B]^2}{s + t[Ax] + u[B] + v[Ax][B] + w[Ax]^2 + x[B]^2 + y[Ax]^2[B] + z[Ax][B]^2} \quad (16)$$

where the constants p–z are combinations of rate constants. At a fixed concentration of one of the substrates, for example B, this fearsome equation simplifies to:

$$v = \frac{\beta_1[Ax] + \beta_2[Ax]^2}{\alpha_0 + \alpha_1[Ax] + \alpha_2[Ax]^2} \quad (17)$$

An equation of this form, containing squared terms in the substrate concentration, can give rise to a variety of curves describing the variation of initial velocity with substrate concentration, as shown in *Figure 9*, depending on the values of the individual rate constants (58). The behaviour of this system has been considered in detail by Ferdinand (59) who pointed out that sigmoidal behaviour would be expected if the rates through the two alternative pathways (through EAx and through EB) leading to the EAxB ternary complex were sufficiently different. Under such conditions a sigmoid dependence of initial velocity upon substrate concentration will result if the concentration of one of the substrates is varied at a fixed, non-saturating concentration of the other, whereas if the concentration of the other substrate is varied at a fixed non-saturating concentration of the first, a curve such as that shown in *Figure 9* line (d) will result. Several other mechanisms in which there are alternative (two or more) ways in which substrates can bind to an enzyme have been shown to give rise to complex steady-state rate equations (60–63).

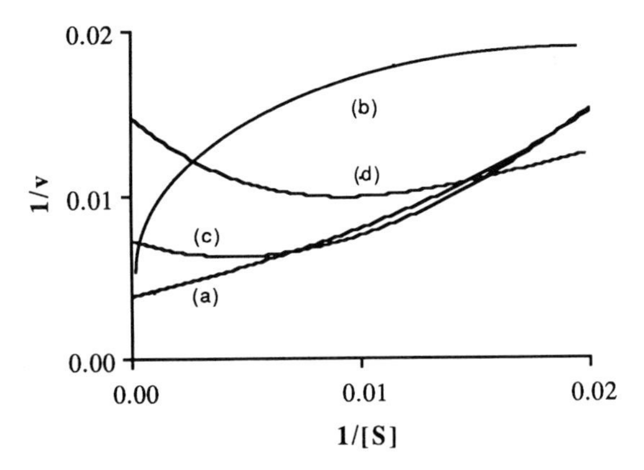


**Figure 9.** Possible initial velocity-substrate concentration relationships that can be obtained with systems obeying Equation 16. The data are shown as double-reciprocal plots.

The random-order mechanism will, of course, yield simple Michaelis–Menten behaviour if the rate of breakdown of the EAxB ternary complex to give products is slow relative to its rate of breakdown to the binary (EAx and EB) complexes so that the system remains in thermodynamic equilibrium (4–7, 48–51).

*iii. The reaction involves two or more molecules of the same substrate*

For a reaction of the type:

$$2\ \mathrm{Ax} \longrightarrow \mathrm{Ax}_2 + \mathrm{A}$$

the initial rate equation for a two-substrate reaction in which the substrates bind randomly under equilibrium conditions, or in an ordered mechanism under steady-state conditions, becomes:

$$v = \frac{V_{\mathrm{max}}}{1 + (K_{\mathrm{m1}}^{\mathrm{Ax}}/[\mathrm{Ax}]) + (K_{\mathrm{m2}}^{\mathrm{Ax}}/[\mathrm{Ax}]) + (K_{\mathrm{s}}^{\mathrm{Ax}} K_{\mathrm{m2}}^{\mathrm{Ax}}/[\mathrm{Ax}]^2)} \qquad (18)$$

which predicts a sigmoidal dependence of initial velocity on substrate concentration. The degree of sigmoidicity will depend on the values of the individual constants. If the two substrate binding steps are separated by an irreversible step, such as in a double-displacement (ping-pong) mechanism, the value of the constant $K_s^{Ax}$ will be zero and thus the equation will simplify to (6, p. 79):

$$v = \frac{V_{max}}{1 + (K_{m1}^{Ax}/[Ax]) + (K_{m2}^{Ax}/[Ax])} \quad (19)$$

and hyperbolic kinetics will result. The enzyme hydroxymethylbilane synthase (EC 4.2.1.24) involves the interaction of four identical substrate molecules to yield the product porphobilinogen. Since this enzyme exhibits simple Michaelis–Menten kinetic behaviour there must be steps that are essentially irreversible between the binding of each of the substrate molecules to the enzyme (64). An alternative way in which an essentially irreversible step may occur is if the binding of the two identical substrate molecules is separated by the binding of a different substrate. In that case very high concentrations of the latter substrate would render its binding essentially irreversible, resulting in an irreversible step between the binding of the two identical molecules, and hyperbolic kinetics would be obtained (65). Determination of the stoichiometry of the reaction catalysed should establish whether sigmoid initial velocity curves are likely to result from this cause; for example, such curves are given by the enzyme carbamoylphosphate synthetase (ammonia) when ATP is the variable substrate because the reaction catalysed involves two molecules of this substrate (65).

*iv. Enzyme isomerization*

Several possible mechanisms in which an enzyme exists in two or more forms which interconvert relatively slowly can give rise to kinetic equations of the form of Equation 17. Different systems of this type have been presented in several publications (66–69). If the rates of the isomerization steps are sufficiently slow hysteric effects (Section 2.4.9) may also be observed. Mechanisms of this type do not require the presence of multiple binding sites for the same substrate and thus may not show co-operative substrate binding.

*v. Failure to determine initial rate*

If an enzyme is assayed by determining the extent of reaction after an arbitrarily fixed time, it may be that the reaction has proceeded to completion, or equilibrium, within that time at the lower substrate concentrations. This would result in an underestimation of the true initial velocity at the lower substrate concentrations, and hence a curve which might appear sigmoid. A similar effect may occur if an enzyme is unstable in the reaction mixture but is stabilized by the binding of substrate. In this case the effectiveness of stabilization will increase with substrate concentration as the enzyme tends to become saturated with substrate. This will lead to an increasing underestimation of the true initial velocity as the substrate concentration decreases below that required to saturate the enzyme. Such an effect, leading to apparently sigmoid curves of velocity against substrate concentration, has been reported for the enzyme threonine deaminase (EC 1.1.1.103) (70).

*vi. Failure to take account of substrate–activator complexes*

The true substrate for a number of enzymes is a complex between substrate and an activator, usually a divalent metal ion. One of the most widely studied cases is the complex between ATP and magnesium ions which is the true substrate for many enzymes catalysing reactions involving ATP. In such cases it is necessary to calculate the concentration of the complexed form, as the true substrate, for kinetic studies. In general, the concentration of the metal–substrate complex, [MS], will be given by the equation:

$$[MS] = 1/2\{([S_T] + [M_T] + K_d) - \sqrt{([S_T] + [M_T] + K_d)^2 - 4[M_T][S_T]}\} \quad (20)$$

where $[S_T]$ and $[M_T]$ are the total concentrations of substrate and metal ion, respectively, and $K_d$ is the dissociation constant of the (MS) complex. This equation does not predict a linear dependence of the concentration of the MS complex upon the concentrations of M plus S if these are mixed together and varied in a fixed ratio. Several kinases have been erroneously reported to show co-operative behaviour because sigmoid kinetics were observed when the ATP and magnesium were mixed together in a fixed ratio and the 'substrate concentration' was varied by the addition of varying amounts of this mixture. In fact, such behaviour is predicted by the relationship shown in the above equation and has nothing to do with co-operativity (65, 71, 72, 73). The simplest way to overcome this problem is to work at such high concentrations of the metal ion, compared to the $K_d$ value, that the substrate remains essentially fully in the complexed form at all concentrations used. However, this may not be possible; for example, if the free metal ion inhibits the enzyme. In such cases it will be necessary to calculate the complex concentration at each concentration used. This is no easy task because the interactions between metals and ligands will be affected by the pH, temperature, and ionic strength (74). A useful procedure and computer program to calculate the binding of metal ions to ATP has been presented by Storer and Cornish-Bowden (53).

It should be remembered that the substrate may not be the only species in solution that binds metal ions. Many buffer species, such as phosphate or citrate buffer, are metal chelators and allowance should be made for this when calculating the amount of complexed species in solution. A number of buffers which do not bind magnesium ions, or bind them only very weakly, are available (75).

The special problems of studying the kinetics of enzyme reactions in which a metal–substrate complex may be involved are beyond the scope of the present account and the reader is referred to several fuller descriptions of the procedures involved (73, 76, 77, 78). Although the above analysis has concentrated on metal–substrate complexes, the possibility that other activator–substrate complexes might be involved in the activity of an enzyme should not be excluded; for example, the true substrate for formaldehyde dehydrogenase (EC 1.2.1.1) is the reversible adduct formed between formaldehyde and glutathione (79). The analysis of the behaviour of such systems would be similar to that discussed above.

### 5.2.3 Apparent negative co-operativity

Negative co-operativity, in which the binding of one molecule of substrate to an enzyme decreases its affinity for binding subsequent molecules of the same substrate, leads to a Hill constant of less than unity and a double-reciprocal plot that curves downward (see *Figure 8*). The curve of velocity against substrate concentration may at first-sight appear to be normal but, as shown in *Figure 8*, this is illusory. As discussed in Section 5.2.2, true negative co-operativity is a substrate-binding phenomenon which should be reflected in similar behaviour if equilibrium binding is studied. There are, however, a number of other possible causes of such effects which may be difficult to distinguish from negative co-operativity.

*i. Alternative pathways in a steady-state system*

As discussed in Section 5.2.2 *ii* mechanisms of this type can yield initial velocity–substrate concentration curves that resemble those seen with negative co-operativity (see *Figures 8* and *9*). However, because this effect arises from steady-state rather than substrate-binding complexities the substrate binding curves determined at equilibrium should be rectangularly hyperbolic.

*ii. The presence of more than one enzyme catalysing the same reaction*

If the same reaction is catalysed by two enzymes which have different $K_m$ values, the rate of the overall reaction will be given by:

$$v = \frac{V^{a}_{max}}{1 + K^{a}_{m}/[S]} + \frac{V^{b}_{max}}{1 + K^{b}_{m}/[S]} \tag{21}$$

where $K_m^{a}$ and $V^{a}_{max}$ are the Michaelis constant and maximum velocity of one enzyme and $K_m^{b}$ and $V^{b}_{max}$ are the corresponding constants for the other. This equation will result in a behaviour similar to that of negative co-operativity. Thus it is always necessary to check for the presence of more than one enzyme if such behaviour is observed. This is particularly important since if the two enzymes had different binding affinities for their substrates, as well as $K_m$ values, similar behaviour would be shown in studies of the equilibrium binding of substrate.

Equation 21 predicts a smooth curve when the data are plotted in double-reciprocal form (see *Figure 8*) without sharp breaks. It is not possible to obtain accurate estimates of the two $K_m$ and $V_{max}$ values by extrapolation of the apparently linear portions of the double-reciprocal plots at very high and very low substrate concentrations. Such an approach does not yield accurate values (80) unless the difference between the $K_m$ values is greater than 1000-fold. It is possible, however, to determine the individual values using an iterative procedure that fits the data points to the sum of two rectangular hyperbolas (e.g. 81, 82). It is always good practice to construct a curve from the calculated values using Equation 21 and compare it with the data points to ensure that a good fit has been obtained.

*iii. Failure to subtract a blank rate*

If a blank rate is not subtracted from the rates observed with the full reaction mixture (Section 2.5) a curve resembling negative co-operativity can result because the rate will appear to be finite when the substrate concentration is zero. A similar distortion can occur in equilibrium studies if there is non-specific binding of the substrate that is not corrected for. If there is a blank rate that is proportional to the concentration of the substrate, failure to subtract it will give rise to a downwardly curved double-reciprocal plot that passes through the origin ($1/v = 0$ when $1/[S] = 0$) since the velocity will tend to become infinite as the substrate concentration is raised towards infinity (83).

*iv. High-substrate activation*

A mechanism in which the binding of substrate to an enzyme–product complex facilities the release of the product, and thus accelerates the reaction, has been proposed to account for the downwardly curving double-reciprocal plots seen with aldehyde dehydrogenase (84). In this case, since the second substrate molecule binds to an enzyme–product complex, normal saturation curves would be expected from equilibrium binding studies with substrate or substrate analogues in the absence of products.

*v. Enzyme isomerizations*

Systems involving isomerization of the enzyme, such as those discussed in Section 5.2.2 *iv*, can give rise to curves resembling negative as well as positive co-operativity.

*vi. Tight binding*

If an enzyme has a very high affinity for its substrate it may be necessary to use substrate concentrations that are similar to those of the enzyme in binding, or initial-rate kinetic, studies. Under these conditions the free substrate concentration will be significantly altered by enzyme–substrate complex formation and the dissociation constant ($K_s$) for the reaction:

$$E + S \rightleftharpoons ES$$

will be:

$$K_s = \frac{([E] - [ES])\,([S] - [ES])}{[ES]} \tag{22}$$

This relationship leads to an equation of the same form as Equation 20 and does not predict a simple binding curve. Methods for analysing the behaviour of such systems have been presented elsewhere (85, 86).

### 5.2.4 Even more bizarre curves

Complexities of the steady-state reaction mechanism (Section 5.2.2 *ii*) can result in multiple inflection points or waves in the curves of initial velocity against substrate concentration (62). True co-operativity in which an enzyme shows a mixture of positive and negative co-operativity for the successive binding of molecules of the same substrate can also give multiple inflection points and plateaus in the saturation or velocity–substrate concentration curve (87). Such behaviour could also arise from a mixture of two enzymes, one exhibiting positive co-operativity and the other no co-operativity. This might come about if the properties of a co-operative enzyme became modified during the purification in such a way that a proportion of the molecules had lost their ability to interact co-operatively with the substrate. Complex negative co-operativity for multiple binding sites can result in curves of velocity against substrate concentration which may appear to be composed of linear sections with apparently sharp breaks between them (88, but see also 89).

# 6. Experimental approaches

## 6.1 Type of assay

Although many enzyme-catalysed reactions result in changes in the properties of the reactants that are relatively easy to measure directly and continuously, others do not; in the latter case it is necessary to use an indirect assay method that involves some further treatment of the reaction mixture. In some cases it may be possible to use such indirect assays to monitor the progress of the reaction continuously, but in others it is necessary to stop the reaction before further treatment of the assay mixture to allow the extent of reaction to be determined. Continuous assays have the advantage of allowing progress curves for the reaction to be followed directly, and should thus make it a relatively simple matter to determine initial rates and see any deviations from the initial linear phase of the reaction or any of the anomalous types of behaviour discussed earlier. Because discontinuous assays will give the extent of a reaction after a chosen fixed time, it is tempting to select a reaction time and assume that initial-rate conditions will hold for that time. I hope that the discussion in the previous sections has made it clear that such assumptions can lead to gross errors and that it is necessary to show that product formation proceeds linearly for the time used in such assays under all conditions employed.

### 6.1.1 Direct continuous assays

Any difference between the properties of the substrates and products that can be directly measured may be used to provide the basis for direct assays. Changes in absorbance, fluorescence, pH, optical rotation, conductivity, enthalpy, viscosity, or volume of the reaction mixture have all been used to assay the activities of individual enzymes. Provided that the sensitivity is sufficiently high and the procedure does not impose undesirable limitations on the assay conditions that can be used, direct continuous assays are always to be preferred, because they allow observation of the progress curve, which simplifies the estimation of initial rates and allows detection of any anomalous behaviour. Some of the most commonly-used procedures are mentioned below (see also references 1–3).

Spectrophotometric assays are probably the most widely-used procedures (2, 90). The high standards of accuracy and reliability of many commercially-available spectrophotometers makes such assays particularly convenient. In cases where the reaction results in a change in fluorescence of one of the substrates, fluorimetric assays can be used. This usually results in a considerable gain in sensitivity and can be particularly valuable when only small amounts of the enzyme are available or if it has a very low $K_m$ value for its substrates. The applications of spectrophotometric and fluorimetric assays, and some of the precautions that should be taken in their use, are discussed in more detail in Chapter 2.

Reactions that involve the release or uptake of hydrogen ions can be assayed directly in unbuffered, or weakly-buffered, solutions by following the change in pH with a glass electrode. The use of such assays should be restricted to a pH range over which the change does not appreciably affect the activity of the enzyme. Changes in hydrogen ion concentrations may also be followed spectrophotometrically by use of an indicator which changes its absorbance with protonation state (Chapter 2). An alternative method which avoids significant change of pH during the assay is to use a pH-stat which titrates the reaction mixture with either acid or alkali to keep the pH constant whilst recording the rate of addition (91; Chapter 7). Ion-sensitive electrodes or gas electrodes can be used for monitoring changes in the concentrations of some other reactants, such as ammonia or $CO_2$ (2). Reactions involving the uptake or output of oxygen can be followed polarographically by means of an oxygen electrode (92; Chapter 6).

### 6.1.2 Indirect assays

These involve some further treatment of the reaction mixture either to produce a measurable product or to increase the sensitivity or convenience of the assay procedure. In some cases it is possible to use indirect assays continuously to monitor the progress of the reaction whereas in others the method can only be used discontinuously.

*i. Discontinuous indirect assays*

These assays, which may also be termed sampling assays, involve stopping the reaction after a fixed time and treating the reaction mixture to separate a product for analysis, or to produce a change in the properties of one of the substrates or products which can then be measured. Examples of the former type of assay include radiochemical assays, which will be discussed in detail in Chapter 3. The development of liquid chromatographic systems for rapid separation and quantitation of reactants has allowed many assays to be devised that are based on this technique. These are discussed in Chapter 4.

Measurement of luminescence can form the basis of highly-sensitive assay procedures. The formation or disappearance of ATP can be determined by measuring the light emission in the presence of fire-fly luciferase which catalyses the reaction:

$$\text{ATP} + \text{luciferin} + O_2 \rightarrow \text{oxyluciferin} + PP_i + CO_2 + \text{AMP} + \text{light}$$

Similarly, NAD(P)H can be determined using the bacterial luciferase system which catalyses the reactions:

$$\text{NADPH} + H^+ + \text{FMN} \rightarrow \text{NADP}^+ + \text{FMNH}_2$$
$$\text{FMNH}_2 + \text{RCHO} + O_2 \rightarrow H_2O + \text{RCOOH} + \text{light}$$

where RCHO represents a long-chain (8–14 carbons) aliphatic aldehyde. This reaction has also been used to determine the oxidation of aliphatic amines to the corresponding aldehydes by the enzyme monoamine oxidase (93). Mutants of the bacteria that require the presence of fatty acids have been used to determine lipase and phospholipase activities (94) and a cyclic AMP requiring mutant can also be used to determine that compound (95).

The earthworm luciferase system requires $H_2O_2$ and may be used as a basis for determining the formation of that compound (96). There are also several chemiluminescence reactions that can be used for the determination of $H_2O_2$ formation. These include the light-emitting reaction with luminol (3-amino-phtalazine-1, 4-dione) which is catalysed by metal-ion complexes at alkaline pH values or by peroxidase (EC 1.11.1.7). A detailed account of the applications of luminescence methods has recently been published (97).

Manipulations to render a product detectable include, for example, adjustment of the pH to alkaline values to allow the *p*-nitrophenol, produced by the action of acid phosphatase on *p*-nitrophenylphosphate, to be detected, and the use of colour reactions to determine inorganic phosphate (e.g. 98). The use of discontinuous assays to increase the sensitivity of detection can be illustrated by the use of strong alkali to convert $NAD^+$ or $NADP^+$ into highly fluorescent derivatives. These products fluoresce at 460 nm when excited at 360 nm with an intensity that is about 10-fold higher than that given by

NAD(P)H (90, 99). Although the reduced coenzymes do not react in this way, it is necessary to remove them to avoid interference from their fluorescence. This may be achieved by prior treatment with 0.2 M HCl which destroys the reduced coenzymes without affecting the oxidized forms. To take advantage of this enhanced sensitivity for the determination of NAD(P)H, the oxidized coenzymes can be destroyed by treatment with dilute alkali and the reduced forms can then be reoxidized with $H_2O_2$ before the strong alkali treatment.

With all discontinuous assays it is important to ensure that the procedure used to terminate the reaction does so instantaneously. Methods involving rapid mixing with acid or alkali to alter the pH to a value where the enzyme is inactive are usually effective, but methods involving transfer of the reaction vessel to an ice or boiling-water bath may be less satisfactory if the volume of the assay mixture is relatively large. It is essential to check that the method used does, in fact, stop the reaction instantaneously. This can often be done by comparing the results given by samples in which the reaction is stopped at zero-time with those given by samples from which either the enzyme or one of the substrates has been omitted, or the enzyme has been inactivated in some way, such as by heat-treatment or incubation with an irreversible inhibitor, before it is added to the assay mixture.

*ii. Continuous indirect assays*

Assays of this type involve carrying out the manipulations necessary to detect product formation, or product remaining within the assay mixture, in such a way as to allow the change to be followed continuously as it occurs. Such assays should allow progress curves to be determined in a single assay and thus they may be less prone to errors arising from the sample manipulations necessary in discontinuous assays.

Reagents that react with one of the products of the reaction to form a detectable compound can be included in the assay mixture. If such an assay is to give valid results the detection reaction must occur so rapidly that the enzyme-catalysed reaction is always rate-limiting, so that the rate determined will correspond to the activity of the enzyme under study. It is also necessary that the reagent used has no effect on the activity of the enzyme and that it does not react with any of the other components of the system. An example of this type of reaction is the detection of carnitine acyltransferases (100). These enzymes will catalyse the transfer of acyl groups from coenzyme A to carnitine. This process results in the liberation of the free sulphydryl group of CoASH which reacts extremely rapidly with the reagent 5,5′-dithiobis-2-nitrobenzoate (DTNB) releasing a yellow-coloured compound whose formation can be followed at 412 nm. A combination of this approach with the use of a synthetic substrate has been used for the assay of acetylcholine esterase (EC 3.1.1.7). In this assay DTNB is included in the assay mixture and acetylcholine is replaced by acetylthiocholine which is hydrolysed by the enzyme to yield acetate plus thiocholine (101).

*iii. Coupled assays*

The most commonly-used assays of this type involve the use of one, or more, additional enzymes to catalyse a reaction of one of the products to yield a compound that can be detected directly. Assays of this type are known as coupled assays and the auxiliary enzymes added are frequently referred to as coupling enzymes. A number of such assays have been listed by Rudolph *et al.* (102). Coupled assays often involve the reduction of $NAD(P)^+$, or the oxidation of the corresponding reduced coenzymes, because these processes can be readily determined spectrophotometrically or fluorimetrically. However, there are many other possibilities; for example, some of the luminescence systems discussed in the previous section may be adapted for continuous use under appropriate conditions (103, 104).

A simple coupled assay for the determination of hexokinase activity by coupling the formation of glucose-6-phosphate to the reduction of $NADP^+$ in the presence of glucose-6-phosphate dehydrogenase (EC 1.1.1.49) is shown in *Figure 10* and a coupled assay for pyruvate carboxylase was discussed in Section 2.5.1 *vi*. It is not necessary for coupled assays to be restricted to a

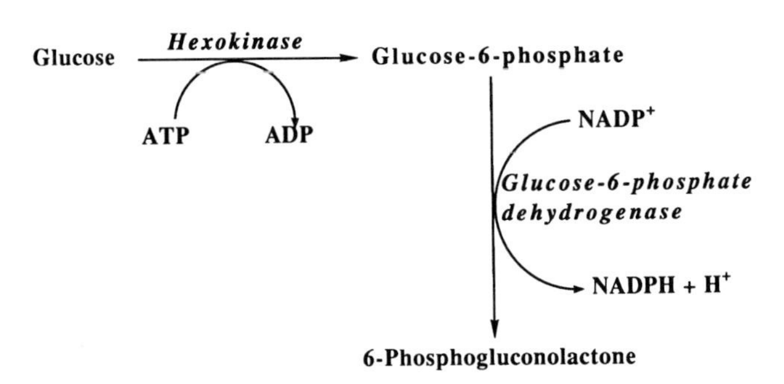


**Figure 10.** A coupled assay for hexokinase.

single coupling enzyme. *Figure 11* shows two alternative assays for phosphofructokinase. That involving the reaction of the ADP produced with phosphoenolpyruvate in the presence of pyruvate kinase (EC 2.7.1.40) has the advantage that the ATP used in the reaction is continuously regenerated, which should prevent any fall-off in the reaction velocity due to depletion of this substrate (Section 2.1.1). However, it has the limitation that phosphofructokinase from some sources is allosterically inhibited by phosphoenolpyruvate (78).

If assays of this type are to yield valid results it is essential that the coupling enzyme(s) used never becomes rate-limiting so that the measured rate is always determined by the activity of the enzyme under study. The velocity of

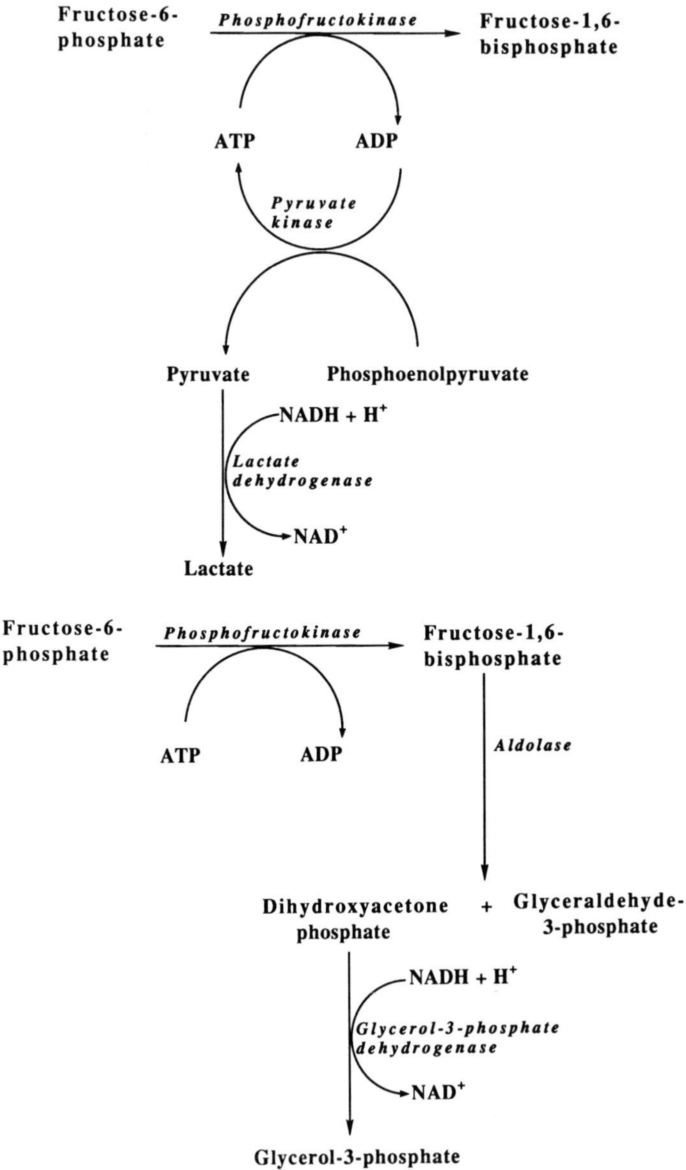


**Figure 11.** Coupled assays for phosphofructokinase based on (a) ADP formation and (b) fructose-1,6-bisphosphate production.

the reaction catalysed by a coupling enzyme will depend upon the substrate concentration available to it. Since this is produced by the activity of the enzyme under study there will be very little available during the early part of the reaction and thus the coupling enzyme will be functioning at only a small fraction of its maximum velocity. As the reaction proceeds the concentration of the intermediate substrate will increase which will, in turn, allow the coupling enzyme to work faster. Thus the rate of the coupling reaction will increase with time until it equals the rate of the reaction catalysed by the first enzyme. At this stage the concentration of the intermediate substrate will remain constant because of a balance between the rate of its formation, by the enzyme under study, and the rate of its removal by the coupling enzyme. This behaviour of coupled enzyme assays is illustrated in *Figure 12*. It results in a lag in the rate of formation of the product of the coupled reaction. It is necessary to minimize this lag period, which is often referred to as the coupling-time, because of the possibility that the reaction catalysed by the first enzyme will have started to slow down (Section 2.1) before the coupling enzyme has reached its steady-state velocity. In that case the coupled assay would never give an accurate measure of the activity of the enzyme under study.

The efficiency with which a coupling enzyme can function will depend on its $K_m$ value for the substrate being formed. The lower its $K_m$ value, the more efficiently it will be able to work at low substrate concentrations. The lag period can also be reduced by increasing the amount of the coupling enzyme present so that it can catalyse the reaction more rapidly at low substrate

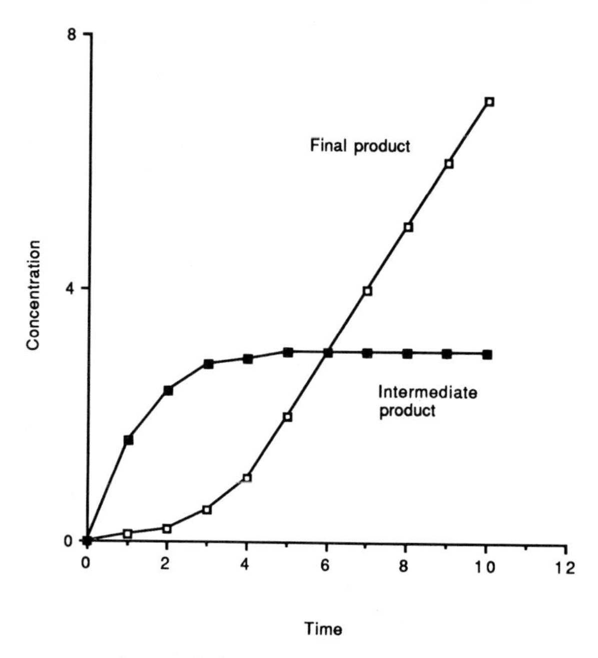


**Figure 12.** Time-course of a coupled enzyme assay involving a single coupling enzyme. The lag in the appearance of the product formed by the coupling enzyme corresponds to the period in which the concentration of the intermediate product, which is the product of the reaction catalysed by the first enzyme and the substrate for the coupling enzyme, rises to a constant (steady-state) concentration.

concentrations. The higher the $K_m$ value of the coupling enzyme, the greater the amount of it will be required to produce the same lag period. Thus it is necessary to characterize the performance of a coupled assay to ensure that it gives an accurate measure of the activity of the enzyme under study. Usually this can be done experimentally by checking that the measured velocity is not increased by increasing the amount of the coupling enzyme present, and that it is proportional to the amount of the first enzyme present at all substrate concentrations and under all conditions that are to be used. Generally, this is achieved by having a very large excess of the coupling enzyme(s) present. It is possible to calculate the amount of a coupling enzyme that must be added to give any given coupling time (102) and such calculations may be useful in saving the expense of adding too much reagent. It must be remembered, however, that it will be necessary to re-check that a coupled assay is performing correctly each time the assay conditions are altered since these may affect the behaviour of the coupling enzyme(s). The purity of the coupling enzyme(s) used should also be checked. Since these are used in relatively high concentrations, even a small degree of contamination with other enzymes or substrates that might affect the reaction under study could become important.

A rather unusual type of coupled assay is the cycling assay for alcohol dehydrogenase activity (105). In this reaction, shown in *Figure 13,* the NADH formed by the oxidation of ethanol is used to reduce lactaldehyde to propanediol in a reaction catalysed by the same enzyme. In this reaction NAD(H) functions catalytically and one mol of propanediol is produced per mol of ethanol oxidized. The rates of such cycling assays can be much faster than conventional assays because the rate-limiting step in the reaction catalysed by this enzyme is the dissociation of the product NADH, which does not need to occur in the cycling system. Because the coenzyme functions catalytically it is only necessary for it to be present at low concentrations and this minimizes competition from other enzymes that use NAD(H] (Section 2.5.3). The cycling assay illustrated in *Figure 13* is performed discontinuously by measuring the amount of propanediol formed chemically after stopping the reaction. However, if the lactaldehyde is replaced by an aldehyde which undergoes an absorbance change on reduction to the corresponding alcohol, a continuous assay is possible (106). Cycling assays of this type have also been used as the basis of sensitive methods for determining metabolite concentrations (90).

Another type of coupled assay is the, so-called, forward-coupled assay. In this system the enzyme that is used to catalyse the reaction leading to a measurable product catalyses a reaction that provides the substrate for the enzyme whose activity is being determined. An example of this is the assay of citrate synthase (EC 4.1.3.7) using malate dehydrogenase (EC 1.1.1.37) as the coupling enzyme (107). This is illustrated in *Figure 14.* If the reaction catalysed by the forward-coupling enzyme, malate dehydrogenase, is at equilibrium, removal of oxaloacetate by citrate synthase should result in an

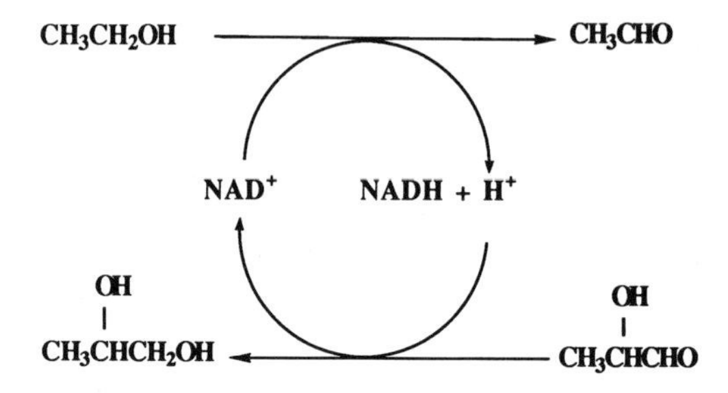


**Figure 13.** A cycling assay for alcohol dehydrogenase.

increase in NADH concentration as the equilibrium is re-established, and this may be used to follow the reaction.

The behaviour of such systems can be complex and the appearance of NADH may not be simply related to rate of the reaction by the enzyme under study (108). This can be illustrated by considering the simple case where the forward coupling enzyme has only a single substrate (A) and product (B) and the latter serves as a substrate for the enzyme under study. This can be represented by the simple scheme:

$$\mathrm{A} \overset{K}{\rightleftharpoons} \mathrm{B} \rightarrow \mathrm{C} \tag{23}$$

where the disappearance of A is used as a measure of the activity of the enzyme that converts B to C. In this system the formation of C will result in a decrease in the concentration of B which will, in turn, be compensated by a decrease in the concentration of A as the equilibrium of the first reaction is maintained. The concentration of C formed will thus be given by the sum of the decreased in the concentrations of A and B.

$$-(\Delta[\mathrm{A}] + \Delta[\mathrm{B}]) = \Delta[\mathrm{C}] \tag{24}$$

Since the equilibrium of the forward coupling reaction will be given initially by $K$=[B]/[A], after the formation of some of the product C this will become:

$$K = \frac{([\mathrm{B}] - \Delta[\mathrm{B}])}{([\mathrm{A}] - \Delta[\mathrm{A}])} \tag{25}$$

Thus the relationship between the formation of C and the decrease in the concentration of A will depend on the value of the equilibrium constant $K$. If $K = 1$, $\Delta[\mathrm{A}] = \Delta[\mathrm{B}]$ and therefore $\Delta[\mathrm{A}] = \Delta[\mathrm{C}]/2$. If the equilibrium position is far in the direction of B the decrease in the concentration of A will only be a small fraction of the amount of C formed; for example if $K = 100$, $\Delta[\mathrm{B}]/\Delta[\mathrm{A}]$ will be equal to this value and the change in the concentration of A will only be 1% of the increase in [C]. Only when the equilibrium position is far over in the direction of A will the decrease in its concentration approach the increase in the concentration of C. At an equilibrium constant of 0.01, for example, the relationship will give $\triangle[C] = 99\% \triangle[A]$.

This analysis may be extended to enzymes involving the interaction between two substrates (108), such as the use of malate dehydrogenase as the forward-coupling enzyme for the assay of citrate synthase (*Figure 14*). In this case the relationship between NADH formation and citrate production will be more complex, since the equilibrium concentration of oxalate initially present will depend on the concentrations of the other reactants. The full equations for this system show that, at any fixed pH value, the formation of NADH will approach that of citrate only when the initial concentration of the former compound is sufficiently high for the initial malate dehydrogenase equilibrium to be far over towards malate (108).

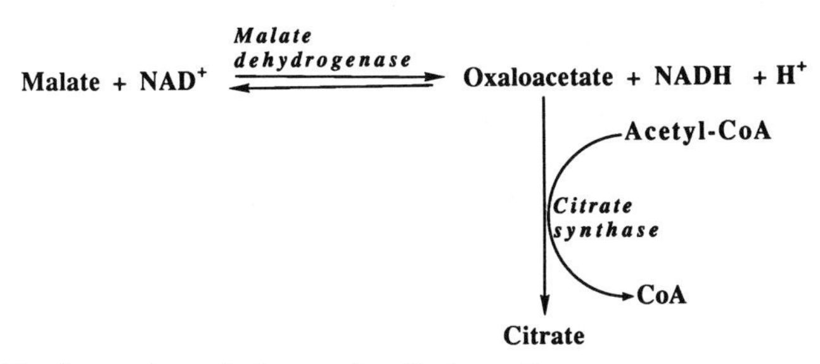


**Figure 14.** The forward-coupled assay for citrate synthase.

An assay system of this type can be useful if the substrate for the enzyme under study is unstable but can be formed in the assay mixture by the action of another enzyme. Such a system in shown in *Figure 15* where the $N^{\tau}$-methylimidazole acetaldehyde is generated as a substrate for aldehyde dehydrogenase by the action of amine oxidase (EC 1.4.3.6) on $N^{\tau}$-methylhistamine (109). If the formation of NADH is monitored under conditions where there is an appreciable lag-phase in the progress curve (see *Figure 12)*, it is possible to determine the $K_m$ and maximum velocities for aldehyde dehydrogenase by

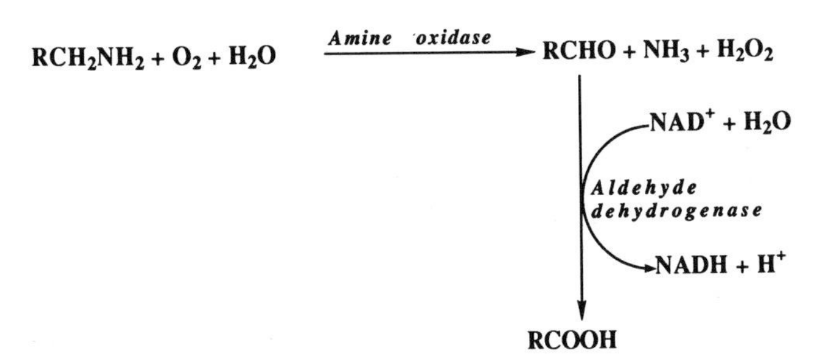


**Figure 15.** The use of an amine oxidase to generate substrate for the reaction catalysed by aldehyde dehydrogenase.

analysis of this approach to steady-state conditions [109]. Under appropriate conditions, the analysis of the progress curves for coupled assay of this type may allow the kinetic parameters of both enzymes involved to be determined [110]. It is important to remember that, as with other procedures for the analysis of reaction progress curves, the precautions and limitations discussed in Section 2.1 must be taken into account if valid results are to be obtained.

### 6.1.3 Automated assay procedures

There are a number of procedures that automate the assay of enzymes in order to allow large numbers of samples to be assayed rapidly and efficiently. Many of these involve the determination of the product formation after a fixed time from the start of the reaction. It will therefore be necessary to ensure that the values obtained represent a true reflection of the initial rate of the reaction. The use of flow-systems involving multi-channel pumps to mix reactants and determine the product formation after a fixed time have been discussed in detail [111] and the basic principles have not changed significantly since then. Immobilized enzymes may also be used to determine product formation by automated procedures (e.g. 112). Flow systems may also be used where the detection system has a slow response. A classical example of this is the work of Roughton and Rossi-Bernardi [113] who were able to make measurements of the extent of a reaction at times considerably shorter than the response-time of a $CO_2$ electrode by using a flow system in which the 'age' of the reaction mixturc was constant as it flowed past the detector.

Many modern spectrophotometers can be equipped with multiple cuvette holders which allow several reactions to be followed by determining the absorbance in each sequentially and repetitively. Typically four or six samples, with appropriate blanks may be studied in this way. Micro-titre plate readers with appropriate wavelength selection and temperature control offer the possibility of assaying many more samples and, by taking rapid multiple readings of the absorbance in each sample well, reaction progress curves may be determined for them all. The centrifugal fast analyser offers an alternative method for determining the time-courses of samples and has been applied to the assay of several enzymes (e.g. 114).

A rather different application of automated assay procedures is to use a pumping system to generate a linearly-increasing gradient of substrate concentration which is then mixed with enzyme and the extent of reaction is measured after a fixed time (115, 116). Such a procedure allows the curve for the dependence of velocity upon substrate concentration to be generated in a single experiment. Since this procedure determines the extent of reaction after a fixed time it is necessary to ensure that true initial-rate conditions apply at all substrate concentrations. However, it is relatively easy to check this by stopping the flow of enzyme and substrates and observing the time-course of the reaction as the mixture 'ages' in the detector (115).

## 6.2 Choice of assay method

There is often a variety of different assay methods available for an enzyme. Provided that adequate controls are carried out to ensure that they do in fact determine the initial velocity of the reaction and that the measurements are free from the potential artefacts discussed earlier, the choice may simply depend on convenience and the availability of the appropriate materials and apparatus. However, there are some general considerations that may be helpful in choosing between alternative methods.

In the assay of enzymes with high $K_m$ values towards their substrates, it may be necessary to work at high substrate concentrations to ensure a sufficient period of linearity of the reaction progress curve. In such cases it may be more accurate to follow the reaction by determining the extent of product formation rather than the disappearane of substrate, since the latter would involve measurement of decreases from very high initial values.

Many assay types impose some limitations on the assay conditions. This may be due to substrate instability under some conditions, such as the breakdown of NAD(P)H at acid pH values, which makes it difficult to estimate the rates of the enzyme-catalysed reaction accurately, or the requirement for alkaline conditions to observe the liberation of *p*-nitrophenol spectrophotometrically. In coupled assays one of the products is continuously removed and thus such methods will not be suitable for studies on the inhibition of the enzyme activity by that product. Other assays that involve determination of the formation of a specific product may become less practicable if large amounts of that product are added for inhibition studies. In coupled assays it must always be remembered that any change in the assay conditions may affect the behaviour of the coupling enzyme(s) as well as those of the enzyme under study.

The above considerations mean that it is often necessary to use more than one type of assay in a complete study of the behaviour of an enzyme. During the purification of an enzyme it is convenient to have a procedure that can rapidly give an estimate of the activity. This may make direct assay procedures more useful than discontinuous ones which involve time-consuming separation or detection procedures. However, the possibility that contaminating activities in impure tissue preparations may prevent the accurate application of some procedures (Sections 2.5.1 *vi* and 2.5.3) can also affect the choice under these conditions.

Many discontinuous assay procedures appear to be particularly attractive because, once a reliable method has been obtained, it is often possible to perform a large number of incubations at the same time. However, it is essential to remember the necessity of ensuring that the procedure is giving a true measure of the initial rate of the reaction, by determining the time-course of the reaction by stopping it at different times, either after removal of samples from a single assay mixture or by treating individual incubation samples, for determination of the extent of the reaction. It will be necessary to repeat such experiments each time the assay conditions are changed.

## 6.3 The effects of pH

Many studies of the behaviour of enzymes involve investigations of the effects of variations of pH on their activities. Correctly designed and analysed studies of this type can yield a great deal of valuable information about the mechanisms involved in the catalytic process. Space does not permit a detailed consideration of the effects of pH on enzyme activity and the reader is referred to more complete treatments (117–119). Other studies are simply aimed at determining the optimum pH of the reaction. In all studies on the effects of pH it is important to show that the effects are reversible. This can usually be achieved by adjusting the pH to different values and incubating under the assay conditions in the absence of one of the substrates for the period of an assay before readjusting to a value where the enzyme is known to be stable and determining the activity. Many enzymes are unstable at the more extreme pH values and some may precipitate. Clearly, any data obtained in regions where the enzyme is rapidly losing activity may be difficult to interpret. A more detailed consideration of buffers and ionic strength is given in Chapter 11.

## 6.4 Practical considerations

It is not possible to cover all the factors that may affect the performance and validity of enzyme assays and this section will be restricted to a general consideration of some of the more important aspects.

### 6.4.1 Enzyme nature and purity

The state of purity of an enzyme preparation affects the ease with which it may be studied. Some optical assays may be difficult to perform accurately with impure preparations which contain large amounts of absorbing, and possibly particulate, material. Furthermore, the presence of contaminating activities (Sections 2.5.1 *vi* and 2.5.3) or substrates may make it difficult to perform certain assays, and in studies on the inhibition by products it is also necessary to ensure the absence of contaminating enzymes that might react with the added product. Such problems often make it necessary to purify an enzyme at least partly in order to obtain reliable data on its activity.

Membrane-bound enzymes pose a particular problem since removal of the enzyme from its membrane environment can lead to changes in properties ranging from complete loss of activity (120) to changes in the kinetic mechanism obeyed (121). It has been pointed out that the surface charge on the membranes can have profound effects on the accessibility of charged substrates to the active sites of enzymes associated with them (122).

If a published assay procedure cannot be made to work, it is possible that the enzyme under study has not been identified correctly. There is considerable scope for confusion since there are many cases where more than one enzyme will catalyse a similar reaction; for example, the amine oxidases EC 1.4.3.4 and EC 1.4.3.6 catalyse similar reactions but have very different substrate specificities, and the two carbamoyl phosphate synthetases (EC 6.3.5.5 and EC 6.4.3.16) differ in the source of ammonia that they can use. There are several examples in the literature where such pairs of enzymes have been confused. The use of the EC nomenclature can help to avoid such confusion, but it should also be remembered that there may be considerable species differences and differences in the specificities of isoenzymes. Thus, for example, the standard assay of alcohol dehydrogenase with ethanol as substrate will not detect some of its isoenzymes, since they have very low activities towards that alcohol (123).

### 6.4.2 Enzyme stability

In any series of studies with an enzyme preparation it is essential to assay its activity at regular intervals under standard conditions to check that it remains constant. If an enzyme is unstable and steadily loses activity with time, it may be possible to correct all the values obtained to those that existed at the start of the studies if a series of standard assays have been carried out at different times to allow a calibration curve of activity against time to be constructed. It is a much better procedure, however, to try to find conditions under which the stock enzyme solutions may be stored without appreciable loss of activity over the time involved (see 124 for a discussion of enzyme stability). Blank rates (Section 2.5) may develop during enzyme storage and it is important to check for their appearance.

### 6.4.3 The substrates

It is essential that pure substrates are used. The presence of contaminants will lead to incorrect estimates of the concentrations of solutions prepared by weight and it will also lead to errors in the determination of kinetic constants if any of the impuritics is inhibitory. If a substrate is contaminated by a competitive inhibitor, the inhibition equation can be written as:

$$v = \frac{V_{max}}{1 + [(K_m/[S])(1 + [I]/K_i)]} \tag{26}$$

where $K_i$ is the inhibitor constant (6). If the inhibitor is present in the substrate solution its concentration will be some fixed proportion (x) of that of the substrate. Thus we can write

$$[I] = x[S]$$

Substituting x[S] for [I] in equation (25) gives:

$$v = \frac{V_{max}}{1 + (K_m/[S])[1 + (x[S]/K_i)]} = \frac{V_{max}}{1 + (K_m/[S]) + xK_m/K_i)} \tag{27}$$

This will result in a proportional decrease in both $K_m$ and $V_{max}$ as shown in *Figure 16,* curve (a) (125, 126). Several stereospecific enzymes are competitively inhibited by other stereoisomers of the substrate and thus the use of racaemic mixtures, rather than the correct enantiomer, may yield incorrect estimates of both $V_{max}$ and $K_m$ values. If a D, L-substrate is used with an enzyme that is only active towards one of the enantiomers, it cannot be assumed that the effective substrate concentration can be taken as one half of that prepared by weight. This analysis has been extended to cases where the inhibitory contaminant is an uncompetitive or a non-competitive (mixed) inhibitor of the enzyme (127). In the uncompetitive case the relationship becomes:

$$v = \frac{V_{max}}{1 + (K_m/[S]) + (x[S]/K_i)} \tag{28}$$

This equation predicts that the $K_m$ and $V_{max}$ values will not be affected by the contaminant, provided that the true substrate concentration is known, but that there will be apparent inhibition at high substrate concentrations, as shown in *Figure 16*, curve (b). Since non-competitive (mixed) inhibition can be regarded as a combination of competitive and uncompetitive effects, the contaminant would result in a decrease in both $K_m$ and $V_{max}$ values as well as the appearance of inhibition at high substrate concentrations (see Equations 27 and 28), as shown in *Figure 16*, curve (c).

The purity of substrate solutions should be checked chromatographically or enzymically (2, 90). If the latter procedure is used a discrepancy between the enzymically-determined substrate concentration and that calculated on a weight basis may indicate whether purification is necessary. The stability of the substrate solutions is also important. Many biochemicals are relatively

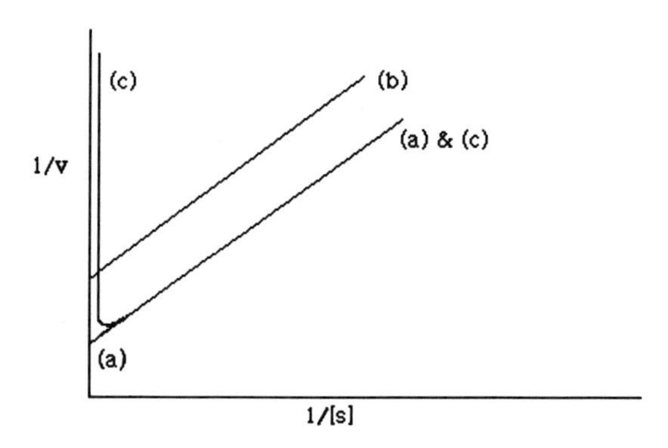


**Figure 16.** The possible effects of inhibitory contaminants in the substrate solution. Curve (a) shows the behaviour in the absence of the contaminant, curve (b) would result if the contaminant were a competitive inhibitor, and curve (c) would result if it were an uncompetitive inhibitor. The behaviour of a mixed or non-competitive inhibitor would have the features of both curves (b) and (c).

unstable and their breakdown would lead to changes in the true substrate concentrations. Thus it is important to use freshly-prepared substrate solutions and to assay their concentrations at intervals to check for any changes. Since the breakdown of substrates in solution may give rise to the formation of inhibitors, it may not be satisfactory simply to correct for changes in substrate concentration with time by using data on the rate of its decomposition. The storage conditions can affect the stabilities of substrate solutions; data on this and the most satisfactory storage conditions for a number of commonly-used substrates have been listed (2, 90) and are also available in the literature from a number of suppliers of biochemicals.

Although it may be relatively simple to purify substrates that are available in large amounts, the small quantities of radioactively-labelled substrates used in many radiochemical assays together with their high cost often leads to the possibility that they may contain impurities being ignored. One way of testing for the presence of kinetically-significant contaminants in radiochemicals is to compare the kinetic behaviour obtained from assays where the radioactive substrate is varied as a fixed proportion of the unlabelled substrate (constant specific radioactivity) with that observed when the amount of radioactivity is kept constant whilst the total substrate concentration is varied (constant radioactivity (128).

### 6.4.4 Solvents and buffers

It is essential to ensure that all solvents used in enzyme assays are free from contaminating material that could affect the activity of the enzyme. Heavy metals are inhibitors of many enzymes and great care is necessary to exclude them; for example, the persistent claim that EDTA was a specific activator of fructose-1, 6-bisphosphatase was eventually shown to be merely an effect of this compound chelating heavy-metal contaminants (129). Some substrates are not readily soluble in water and are added to the enzyme assay in solution in an organic solvent. It is necessary in such cases to carry out adequate control experiments to check that the solvent itself has no effect on the activity of the enzyme under study or on the assay method being used. In any enzyme assay it is important to check that the addition of the other components does not affect the pH of the assay. It is often convenient to adjust the pH of enzyme and substrate solutions to the same value as that of the assay buffer, but this may not always be possible if any of these are not stable for long periods at that value. As discussed in Chapter 11, the choice of buffer solution may be of great importance.

### 6.4.5 Assay mixtures

For assay mixtures that contain a number of different components it may be possible to make up a bulk mixture containing all of them except the enzyme to be assayed. Such a 'cocktail' can be particularly useful if a number of assays are to be performed under identical conditions, such as during enzyme purification. If such a cocktail gives rise to a blank rate, it will of course be necessary to find out the components responsible by using incomplete mixtures. More seriously, the use of such cocktails does not allow one to check for blank rates involving the enzyme in an incomplete reaction mixture (Section 2.5). Thus it is necessary to carry out adequate controls to test for this before relying on such an assay cocktail.

Although the use of assay cocktails can save time, it must be remembered that the different components may differ in their stabilities; for example, one of the substrates or an enzyme involved in a coupled assay may decay more rapidly than the other components thus limiting the useful lifetime of the entire mixture. The decay of one of the components of an assay cocktail may lead to the formation of inhibitory products so it may not be possible to 'revitalize' the mixture by adding more of the component that has degraded. Thus, if a cocktail is to be used, it is preferable to pre-mix only those components that are stable under the storage conditions in order to avoid the expense of having to discard it before it has all been used. In the case of coupled assays, if no activity of the enzyme under study can be detected it will be possible to check that the coupling system is working adequately by seeing whether there is a response when the product that is being determined is added to the system.

### 6.4.6 Mixing

If there are no significant blank rates in any of the incomplete reaction mixtures, the choice of whether to start the assay by the addition of enzyme or one of the substrates may not be important. If the enzyme is unstable in the assay mixture it would be preferable to start the reaction by adding it last. In other cases, however, it might be preferred to pre-incubate the enzyme in the incomplete assay mixture and start by the addition of one of the substrates. This, for example, might be the case if there were hysteretic effects resulting from the dilution of the enzyme. It is often necessary to store the enzyme preparation and substrates in ice to ensure their stability. Thus, it is important to keep the final volume of material used to start the reaction as small as possible to avoid too great a fall in the temperature of the assay mixture when it is added (Section 2.4.1). If the effects of varying the concentrations of enzyme are to be studied it may be possible to keep the volumes of these added so low in relation to that of the total assay mixture that the differences between the volumes added are negligible. However, if this is not possible, it will be necessary to adjust the volume of the assay mixture to a constant final value in each case using the same buffer as that in which the varied enzyme or substrate is contained.

It is essential to ensure that the components are properly mixed in the assay mixture. However, some enzymes are denatured by too vigorous shaking. With an assay vessel that is not mechanically stirred or shaken it is usually possible to ensure adequate mixing by covering the top with parafilm and inverting the tube or cuvette a few times. This can be achieved relatively quickly, but more rapid and just as effective mixing can be achieved by using a stirrer made from a small piece of plastic that has been bent over and flattened at one end. A few vertical strokes of the stirrer should be adequate and in spectrophotometric or fluorimetric assays it is possible to do this without removing the cuvette from the apparatus. If the volume of the material used to start the reaction is relatively small it may be possible to place it on the flattened end of the stirrer so that addition and stirring can be performed in a single operation. It is a simple matter to make such stirrers in the laboratory although they are also available commercially. If it is necessary to start a reaction by adding a component to a blank and reaction cuvette simultaneously, it is possible to mix them both at once using a stirrer made from a U-shaped plastic rod flattened at both ends. Particular care in mixing is necessary if the enzyme is in a dense medium, such as glycerol or sucrose solution, which are sometimes used to enhance stability (124).

## 6.5 Conclusions

As stated earlier, it is not possible to recommend the ideal type of assay method to use since this will depend on the nature of the enzyme being studied, its degree of purity, and the purpose of the assays. Indeed, it may prove necessary to use more than one assay method in a complete study of an enzyme if, for example, product inhibition is to be studied and a coupled assay involving that product has been routinely used, or if the assay of choice cannot be used during the early stages of the purification.

It has not been possible to cover all aspects of enzyme assay and kinetics in the space available for this review, but references given in the appropriate section of the text should fill any gaps. The subject of enzyme inhibition has been treated in detail in several works (4, 6) and an account of some of the pitfalls and problems in such studies has been presented (130). This account has concentrated on the difficulties that can be encountered with enzyme assays, because the results from invalid determinations are of no use to anyone. By taking care to validate the assay procedures used, under all conditions of the studies, it is possible to ensure that meaningful results are obtained.


# Acknowledgements

I am grateful to Dr Michael Danson and Dr Robert Eisenthal for their critical comments and many helpful suggestions.


# References


1. Colowick, S. P., Kaplan, N. O., and others (1955 *et seq.*). *Methods in Enzymology*. Vol. 1 and continuing, Academic Press Inc., New York.

2. Bergmeyer, H. U. (1983). *Methods of Enzymatic Analysis*, Vol. 1–10, (3rd edn). Verlag Chemie, Weinheim.
3. Barman, T. E. (1969 and 1974). *Enzyme Handbook*. Springer-Verlag, Heidelberg.
4. Segel, I. H. (1975). *Enzyme Kinetics*. Wiley-Interscience Inc., New York.
5. Tipton, K. F. (1974). In: *Companion to Biochemistry* (ed. A. T. Bull, J. R. Lagnado, J. O. Thomas, and K. F. Tipton), p. 327. Longman Ltd, London.
6. Dixon, M. and Webb, E. C. (1979). *Enzymes* (3rd edn). Longman Ltd, London.
7. Cleland, W. W. (1970). In *The Enzymes*, (ed. P. D. Boyer), Vol. 2, p. 1. Academic Press Inc., New York.
8. Seiler, N., Jung, M. J., and Koch-Weser, J. (1978). *Enzyme Activated Irreversible Inhibitors*. Elsevier–North Holland BV, Amsterdam.
9. Palfreyman, M. G. (1987). In *Essays in Biochemistry*, (ed. R. D. Marshall and K. F. Tipton), Vol. 23, p. 28. Academic Press Ltd, London.
10. Tipton, K. F. (1980). In: *Enzyme Inhibitors as Drugs*, (ed. M. Sandler), p. 1. Macmillan Ltd, London.
11. Silverman, R. B. and Hoffman, S. J. (1984). *Med. Res. Rev.*, **44,** 415.
12. Tipton, K. F. (1989). In: *Design of Enzyme Inhibitors as Drugs*, (ed. M. Sandler and H. J. Smith), p. 70. Oxford University Press.
13. Waley, S. G. (1985). *Biochem. J.*, **277,** 843.
14. Tatsunami, S., Yago, M., and Hosoe, M. (1981). *Biochim. Biophys. Acta,* **662,** 226.
15. Tipton, K. F., Fowler, C. J., McCrodden, J. M., and Strolin Benedetti, M. (1983). *Biochem. J.*, **209,** 235.
16. Tipton, K. F., McCrodden, J. M., and Youdim, M. B. H. (1986). *Biochem. J.*, **240,** 379.
17. Kinemuchi, H., Arai, Y., Oreland, L., Tipton, K. F., and Fowler, C. J. (1982). *Biochem. Pharmacol.*, **31,** 959.
18. Donovan, J. W. (1973). In: *Methods in Enzymology*, (ed. C. W. Hirs and S. N. Timasheff), Vol. 27, p. 497. Academic Press Inc., New York.
19. Undenfriend, S. (1962). *Fluorescence Assay in Biology and Medicine*. Academic Press Inc., New York.
20. Wharton, C. W. (1983). *Biochem. Soc. Trans.*, **11,** 817.
21. Waley, S. G. (1981). *Biochem. J.*, **193,** 2009.
22. Nicholls, R. G., Jerfy, M., and Roy, A. B. (1974). *Anal. Biochem.*, **61,** 93.
23. Orsi, B. A. and Tipton, K. F. (1979). In: *Methods in Enzymology*, (ed. D. L. Purich), Vol. 63, p. 159. Academic Press Inc., New York.
24. Wharton, C. W. and Szawelski, R. J. (1982). *Biochem. Soc. Trans.*, **10,** 233.
25. Selwyn, M. J. (1965). *Biochim. Biophys. Acta*, **105,** 193.
26. Tipton, K. F. and Fowler, C. J. (1984). In: *Monoamine Oxidase and Disease* (ed. K. F. Tipton, P. Distert, and M. Strolin Benedetti), p. 27. Academic Press Ltd, London.
27. Cha, S. (1976). *Biochem Pharmacol.*, **25,** 1561 and 2695.
28. John, R. A. (1985). In: *Techniques in the Life Sciences,* (ed. K. F. Tipton). Vol. BI/II, Supplement BS 118, p. 1. Elsevier Ltd, Ireland.
29. Hiromi, K. (1980). In: *Methods of Biochemical Analysis*, (ed. D. Glick), Vol. 26, p. 137. Wiley, New York.
30. Hartley, B. S. and Kilby, B. A. (1954). *Biochem. J.*, **56,** 288.

31. O'Fagain, C., Bond, U., Orsi, B. A., and Mantle, T. J. (1982). *Biochem. J.*, **201,** 345.
32. Söling, H. J., Bernhard, G., Kuhn, A., and Luck, H. J. (1977). *Arch. Biochem. Biophys.*, **182,** 563.
33. Cronin, C. N. and Tipton, K. F. (1985). *Biochem. J.*, **227,** 113.
34. Rauschel, F. M. and Cleland, W. W. (1977). *Biochemistry,* **16,** 2169.
35. Frieden, C. (1970). *J. Biol. Chem.*, **245,** 5788.
36. Kurganov, B. I., Dorozhko, A. I., Kagan, Z. S., and Yakovlev, V. A. (1976). *J. Theor. Biol.*, **60,** 247, 271, and 287; and **61,** 531.
37. Neet, K. E. and Ainslie, G. R. (1980). In: *Methods in Enzymology,* (ed. D. L. Purich), Vol. 64, p. 192. Academic Press Inc., New York.
38. Frieden, C. (1979). *Ann. Rev. Biochem.*, **48,** 471.
39. Williams, D. C. and Jones, J. G. (1976). *Biochem. J.*, **155,** 661.
40. Nimmo, G. A. and Tipton, K. F. (1981). *Biochem. Pharmacol.*, **30,** 1635.
41. Allanson, S. and Dickinson, F. M. (1984). *Biochem. J.*, **223,** 163.
42. Dickinson, F. M. and Allanson, S. (1985). In: *Enzymology of Carbonyl Metabolism*, (ed. T. G. Flynn and H. Weiner), Vol. 2, p. 71. A. R. Liss Inc., New York.
43. Pietruszko, R., Ferenca-Biro, K., and McKerrell, A. D. (1985). In: *Enzymology of Carbonyl Metabolism*, (ed. T. G. Flynn and H. Weiner), Vol. 2, p. 29. A. R. Liss Inc., New York.
44. Bates, D. J. and Frieden, C. (1973). *J. Biol. Chem.*, **248,** 7878 and 7885.
45. Duncan, R. J. S. and Tipton, K. F. (1971). *Eur. J. Biochem.*, **22,** 257.
46. Estabrook-Smith, S. B., Hudson, P. J., Gross, N. H., Keech, D. B., and Wallace, J. C. (1976). *Arch. Biochem. Biophys.*, **171,** 709.
47. Hulme, E. C. and Tipton, K. F. (1971). *FEBS Lett.*, **12,** 197.
48. Cornish-Bowden, A. (1979). *Fundamentals of Enzyme Kinetics*, Butterworths Ltd, London.
49. Wong, J. T-F. (1975). *Kinetics and Enzyme Mechanism,* Academic Press Inc., New York.
50. Fromm, H. J. (1975). *Initial Rate Enzyme Kinetics*, Springer-Verlag, Heidelberg.
51. Roberts, D. V. (1977). *Enzyme Kinetics*, Cambridge University Press, Cambridge.
52. Dalziel, K. (1957). *Acta Chem. Scand.*, **11,** 1706.
53. Storer, A. C. and Cornish-Bowden, A. (1976). *Biochem. J.*, **159,** 1.
54. Monod, J., Wyman, J., and Changeux, J. P. (1966). *J. Mol. Biol.*, **12,** 88.
55. Whitehead, E. P. (1970). *Prog. Biophys. Mol. Biol.*, **21,** 323.
56. Tipton, K. F. (1979). In: *Companion to Biochemistry*, (ed. A. T. Bull, J. R. Lagnado, J. O. Thomas, and K. F. Tipton), p. 327. Longman Ltd, London.
57. Hill, A. V. (1910). *J. Physiol.* (London), **40,** 4.
58. Hearon, J. Z., Bernhard, S. A., Friess, S. L., Botts, D. J., and Morales, M. F. (1959). In: *The Enzymes* (ed. P. D. Boyer, H. A. Lardy, and K. Myrbäck), (2nd edn), p. 49. Academic Press Inc., New York.
59. Ferdinand, W. (1966). *Biochem. J.*, **98,** 273.
60. Sweeny, J. F. and Fisher, J. R. (1968). *Biochemistry,* **7,** 561.
61. Whitehead, E. P. (1976). *Biochem. J.*, **159,** 449.
62. Bardsley, W. G. and Childs, R. E. (1975) *Biochem. J.*, **149,** 313.
63. Wong, J. T-F., Gurr, P. A., Bronskill, P. M., and Hanes, C. S. (1972). In:

*Analysis and Simulation of Biochemical Systems*, (ed. H. C. Hemker and B. Hess), p. 327. North Holland BV, Amsterdam.
64. Williams, D. C., Morgan, G. S., McDonald, E., and Battersby, A. R. (1981). *Biochem. J.,* **193,** 301.
65. Elliott, K. R. F. and Tipton K. F. (1974). *Biochem. J.,* **141,** 807.
66. Rabin, B. R. (1967). *Biochem. J.,* **102,** 22C.
67. Frieden, C. (1967). *J. Biol. Chem.,* **242,** 4045.
68. Ainslie, G. R., Shill, J. P., and Neet, K. E. (1972). *J. Biol. Chem.,* **247,** 7088.
69. Ricard, J., Meunier, J. C., and Buc, J. [1974). *Eur. J. Biochem.,* **49,** 195.
70. Harding, W. M. (1969). *Arch. Biochem. Biophys.,* **129,** 57.
71. Blair, J. McD. (1969). *FEBS Lett.,* **2,** 245.
72. Purich, D. L. and Fromm, H. J. (1972). *Biochem. J.,* **130,** 63.
73. Ainsworth, S. (1977). *Steady-state Enzyme Kinetics*, p. 62. Macmillan, London.
74. O'Sullivan, W. J. and Smithers, G. W. (1979). In: *Methods in Enzymology,* (ed. D. L. Purich), Vol. 63, p. 294. Academic Press Inc., New York.
75. Good, N. E. and Izawa, S. (1972). In: *Methods in Enzymology* (ed. A. San Pietro), Vol. 24, p. 53. Academic Press, New York.
76. Tipton, K. F. (1985). In: *Techniques in the Life Sciences*, (ed. K. F. Tipton), Vol. B1/II, Supplement BS 113, p. 1, Elsevier Ltd, Ireland.
77. Morrison, J. F. (1979). In: *Methods in Enzymology*, (ed. D. I. Purich), Vol. 63, p. 257. Academic Press Inc., New York.
78. Cronin, C. N. and Tipton, K. F. (1987). *Biochem. J.,* **247,** 41.
79. Uotila, L. and Mannervik, B. (1979). *Biochem. J.,* **177,** 869.
80. Dixon, H. B. F. and Tipton, K. F. (1973). *Biochem. J.,* **133,** 837.
81. Burns, D. J. W. and Tucker, S. A. (1977). *Eur. J. Biochem.,* **81,** 45.
82. Spears, G., Sneyd, J. T., and Loten, E. G. (1971). *Biochem. J.,* **125,** 1149.
83. Denizeau, F., Wyse, J., and Sourkes, T. L. (1976). *J. Theor. Biol.,* **63,** 99.
84. Dickinson, F. M. (1986). *Biochem. J.,* **238,** 75.
85. Williams, J. W. and Morrison, J. F. (1979). In: *Methods in Enzymology,* (ed. D. L. Purich), Vol. 63, p. 437. Academic Press Inc., New York.
86. Henderson, P. J. F. (1973). *Biochem. J.,* **135,** 101.
87. Teipel, J. and Koshland, D. E. (1989). *Biochemistry,* **8,** 4656.
88. Dalziel, K. and Engel, P. C. (1968). *FEBS Lett.,* **1,** 349.
89. Cornish-Bowden, A. (1988). *Biochem. J.,* **250,** 309.
90. Lowry, O. H. and Passonneau, J. V. (1972). *A Flexible System of Enzymatic Analysis*. Academic Press Inc., New York.
91. Jackobson, C. F., Leonis, J., Linderstrom-Lang, K., and Ottesen, M. (1957). *Meth. Biochem. Anal.,* **4,** 171.
92. Lessler, M. A. and Vrierley, G. P. (1969). *Meth. Biochem. Anal.,* **17,** 1.
93. Youdim, M. B. H. and Tenne, M. (1987). In: *Methods in Enzymology* (ed. S. Kaufman), Vol. 142, p. 617. Academic Press Inc., New York.
94. Ulitzer, S. and Hastings, J. W. (1978). In: *Methods in Enzymology* (ed. M. A. De Luca), Vol. 57, p. 189. Academic Press Inc., New York.
95. Hastings, J. W. (1978). In: *Methods in Enzymology*, (ed. M. A. De Luca), Vol. 57, p. 125. Academic Press Inc., New York.
96. Mulkerrin, M. G. and Wampler, J. E. (1978). In: *Methods in Enzymology* (ed. M. A. De Luca), Vol. 57, p. 375. Academic Press Inc., New York.

97. Campbell, A. K. (1989). In: *Essays in Biochemistry*, (ed. R. D. Marshall and K. F. Tipton), Vol. 24, p. 41. Academic Press Ltd, London.
98. Lebel, D., Poirier, G. G., and Beaudoin, A. R. (1978). *Anal. Biochem.*, **85,** 86.
99. Kaplan, N. O., Colowick, S. P., and Barnes, C. C. (1951). *J. Biol. Chem.*, **191,** 461.
100. Fritz, I. B., Schultz, S. K., and Srere, P. A. (1963). *J. Biol. Chem.*, **238,** 2509.
101. Ellman, G. L., Courtney, K. D., Andres, V., and Featherstone, R. M. (1961). *Biochem. Pharmacol.*, **7,** 88.
102. Rudolph, F. B., Baugher, B. W., and Beissner, R. S. (1979). In: *Methods in Enzymology*, (ed. D. L. Purich), Vol. 63, p. 22. Academic Press Inc., New York.
103. Wulff, K. (1983). In: *Methods of Enzymatic Analysis*, (ed. H-U. Bergmeyer), Vol. 1, p. 340. Verlag-Chemie, Weinheim.
104. De Luca, M. A. (ed.) (1978). *Methods in Enzymology*, Vol. 57. Academic Press Inc., New York.
105. Raskin, N. H. and Sokoloff, L. (1979). *J. Neurochem.*, **17,** 1677.
106. Dunn, M. F. and Bernhard, S. A. (1971). *Biochemistry,* **10,** 4569.
107. Tubbs, P. K. and Garland, P. B. (1964). *Biochem. J.*, **93,** 550.
108. Pearson, D. J. (1965). *Biochem. J.*, **95,** 23C.
109. Gitomer, W. L. and Tipton, K. F. (1983). *Biochem. J.*, **211,** 277.
110. Duggleby, R. G. (1983). *Biochim. Biophys. Acta,* **744,** 249.
111. Roodyn, D. B. (1970). *Automated Enzyme Assay*, Elsevier BV, Amsterdam.
112. Wienhausen, G. and De Luca, M. (1986). In: *Methods in Enzymology*, (ed. M. A. De Luca and W. D. McElroy), Vol. 133, p. 198. Academic Press Inc., New York.
113. Roughton, F. J. W. and Rossi-Bernardi, L. (1964). *Proc. Roy. Soc.*, Ser. B, **164,** 381.
114. Boghosian, R. A. and McGuinness, E. T. (1981). *Int. J. Biochem.*, **13,** 909.
115. Illingworth, J. A. and Tipton, K. F. (1969). *Biochem. J.*, **115,** 511.
116. Gurr, P. A., Wong, J. T-F., and Hanes, C. S. (1973). *Anal. Biochem.*, **51,** 584.
117. Tipton, K. F. and Dixon, H. B. F. (1979). In: *Methods in Enzymology*, (ed. D. L. Purich), Vol. 63, p. 183. Academic Press Inc., New York.
118. Laidler, K. J. and Bunting, P. S. (1973). *The Chemical Kinetics of Enzyme Action*. Oxford University Press, Oxford.
119. Cleland, W. W. (1982). In: *Methods in Enzymology,* (ed. D. L. Purich), Vol. 87, p. 390. Academic Press Inc., New York.
120. Bock, H-G. O. and Fleischer, S. (1974). In: *Methods in Enzymology*, (ed. S. Fleischer and L. Packer), Vol. 32, p. 374. Academic Press Inc., New York.
121. Houslay, M. D. and Tipton, K. F. (1975). *Biochem. J.*, **145,** 311.
122. Wojtczak, L. and Nalecz, M. J. (1979). *Eur. J. Biochem.*, **94,** 99.
123. Boleda, M. D., Pere, J., Moreno, A., and Pares, X. (1989). *Arch. Biochem. Biophys.*, **274,** 74.
124. Scopes, R. K., (1982). *Protein Purification. Principles and Practice*, p. 185. Springer-Verlag, New York.

125. Tubbs, P. K. (1962). *Biochem. J.,* **82,** 36.
126. Dalziel, K. (1962). *Biochem. J.,* **83,** 28P.
127. Houslay, M. D. and Tipton, K. F. (1973). *Biochem. J.,* **135,** 735.
128. Tipton, K. F. (1977). *Biochem. Pharmacol.,* **26,** 1525.
129. Nimmo, H. G. and Tipton, K. F. (1982). In: *Methods in Enzymology*, (ed. W. A. Wood), Vol. 90, p. 330. Academic Press Inc., New York.
130. Tipton, K. F. (1973). *Biochem. Pharmacol.,* **22,** 2933.

# 2

# Photometric assays

ROBERT A. JOHN

## 1. Introduction

Photometric methods are undoubtedly the most frequently used of all kinds of enzyme assay. They are convenient to use and are capable of rapidly providing accurate and reproducible results on large numbers of samples.

Useful changes in the optical properties of the system under analysis frequently arise directly from the chemical transformations accompanying the enzyme-catalysed conversion of substrate to product. When the reaction catalysed by the enzyme under assay does not itself produce a useful change in optical properties, incorporation of appropriate additional reagents often allows the reaction to be monitored photometrically. Enzyme assays based on changes in the light *absorbed* by the solution as the reaction proceeds are more frequently used than other photometric methods. However, changes in *fluorescence* and changes in *turbidity* of the solution also provide the bases of useful assay methods.

## 2. Absorption

Absorption of light occurs when electrons of the absorbing molecules are promoted to a higher energy level as a result of an electronic transition. Visible light is electromagnetic radiation of wavelength between 400–750 nm and materials that absorb light in this wavelength range are coloured. Electronic transitions are also brought about by the absorption of ultraviolet light in the range 200–400 nm. The amount of energy required for an electronic transition has a precise value which is proportional to the frequency of oscillation of the electrons undergoing that transition and is determined by the structure of the absorbing molecule. The probability of the transition is greatest when the frequencies of the electronic oscillation and of the irradiating light coincide. It is conventional to refer to ultraviolet (UV) and visible light by its wavelength rather than by its frequency to which it is inversely related.

### 2.1 Terminology

Although strictly speaking the term *chromophore* describes a chemical group that brings colour by absorbing visible light, in this chapter the term will also be applied to groups that absorb in the UV part of the electromagnetic spectrum. Whereas *absorption* may be used to describe the process by which light is absorbed, the word *absorbance* refers to a quantity and has a precisely defined meaning. *Optical density* and *extinction* are synonymous with absorbance but these terms are no longer acceptable by scientific journals.

## 2.2 Absorbance

Instruments designed to measure absorbance invariably make their measurements by determining the amount of light that is not absorbed, i.e. the light which is transmitted by the solution. Naturally, the light transmitted by a solution of a chromophore decreases as the concentration of the chromophore increases. Furthermore, the proportionality between transmitted light and concentration of the chromophore is not linear but logarithmic. The quantity known as absorbance was deliberately designed to increase linearly as the concentration of the chromophore increases but its value must always be derived from measurements of the transmitted light no matter what the design of instrument. Thus any instrument that measures absorbance must in some way make a comparison of the light transmitted by a solution not containing the chromophore ($I_0$) with that of a solution that does contain the chromophore ($I$). The fraction of incident light that is transmitted by the solution is $I/I_0$ and Equation 1 shows the relationship between absorbance ($A$) and this fraction.

$$A = -\log_{10}(I/I_0) \quad (1)$$

Absorbance, being derived from a ratio, does not have units. However, the expressions 'absorbance units' or simply 'A' are often used.

### 2.2.1 Conversion of absorbance to concentration

In the absence of unusual complicating factors such as association of the chromophore into polymers, and at the relatively low concentrations almost invariably used in biochemical experiments, absorbance is directly proportional to concentration. Absorbance is always directly proportional to path-length. These facts are combined in the Beer–Lambert relationship (Equation 2) relating absorbance ($A$) to concentration ($c$) and path-length ($l$).

$$A = \varepsilon c l \quad (2)$$

Although absorbance is now the term recommended by scientific journals, the Greek letter ε, (epsilon, for extinction) is still used as the proportionality constant relating absorbance to concentration. The expression extinction coefficient is still widely used in the scientific literature for absorbance coefficient and this is apparently acceptable.

The most widely-used unit for absorbance coefficient in the biochemical literature is $l\ mol^{-1}\ cm^{-1}$. This is particularly convenient when, as is normal, solutions are contained in cuvettes of path-length 1 cm, because the concentration may be determined simply by dividing the measured absorbance by the relevant extinction coefficient to give the concentration in $mol\ l^{-1}$ (Equation 3).

$$c = A/\varepsilon \tag{3}$$

Another way of looking at absorbance coefficient expressed in these units is to consider it to be the absorbance in a 1-cm path-length cuvette of a one-molar solution of the chromophore. The unit $cm^2\ mol^{-1}$ is used occasionally and, expressed in these units, the values are numerically 1000 times greater than those expressed in $l\ mol^{-1}\ cm^{-1}$, representing the absorbance that would be obtained for a 1000 $mol\ l^{-1}$ solution of the chromophore.

### 2.2.2 Instrumentation

*Figure 1* shows a schematic illustration of an instrument designed to measure absorbance. All instruments operate by the same basic principle in that the wavelength range of light from a source is first restricted appropriately before being passed through the solution containing the chromophore and thence to a photoelectric detector.

This basic format is common to a very wide variety of instruments ranging from the simplest colorimeter to the highest precision spectrophotometer. However, the quality of the result obtained depends very much on the detailed configuration of the instrument used.

*i. Light source*

The main consideration here is the wavelength at which measurements are to be made. The most commonly-used light sources are the glass enveloped filament bulb, the quartz enveloped tungsten halogen lamp, and the deuterium

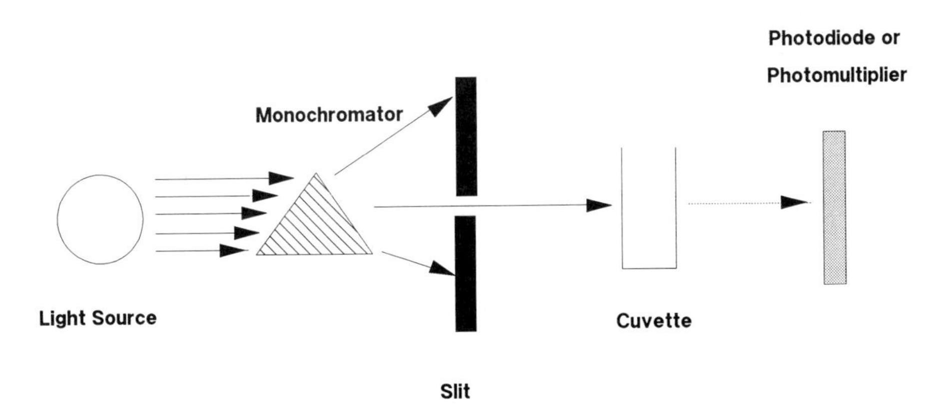


**Figure 1.** Essential features of an instrument designed for measuring absorbance.

lamp. To cover the useful UV/visible range (200–700 nm) an instrument using both deuterium and quartz halogen lamps is necessary. To make measurements in the visible and near UV, including the 340 nm required for NAD(P)/NAD(P)H-based measurements, an instrument using a quartz–halogen lamp as sole light source is adequate. At 340 nm the energies of deuterium and quartz–halogen lamps are approximately equal and either may be used. The glass enveloped filament lamp can only be used in the visible region and its value in enzyme assays is very limited.

*ii. Wavelength selector*

The system used to restrict the wavelength of light to that most appropriate to the relevant chromophore is very important in determining the quality of the measurements made. The colorimeter simply uses a series of filters that have different absorbance characteristics. If, for example, the reading is to be made at 410 nm, the filter chosen is one that absorbs light at wavelengths higher and lower than 410 nm. However, given that the range of available filters is fairly small, the light incident on the cuvette will consist mainly of wavelengths well away from 410 nm where the chromophore absorbs little or not at all. The absorbance readings obtained from such an instrument will be less, sometimes very much less, than those predicted from the extinction coefficient.

The modern spectrophotometer uses a holographic diffraction grating to disperse the light from the source into a spectrum. Rotation of the diffraction grating varies the wavelength of emergent light passing through a slit, and the extent to which this approximates to monochromatic light depends on the width of the slit and is described in the specification of an instrument as the *bandwidth*. It is useful to be able to increase this slit width in situations where increased sensitivity of an assay is required, but on many instruments the bandwidth is predetermined. Photometric accuracy and precision can be increased by using a high bandwidth because the incident beam has more energy. However, this produces a beam that is less-nearly monochromatic, a fact that should be kept in mind when comparing instruments.

*iii. Determination of absorption spectra*

The most versatile spectrophotometers are capable of rapid determination of complete absorbance spectra. Despite the undoubted value of such instruments in the biochemical laboratory, the assay of enzyme activity *per se* does not require a wavelength scanning instrument. Nevertheless, many problems experienced with photometric enzyme assays can be solved more rapidly if the absorbance spectrum of the solutions involved can be determined rapidly.

*iv. Cuvette holder*

For continuous assays it is essential that the temperature of the cuvette holder can be controlled. Electrical heating is superior to circulating water because of the danger of flooding the instrument with the latter. Peltier constant temperature systems, which allow the temperature to be maintained below as well as above ambient, are available for some instruments and are undoubtedly useful in the investigation of enzyme kinetics.

Many instruments have multiple cuvette holders so that several continuous assays can be conducted simultaneously by automatic mechanical movement of the carriage between measurements. This is a considerable advantage when multiple slow reactions have to be followed. Such additional compartments are also useful for holding cuvettes containing assay solution at the right temperature while a measurement on another cuvette is in progress. There is some occasional advantage in having a cuvette holder that can accept cells of path-length longer than 1 cm.

## 2.3 Limitations and sources of error

It is well worth considering the factors that determine the quality of an absorbance measurement. The absorbance measurement reported by a spectrophotometer will differ from the true value due to three main causes.

### 2.3.1 Instrumental noise

This is a random fluctuation in the output from the photodetector which originates within the instrument and does not arise from the solution.

### 2.3.2 Zero drift

This is a slow, steady rise in apparent absorbance observed in single-beam instruments and it is caused by time-dependent changes in the photodetector and in the lamp. Changes are large and rapid for the few minutes immediately after switching the instrument on (*Figure 2*). However, slow changes continue over a very much longer period and 'zero drift' never disappears completely.

### 2.3.3 Non-linearity

This problem (Section 3) usually arises from light that originates within the instrument but reaches the photodetector without passing through the cuvette.

## 2.4 Absorbance range

The fact that determination of absorbance depends upon measurements of transmitted light means that the effects of these sources of error become increasingly important at high absorbance, as illustrated in *Figure 3*. Very low absorbance readings will be difficult to estimate because there is very little difference between the amounts of incident and transmitted light. The range of absorbance values that can be realistically measured varies greatly according to the quality of the instrument and, in the absence of the relevant information for a particular instrument, readings should be kept well within the range 0.01–1.0.

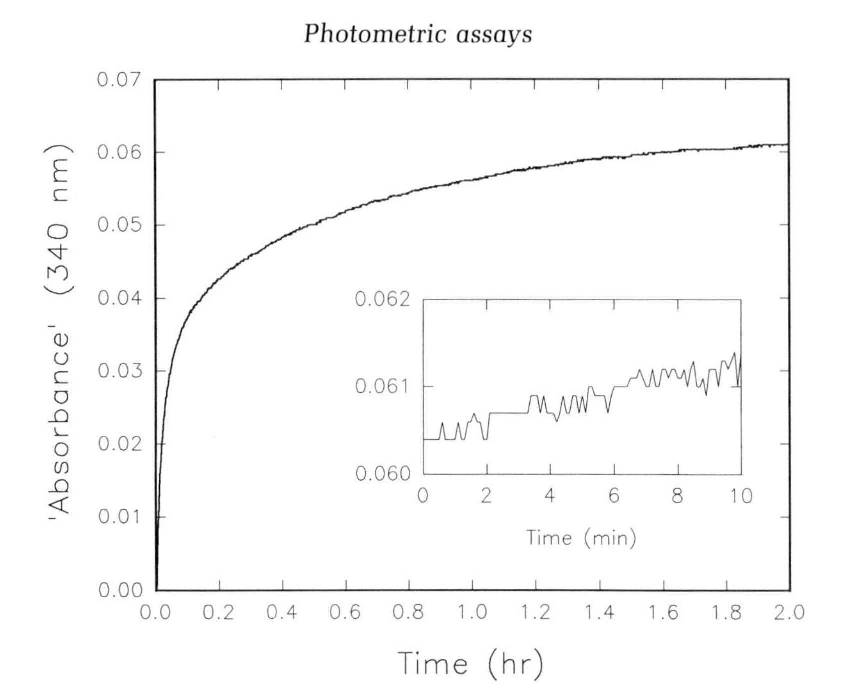


**Figure 2.** Readings of apparent absorbance after switching on tungsten filament lamp. The inset shows the slow continuous increase that persists after the instrument has been turned on for nearly 2 h.

## 2.5 Measurement of low rates of absorbance change

Continuous improvements in instrumentation mean that the lower limit of detectability attainable by absorbance assay has fallen progressively. Because of wavelength-dependent variations in both the energy output from the source and the response of the detector, this limit depends on the wavelength at which the assay is conducted. The least favourable wavelength from this point of view is close to 340 nm which coincides with the most useful of all the chromophores used in the assay of enzymes, namely NADPH and NADH. The lower limit of detectability depends very much upon the absolute value of the absorbance at which the assay takes place, small changes in absorbance being less easy to detect when the absolute absorbance is high. Spectrophotometers intended for making routine continuous enzyme assays typically have absorbance noise values of about 0.001 when the absolute value of absorbance is close to zero. This figure rises with increasing absorbance. In addition, the apparent absorbance reported by single-beam instruments changes continuously by about 0.003 $h^{-1}$ because of time-dependent changes in the photodetector (*Figure 2*). The lower limit of detection is determined by these factors and is illustrated in *Figure 4*. At rates of absorbance change below 0.001 $min^{-1}$ zero drift becomes significant.

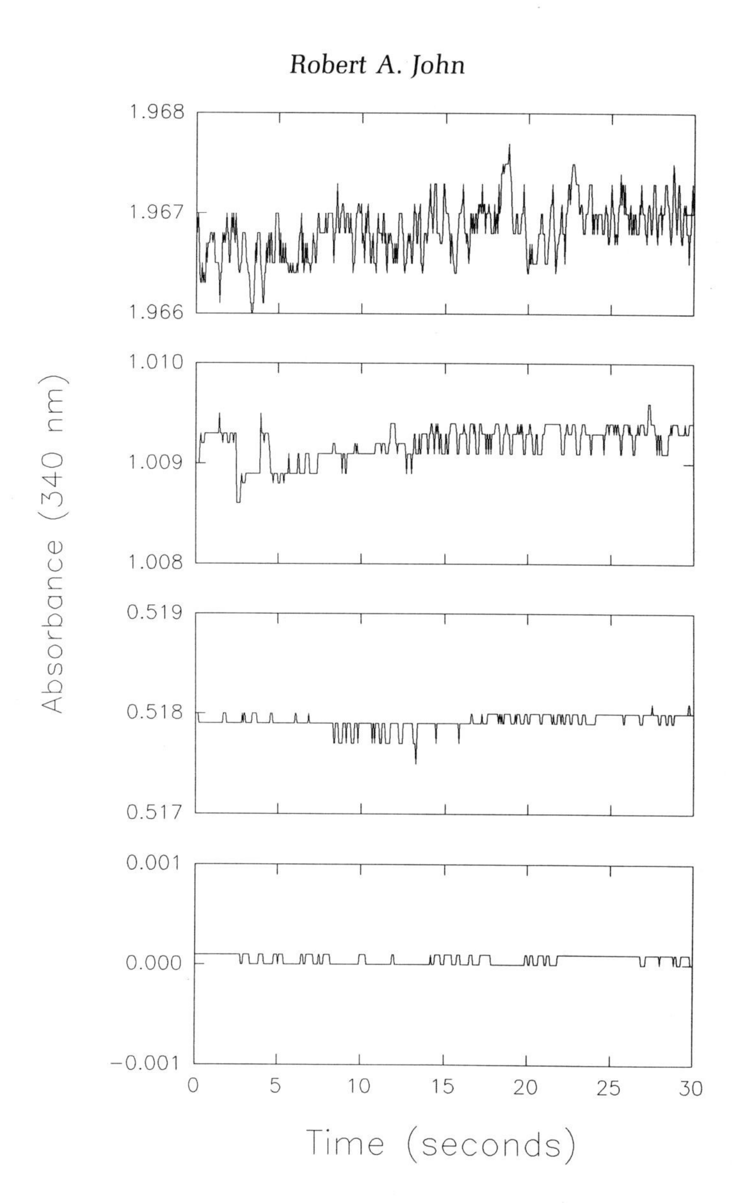


**Figure 3.** Instrument noise as a function of absorbance value. The traces show the level of instrument noise in the region of 0, 0.5, 1.0, and 2.0. The absorbance value of each solution was constant and observations were made at 340 nm.

## 2.6 Stray light

Although the Beer–Lambert law predicts linearity between concentration and absorbance, the relationship frequently appears not to be obeyed. Most often this is due to inadequacy of the instrument rather than to complex behaviour of the chromophore. The underlying reason for the non-linearity

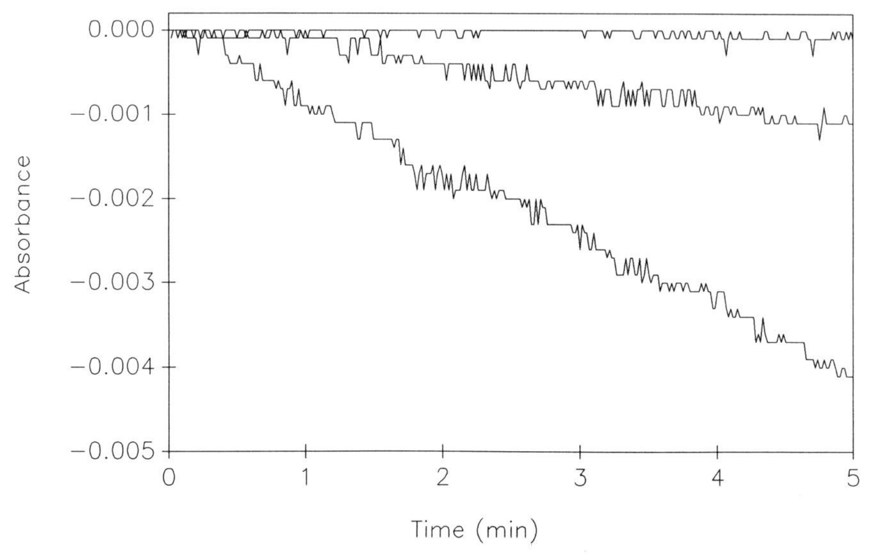


**Figure 4.** Continuous absorbance assays conducted near limits of detection. Lactate dehydrogenase was assayed using pyruvate and NADH as substrates as described in Section 2.10.1, using two concentrations of the enzyme to produce rates of absorbance change of 0.0003 $min^{-1}$ and 0.001 $min^{-1}$, respectively. The line observed with no added enzyme is included for comparison and the initial absorbance was set to zero in each case.

lies in the instruments' determination of values $I_0$ and $I$ for the transmitted light. Despite all efforts it is impossible to prevent some light from reaching the photodetector by routes other than that intended. This 'stray light' is always a small but constant component of the light detected. Thus, whereas the true absorbance value is given by Equation 1, the instrument has to operate with an additional term $I_s$, for stray light, which increases the apparent values of $I$ and $I_0$ by the same amount (Equation 4).

$$A_{app} = -\log_{10} [(I + I_s)/(I_0 + I_s)] \tag{4}$$

It is easy to see that as $I$ becomes small, $I_s$ becomes more and more significant and the apparent value of the absorbance ($A_{app}$) falls below the true value. The result is an apparent deviation from the Beer–Lambert relationship of the type shown in *Figure 5*.

## 2.7 Use of extinction coefficient

So long as the instrument is used within its capabilities, the Beer–Lambert relationship will apply and rates of concentration change can be determined directly using the extinction coefficient for the compound. It should be remembered that published values for extinction coefficients are determined with narrow bandwidths, in other words, with very nearly true monochromatic light. Errors will enter the determination of concentration if bandwidths

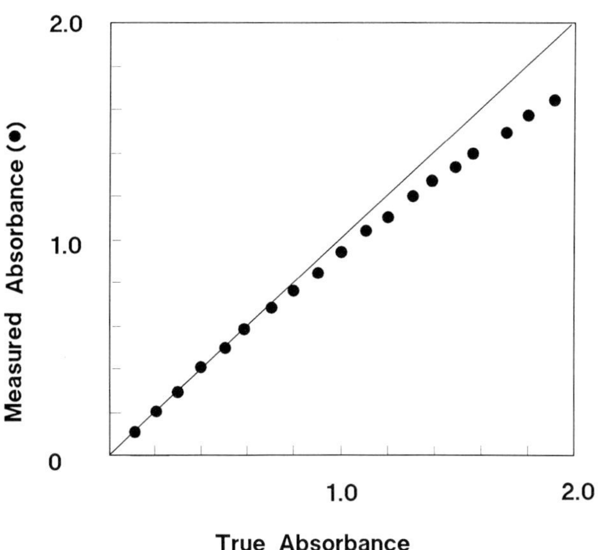


**Figure 5.** Effect of stray light on relationship between true and measured absorbance. The experimental points show a deviation from linearity which is produced by approximately 1% stray light.

are broad relative to the absorbance peak of the chromophore being determined. *Figure 6* illustrates this point. The problem is particularly acute when a colorimeter is used to measure a chromophore that is not well matched by any of the available filters. In such a case, it is not exceptional to obtain readings an order of magnitude lower than those expected from the published absorbance coefficient.

## 2.8 Continuous assays

As has been explained in Chapter 1, a continuous assay, in which the enzyme-catalysed reaction is monitored as it proceeds, is very much to be preferred over one in which the enzyme reaction is run for a fixed time and then stopped before measuring products formed. Because the output of the spectrophotometer can readily be coupled to a chart recorder, *x,y*-plotter, or printer, photometric methods are well suited to continuous assay. Some elementary but important practical points are worth considering.

### 2.8.1 Choice of solution for setting zero absorbance

Regardless of the instrument used and the system employed, at some stage the operator must make a choice as to what is in the cuvette holder when zero absorbance is set. Amongst various possibilities, this 'blank' may be a cuvette containing all of the ingredients except one, or a cuvette containing just water. Alternatively, the cuvette holder can be left empty. Absolute values of

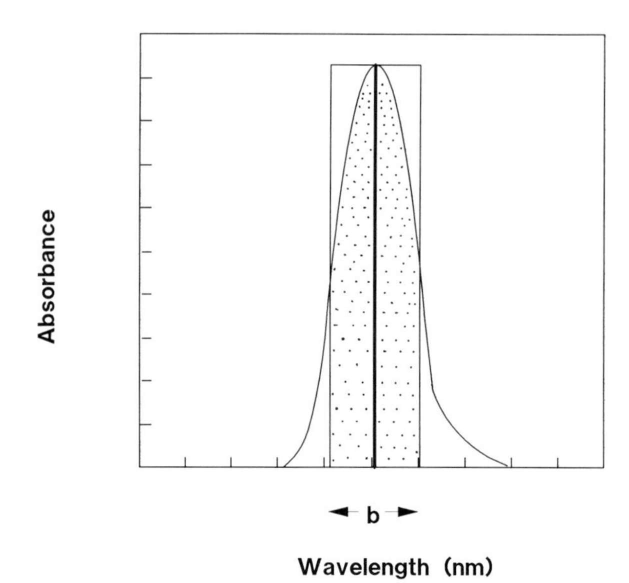


**Figure 6.** Dependence of apparent extinction coefficient on band width. Absorbances measured at broad bandwidth **b** will be lower than those predicted by an extinction coefficient determined at the narrow bandwidth indicated by the solid line. The discrepancy will be in the ratio of the shaded and unshaded portions of the rectangle.

absorbance reported by the instrument during subsequent measurements will be reduced by the absorbance of the blank but rates of absorbance change will be unaffected. If the instrument used sets zero absorbance by attenuation of the light beam, the choice of 'blank' influences the noise level of the measurement. There should be an improvement in noise levels if the high-absorbing solution is used to set zero. Clearly, for continuous methods producing a fall in absorbance, a solution containing the relevant chromophore can only be used to set zero absorbance if the instrument is capable of reading negative values. *Figure 7* shows the assay of LDH conducted (a) by setting the zero with an assay solution containing no NADH and (b) by setting the zero with a cuvette containing NADH at its starting concentration.

### 2.8.2 Temperature control

Photometric assays of an enzyme by a continuous method can only be achieved satisfactorily at a constant known temperature. It is particularly important to ensure that the contents of the cuvette are at the required temperature. The time taken for the contents of a cuvette to reach temperature depends on the nature and volume of contents. *Figure 8* shows how a solution approaches the required temperature when quartz or plastic cuvettes are used. Clearly, there is a need for the cuvettes to be brought to the correct temperature before the reaction is started. The inconvenience of waiting

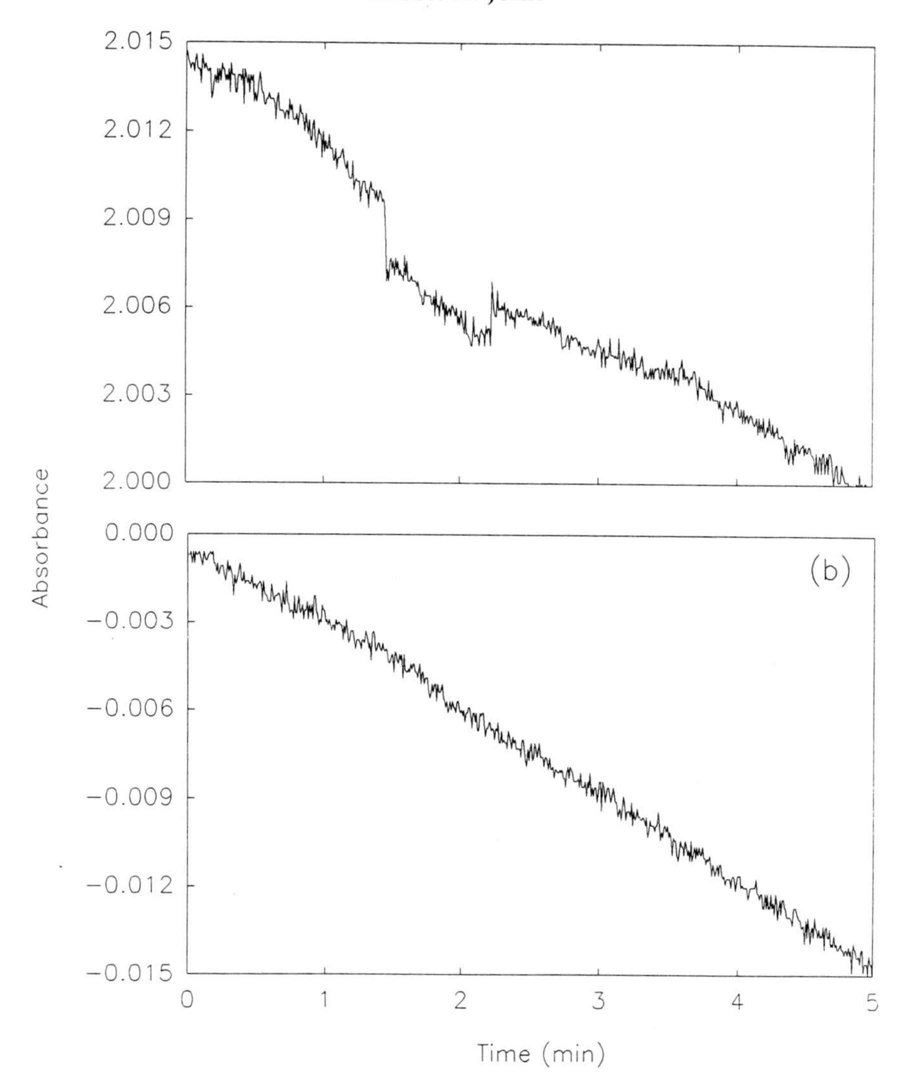


**Figure 7.** Continuous absorbance measurement obtained after setting zero with high absorbing solution. The NADH concentration in the assay was set so that the absorbance was 2.0 and the course of reaction followed after (a) setting zero on a control solution containing no NADH, and (b) setting zero with the high-absorbing NADH solution.

several minutes for temperature to be reached may be overcome in several ways. If the assay is rapid, i.e. complete within a few minutes, then the unused positions in a multiple cuvette holder provide convenient repositories in which successive cuvettes may be held prior to assay. For this system to work, the enzyme, or whatever solution is used to start the reaction, must be only a small part of the total volume. If addition of a relatively large volume

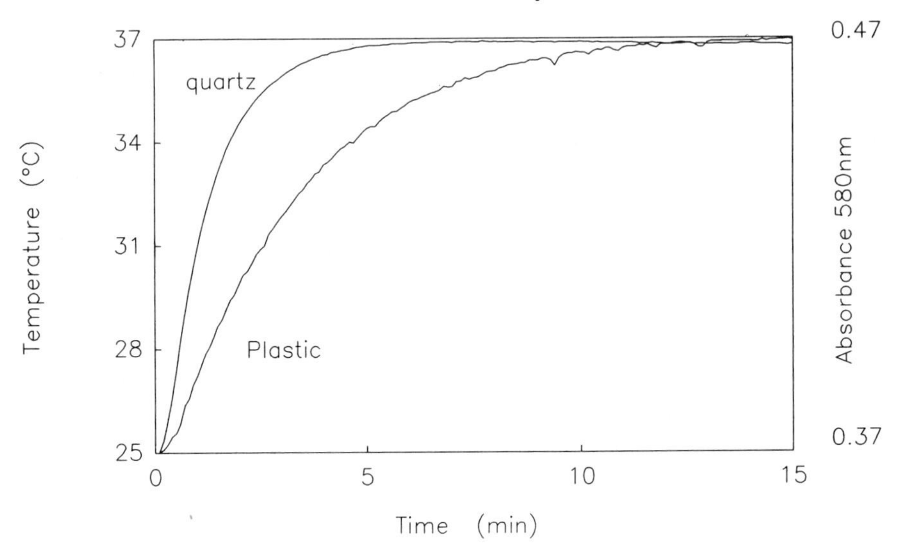


**Figure 8.** Temperature equilibration of cuvettes. Plastic and quartz cuvettes containing phenol red (6 mg/l) in 0.1 M Tris–HCl, pH 7.8, were transferred from room temperature (15 °C) to a cuvette holder at 37 °C, and absorbance at 560 nm was monitored continuously.

to start the reaction is inescapable, then this must be brought separately to the assay temperature before it is added. For continous assays lasting significantly longer than the warming-up period it is more convenient to use all positions in the cuvette holder for assays but to delay starting the reaction for several minutes until assay temperature has been reached.

### 2.8.3 Starting the reaction

The operations involved in starting the reaction, i.e. adding the final 'starter' ingredient, mixing adequately, replacing the cuvette in its holder, closing the lid, and beginning the recording, necessarily take time. Even the most skilled practitioner is unlikely to take much less than 10 s. It is important to arrange things so that the amount of enzyme added does not produce such a rapid reaction that a significant part is over before recording begins. It is not unusual to hear experimentalists, unfamiliar with this sort of system, reporting that there is no enzyme in a sample, whereas there is so much that the reaction is complete by the time the recording is begun.

A convenient method for starting the reaction is as follows:

---

**Protocol 1.** Starting the reaction

1. Add all the ingredients except the 'starter' to the cuvette and allow it to attain the correct temperature. In the case of photometric assays, as in other enzyme assays, it is always best, if possible, to mix all the ingredients

except one in a stock solution and to start the reaction by adding a small (but not too small) volume of the omitted ingredient. (See Chapter 1 on the use of assay 'cocktails'.)

2. Add the 'starter' ingredient in a volume of 10–50 μl per ml of final assay mixture. This may be added directly into the solution. However, the time between mixing and measurement can be shortened if it is added as a 'hanging drop' on the side of the cuvette, so that it does not mix until the cuvette is inverted in the next step.
3. Use parafilm and your thumb to close the cuvette and mix the contents by gently inverting the cuvette twice. Ensure that there is enough space at the top of the cuvette for the solution to mix adequately and that the solution really does mix by falling into this space (with semi-micro cuvettes the solution can remain in place even though the cuvette is inverted). The contents of the cuvette should not be mixed by shaking vigorously as this will introduce bubbles which will interfere with the absorbance measurement and may possibly denature the enzyme.
4. Place the cuvette in the instrument and start the recording.

---

An alternative and equally good method of starting the reaction is to stir the contents of the solution with a commercially-available plastic spatula shaped like a ladle. The operations of adding and mixing may be combined if the solution used to start the reaction is pipetted on to the spatula rather than directly into the cuvette (Chapter 1).

If the rate of reaction is fast so that the process of mixing has to be hurried it is undoubtedly better to dilute the enzyme sample appropriately.

### 2.8.4 Volume needed in cuvette

When reagents and enzyme sample are not in short supply there is much to be said for using full-size (3-ml) cuvettes rather than semi-micro cuvettes. Pipetting errors are somewhat reduced, especially because a larger volume of enzyme can be used to start the reaction without lowering the temperature of the solution in the pre-incubated cuvette. Also, with very many spectrophotometers, problems arise because the beam width is greater than the optical face of the semi-micro cuvette so that unacceptable non-linearity is experienced often at absorbance values well below 1. However, economy of reagents and enzyme frequently requires that the volume used be as small as possible. The dimensions of the beam at the point at which it passes through a semi-micro cuvette may be examined as follows.

(a) Set the wavelength to something clearly visible (~500 nm) and increase the bandwidth to its maximum. This will make the beam sufficiently intense to be visible in a darkened room or under a black cloth.

(b) Insert a piece of white paper into the cuvette holder. The illumination of the paper will give a clear indication of the beam dimensions.

If some of the light reaching the photodetector has passed through the cuvette walls rather than the solution the effect will be the same as that produced by stray light; absorbance values will be underestimated and the extent of the underestimation will increase with increasing absorbance. Light can be prevented from passing through the walls by masking the cuvette with black paint or tape.

It may be that the cuvette need not be pushed all the way into the holder for the beam to be completely contained within the solution. *Figure 9* shows the effect on the absorbance value reported by the spectrophotometer, of altering the volume in a full-size (3-ml) cuvette. With this particular instrument 1.5 ml is a 'safe' volume to work with. A similar effect is observed with semi-micro cuvettes. Frequently 0.4 ml or even less of a solution in a semi-micro cuvette is adequate to contain the beam completely. However, when working close to the limit in this way it is important to avoid bubbles trapped in the surface layer as these may place a variable air–liquid interface within the beam.

### 2.8.5 Calculation of enzyme activity

If a chart recorder is being used to record the results its speed should be set so as to produce a line with a slope that is neither too steep nor too shallow (the most accurate determination of gradient comes from a line of slope 45°). The line should be used to express the rate of change of absorbance with time (d$A$/

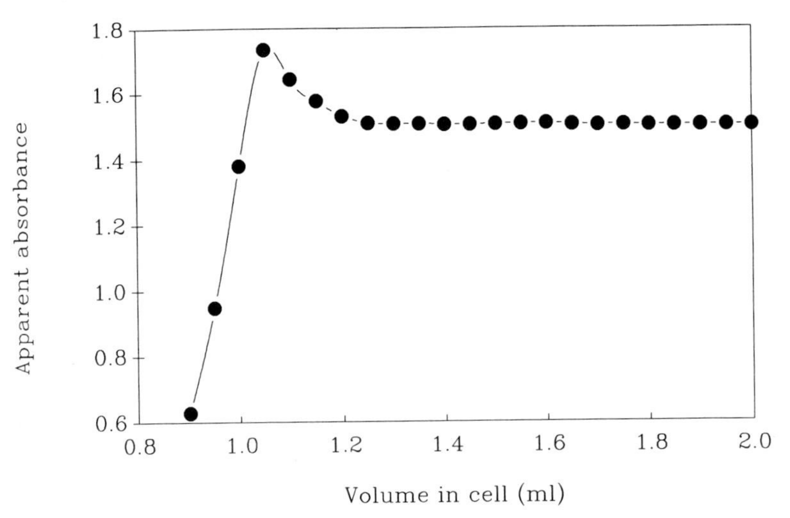


**Figure 9.** Effect of decreasing the volume of solution in cuvette. In this instrument, a constant reading is observed until the volume is reduced to about 1.3 ml. The apparent absorbance then rises as the region of the meniscus comes to lie in the beam. Thereafter apparent absorbance falls sharply. These measurements were obtained by removing successive 50 µl volumes from a cuvette with a total capacity of 3 ml.

d$t$). Division by the molar extinction coefficient ($\varepsilon$) gives the rate of change of concentration (d$c$/d$t$) in mol $l^{-1}$ $min^{-1}$.

The number of units (μmol/min) of enzyme present in the cuvette depends on the final volume, $v_f$ (ml), it contains. Almost invariably the assay uses a very small part of a much larger sample. Suppose that $v_s$ ml of enzyme was taken from a sample of total volume $v_t$ ml and added to make a final volume of $v_f$ ml in the cuvette. All three of these volumes must be used in determining the total number of units in the sample.

The overall calculation is:

$$\text{Units of activity in sample} = [(dA/dt)/\varepsilon] \times v_f \times 1000 \times (v_t/v_s)$$

(The factor of 1000 arises because the units of activity are in μmol $min^{-1}$; use of the extinction coefficient gives the concentration in mol $l^{-1}$ and cuvette volume is in ml.)

### 2.8.6 Causes of artefactual non-linearity

The best continuous assay methods should give linear initial-reaction rates from the time that recording begins until sufficient reaction has been recorded to establish the slope of the line clearly. However, even when the reactant solutions are correctly made, deviations from linearity may be observed because of errors in the use of the spectrophotometer or in the instrument itself.

Upward curvature of the line as in *Figure 10a* indicates an accelerating reaction and suggests inadequate temperature pre-equilibration. A random wavy line indicates inadequate mixing (*Figure 10b*) or particles in solution moving up and down through the beam (*Figure 10c*).

Non-linearity will also be evident if the spectrophotometer does not give a linear response to concentration as discussed above in the section on stray light.

## 2.9 Discontinuous assays

Undoubtedly the great value of absorbance measurement in the assay of enzymes is that the instrumentation is ideal for continuous assay. However, despite the fact that the eventual measurement is of absorbance, some systems cannot be assessed continuously. This may be because the conditions required for the enzyme-catalysed reaction are not the same as those for colour development. Alternatively, interfering background absorbance from, for example, crude tissue samples may be so high that deproteinization must precede absorbance measurement. In these circumstances the reaction is stopped at various intervals and the colour developed in a separate reaction.

### 2.9.1 Sources of error

As with any discontinuous assay there is a temptation to equate the concentration of product formed in a fixed time with the rate of the reaction. This is

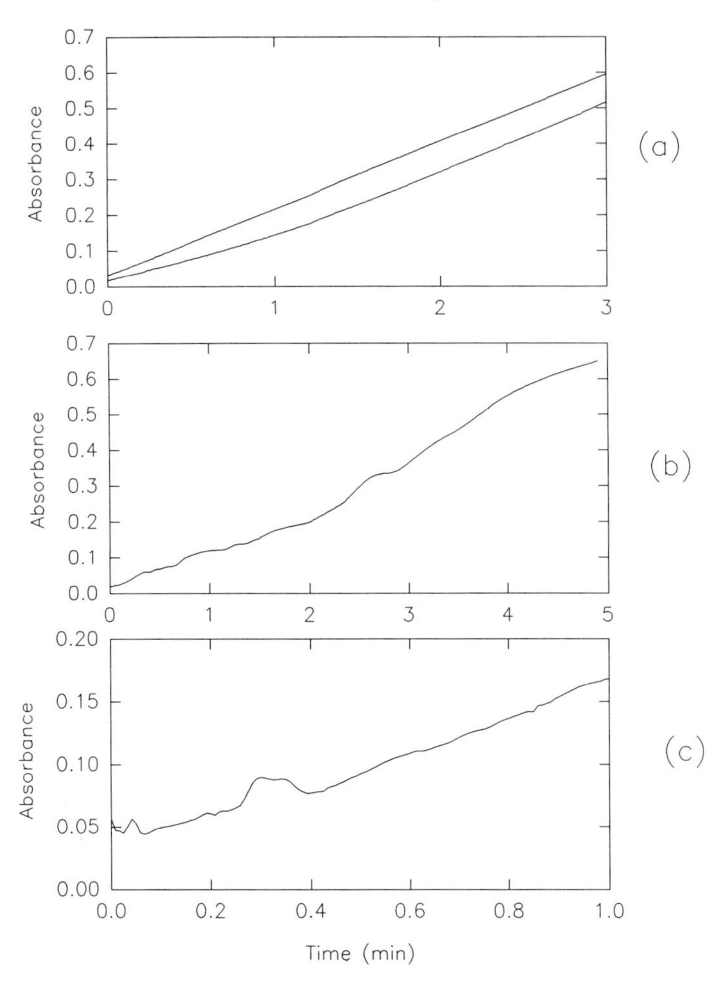


**Figure 10.** Some artefacts produced in continuous absorbance measurement. (a) Inadequate temperature equilibration (lower line) compared with adequate equilibration (upper line); (b) inadequate mixing; (c) dust particles in assay solution.

only true if the reaction runs at constant rate over the period of the assay and this cannot be established with measurements made after only one time interval. The most satisfactory way of overcoming this problem is to make measurements at a minimum of two time intervals in addition to one for zero time. Alternatively, measurements at fixed time can be made using different volumes of enzyme sample (in the same total volume of assay). If the fixed-time absorbance value is linearly related to the volume added, then this provides good evidence that linearity of absorbance change with time is maintained for the duration of the assay. The necessity to make measurements at multiple time-intervals may justifiably be avoided in routine measurements if the conditions of the assay system, such as substrate concentration, are kept constant and the amounts of enzyme measured are well within already established bounds of linearity. For a detailed discussion of this point, see Chapter 1.

### 2.9.2 Use of microtitre-plate readers

The availability of plastic micro-titre plates and associated plate readers considerably reduces the effort involved in obtaining absorbance measurements for fixed-point assays. The reproducibility of measurements made in this way is undoubtedly perfectly adequate for many types of analysis in which large numbers of samples need to be processed. Linearity is maintained to high absorbance readings (*Figure 11a*) and standard deviations in absorbance readings are constant at about 0.01 over this range (*Figure 11b*). Thus, if one is not prepared to accept errors greater than 10% from readings made in duplicate, minimum absorbance values of 0.2 should be used. The variance in the measurement appears to be associated with the fact that one optical face is the meniscus at the air–liquid interface. Consequently, to maximize the absorbance reading, it is best to fill the wells close to the top using a total volume of about 0.25 ml, which gives a pathlength of about 1 cm. Because the instruments use filters, the absorbances measured will be lower than those expected on the basis of published extinction coefficients, and concentrations should be determined by using an appropriate standard.

## 2.10 Examples of enzymes assayed by absorbance change

The number of enzymes that can be assayed spectrophotometrically is so large that only a limited number of examples can be presented here in detail. For experimental detail of enzymes not mentioned here the reader is referred to two major collections of such assay methods, namely *Methods in Enzymology* (1) and *Methods in Enzymatic Analysis* (2). Much ingenuity has been applied to devising photometric assays for enzymes that catalyse reactions which do not themselves give rise to any direct absorbance change. The following examples are chosen, in part, to illustrate this experimental ingenuity in the hope that consideration of these successful methods will aid the design of new assays for other enzymes.

### 2.10.1 Direct observation of the natural reaction

A limited number of enzyme-catalysed reactions are themselves accompanied by a useful change in absorbance. *Table 1* shows the absorbance properties of some substrates that can be employed directly in the assay of the enzymes indicated.

Oxidation of NADH (and NADPH) as well as the reaction in the opposite direction is accompanied by a large change in absorbance at 340 nm ($\varepsilon_{340}$ = 6200 l $mol^{-1}$ $cm^{-1}$). This makes the direct, continuous absorbance assay of a large and important group of enzymes, the dehydrogenases, very simple.

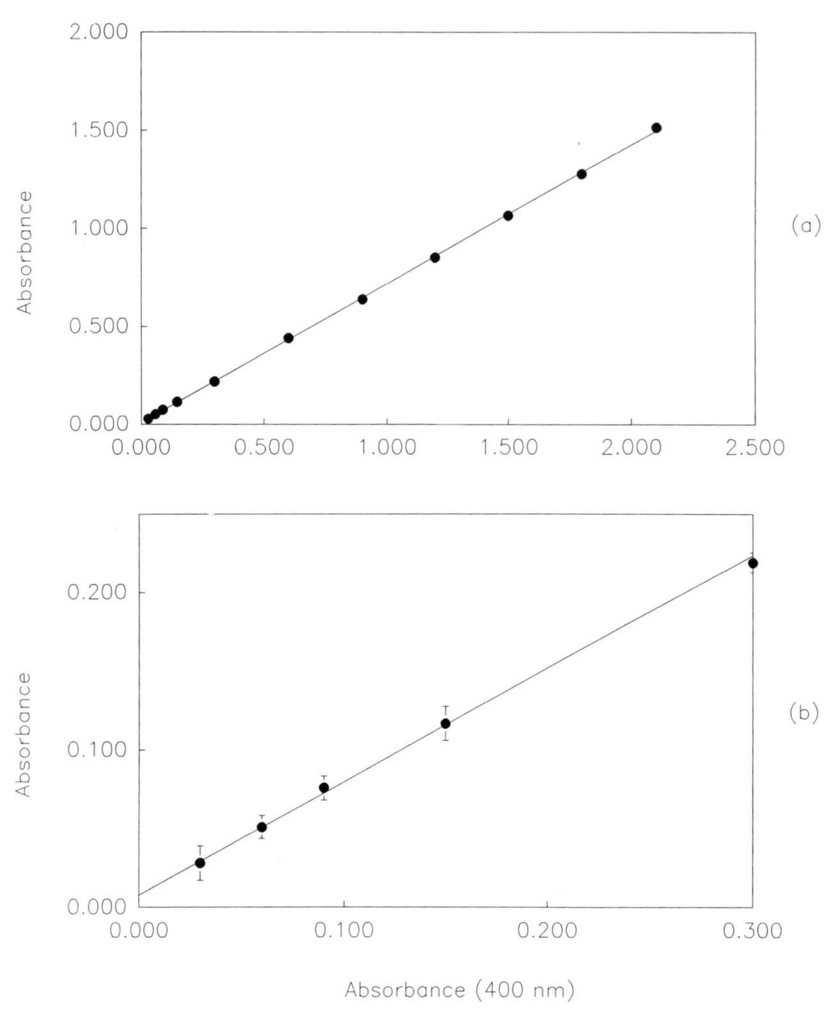


**Figure 11.** Errors associated with absorbance readings made using micro-titre plates. A solution of nitrophenol in 0.1 M NaOH was used to compare readings of true absorbance with values determined by a micro-titre plate reader. Wells were filled to a depth of 1 cm by adding 0.27 ml of solution. (a) How linearity is maintained to high absorbance values. (b) Some of the same data as in (a) but on an expanded scale so that the standard deviation of the measurements can be appreciated.

### *i. Lactate dehydrogenase*

Lactate dehydrogenase (LDH) is one of the most frequently assayed of all enzymes because its presence in serum after tissue damage provides an important aid to clinical diagnosis. It is worth considering the factors that have led to the choice of assay conditions.

Because the reaction catalysed is freely reversible the assay can be made either in the direction of oxidation of lactate by $NAD^+$ or in the reverse

**Table 1.** Absorbance characteristics of some naturally-occurring, UV-absorbing substrates

| Compound | Absorbance characteristics | Enzymes |
|---|---|---|
| Acetyl coenzyme A 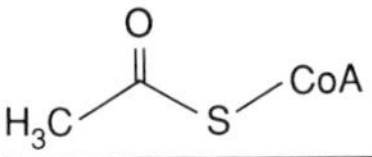 | $\lambda_{max} = 232$ nm<br>$\varepsilon = 4500$ l mol$^{-1}$ cm$^{-1}$ | Choline acetyltransferase and other acetyltransferases |
| Fumarate 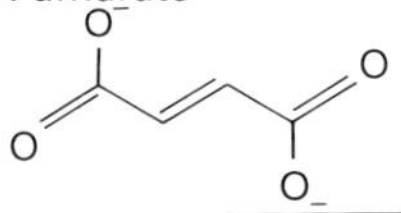 | $\lambda_{max} = 240$ nm<br>$\varepsilon = 2440$ l mol$^{-1}$ cm$^{-1}$ | Fumarase<br>Argininosuccinase |
| Oxaloacetate 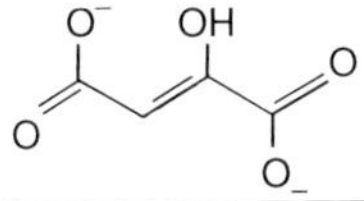 | $\lambda_{max} = 265$ nm<br>$\varepsilon = 950$ l mol$^{-1}$ cm$^{-1}$ | Oxaloacetate decarboxylase |
| Inosine 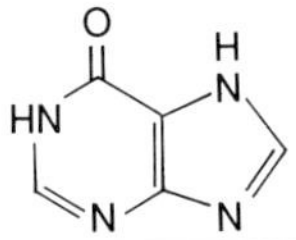 | $\lambda_{max} = 265$ nm<br>$\varepsilon = 8600$ l mol$^{-1}$ cm$^{-1}$ | Adenosine deaminase |
| Uric acid 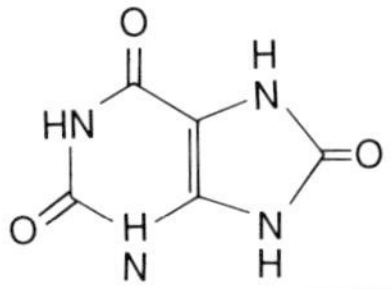 | $\lambda_{max} = 232$ nm<br>$\varepsilon$ 12 200 l mol$^{-1}$ cm$^{-1}$ | Xanthine oxidase |

direction in which pyruvate is reduced by NADH. Methods for assaying the enzyme based on both reaction directions are available but that which is almost invariably used begins with pyruvate and NADH and measures the fall in absorbance at 340 nm that accompanies the reaction.

$$\underset{\text{(Pyruvate)}}{CH_3COCOO^-} + NADH + H^+ \rightleftharpoons \underset{\text{(Lactate)}}{CH_3CHOHCOO^-} + NAD^+$$

Several reasons combine to make reduction of pyruvate the preferred direction for the assay:

(a) The equilibrium lies very much in this direction. Thus, only an insignificant fraction of the complete reaction need occur to give sufficient absorbance change for the measurement.

(b) The maximal velocity is higher in this direction. This makes the assay more sensitive.

(c) NADH can be kept in a more stable condition at the pH of the assay than can $NAD^+$.

The enzyme may be satisfactorily assayed at 30°C, pH 7.2 in 50 mM Tris, with substrate concentrations of 0.15 mM NADH and 1.2 mM sodium pyruvate. The reasoning behind the choice of concentrations is as follows:

(a) Pyruvate. Besides being the substrate, this compound is also an inhibitor because it binds to enzyme–$NAD^+$ to make an 'abortive complex'. Thus the concentration is deliberately kept low so that inhibition is avoided. The conditions given are intended for assay of the enzyme in serum which is likely to contain two isoenzymes of LDH differing in their sensitivity to inhibition by pyruvate.

(b) NADH. The initial concentration gives a convenient absorbance measurement of 0.9 when the solution is used in a cuvette of pathlength 1 cm. A significant fall in this absorbance is essential for the assay but the low $K_m$ for NADH ensures that the velocity does not fall significantly.

### 2.10.2 Indirect assay by coupling with a dehydrogenase

The convenience of the change in absorbance when the nicotinamide coenzymes undergo oxidation or reduction, together with the large number of dehydrogenases having a wide range of specificities, mean that these enzymes frequently provide the final step of coupled enzyme assays (Chapter 1).

*i. Aspartate aminotransferase*

The clinical importance of measuring aspartate aminotransferase means that like lactate dehydrogenase it is one of the most frequently measured of enzymes. Usually known in the clinical biochemistry laboratory as GOT (Glutamate Oxaloacetate Transaminase) its measurement in serum is determined by a standardized method approved by the International Federation of Clinical Chemists who recommend that the assay be conducted at 30°C, pH 7.8, in 80 mM Tris. Recommended substrate concentrations are 240 mM L-aspartate and 12 mM 2-oxoglutarate. In order to activate the significant amounts of apo-enzyme in serum, 0.1 mM pyridoxal phosphate is added. The coupling step and removal of interfering pyruvate require the addition of 0.18 mM NADH, 0.42 units of malate dehydrogenase/ml and 0.6 units of lactate dehydrogenase/ml.

$$\text{2-oxoglutarate} + \text{aspartate} \xrightleftharpoons{\textit{aspartate aminotransferase}} \text{glutamate} + \text{oxaloacetate}$$

$$\text{oxaloacetate} + \text{NADH} + H^+ \xrightleftharpoons{\textit{malate dehydrogenase}} \text{malate} + NAD^+$$

When the assay method is used for measuring the activity of the pure enzyme, all the ingredients can be combined in one solution and the assay started by adding an appropriate amount of the enzyme in a small volume. However, when the enzyme is being measured in an unpurified sample such as serum or tissue homogenate, precautions must be taken to avoid artefacts arising from the presence of varying amounts of pyruvate and lactate dehydrogenase which themselves oxidize NADH and give rise to a falsely high and non-linear initial rate. With such samples, the 2-oxoglutarate is left out of the assay mixture and the serum or other sample is added together with lactate dehydrogenase to remove pyruvate. The mixture is left until the absorbance is constant and the reaction is then started by adding 2-oxoglutarate.

*ii. Triose phosphate isomerase*

This enzyme catalyses the isomerization between glyceraldehyde-3-phosphate and dihydroxyacetone phosphate. It is normally assayed in the direction of dihydroxyacetone phosphate synthesis. Rabbit muscle α-glycerophosphate dehydrogenase is used to couple the reaction to the oxidation of NADH.

$$\text{Glyceraldehyde-3-phosphate} \underset{}{\overset{\textit{Triose phosphate isomerase}}{\rightleftharpoons}} \text{dihydroxyacetone phosphate}$$

$$\text{Dihydroxyacetone phosphate} + \text{NADH} + H^+ \overset{\alpha\textit{-glycerophosphate dehydrogenase}}{\rightleftharpoons} \alpha\text{-glycerophosphate} + NAD^+$$

Conditions for the assay are 30°C, pH 7.9 in 0.1 M triethanolamine containing 0.14 mM NADH, 0.4 mM D-L-α-glyceraldehyde 3-phosphate, α-glycerophosphate dehydrogenase (2 μg/ml) and 5 mM EDTA.

*iii. Enzymes catalysing production of $CO_2$*

A recently-designed system allows photometric assay of enzymes producing $CO_2$ to be achieved by coupling with wheat phosphoenolpyruvate (PEP) carboxylase, an enzyme that produces oxaloacetate. A third enzyme, malate dehydrogenase, is necessary to complete the linkage, with oxidation of NADH producing a fall in 340 nm absorbance.

Decarboxylases functioning in the pH range 6–8 can be assayed in this way. An example is the assay of lysine decarboxylase (3).

$$\text{Lysine} \xrightarrow{\textit{decarboxylase}} \text{cadaverine} + CO_2$$

$$CO_2 + \text{PEP} \overset{\textit{PEP carboxylase}}{\rightleftharpoons} \text{oxaloacetate}$$

$$\text{oxaloacetate} + \text{NADH} + H^+ \overset{\textit{malate dehydrogenase}}{\rightleftharpoons} \text{malate} + NAD^+$$

Assay of lysine decarboxylase is conducted using a solution consisting of 0.1 M Tris–HCl pH 6.0, 8 mM $MgCl_2$, 10 μM pyridoxal phosphate, 10 Units malate dehydrogenase/ml, 1 Unit PEP carboxylase/ml, 45 mM PEP, 0.4 mM NADH, and 0.01% Nonidet detergent. The detergent is included to prevent aggregation of the enzymes. The solution is degassed by evacuation to remove dissolved $CO_2$.

### 2.10.3 Problems with nicotinamide nucleotides

Preparations of these coenzymes have in the past been contaminated with inhibitors, and their storage in aqueous solution under the wrong conditions also allows inhibitors to form rapidly. Clearly, the use of such contaminated preparations is to be avoided. The clinical importance of assays based on these coenzymes means that commercial suppliers are very conscious of the need to provide high-quality reagents. After opening, the preparations should be stored dry at 0–4°C and protected from light. There is the additional problem of 'NADH oxidase' referred to in Chapter 1.

### 2.10.4 Artificial chromogenic substrates

Where the natural reaction is not accompanied by a useful absorbance change it is commonplace to use a synthetic substrate. Ideally, the enzyme should have the same specificity for both synthetic and natural substrates. When synthetic substrates are used to measure the enzyme in a crude mixture there is a very real risk of measuring an enzyme entirely different from that intended.

In the case of *p*-nitrophenol, one of the most commonly used chromogenic groups, the coloured species released by hydrolytic reactions is the nitrophenolate anion which protonates to the colourless acid form with a pK of 7.15. The anion has an extinction coefficient of 18 000 l $mol^{-1}$ $cm^{-1}$ (4). Care must be taken with such assays to ensure that the value used as extinction coefficient takes pH into account. The extinction coefficient that should be applied at different pH values takes the pK of the ionization into account according to Equation 5.

$$\varepsilon_{pH} = 18\,000 \times [10^{(pH-7.15)}/(10^{(pH-7.15)} + 1)] \quad (5)$$

*i. β-galactosidase*

The 'normal' reaction for this enzyme is the hydrolysis of the naturally-occurring β-galactoside, lactose, into galactose and glucose, a reaction which is not accompanied by an absorbance change. Assay of β-galactosidase is made extremely simple by using the synthetic substrate nitrophenyl-galactoside. The simplicity of this assay greatly assisted the classical work that determined the control of gene expression via the *lac* operon. The enzyme is very simply assayed in 0.1 M Tris–HCl pH 7.6, at a nitrophenyl-galactoside concentration of 0.1 mM. At the pH of the assay the effective extinction coefficient of the product nitrophenol is 13 300 l $mol^{-1}$ $cm^{-1}$.

*ii. α-amylase*

A chromogenic substrate, namely 4-nitrophenylmaltoheptoside, provides the basis of an assay for this enzyme. The reaction is linked to the release of nitrophenol by α-glucosidase. Conditions for the assay are 30°C, 0.1 M sodium phosphate, pH 7.1, 0.05 M NaCl, 30 units of α-glucosidase/ml and 5 mM 4-nitrophenyl-maltoheptoside. The use of this concentration of the linking enzyme, α-glucosidase, results in a lag of about 3 min.

*iii. Serine proteases*

Considerable success has been achieved in assaying different serine proteases using synthetic substrates in which the acyl moiety is an oligopeptide, the sequence of which is intended to give specificity. Trypsin can be assayed with benzoyl-argininine-4-nitrophenylalanine, but this substrate is hydrolysed at similar rates by other serine proteases with a preference for basic amino acids as the acyl donor in the scissile bond. The synthetic tetrapeptide benzoyl–Ile–Glu–Gly–Arg–4-nitroaniline is a better substate for the assay of trypsin in that sensitivity is increased 10-fold and specificity for trypsin is higher. Conditions for the assay are 40 mM Tris–HCl, pH 8.2, 16 mM $Ca^{2+}$, 0.1 mM benzoyl–Ile–Glu–Gly–Arg–4–nitroaniline.

*iv. Carboxypeptidase A*

A furoylacryloyl group incorporated at the amino terminus of a synthetic peptide provides an assay for this enzyme (5). Hydrolysis of the carboxy-terminal residue from the substrate furanacryloyl–Phe–Phe (1 mM) gives a decrease in absorbance of the chromophore. In order to avoid high absorbance of the substrate, the assay is best conducted at a wavelength where the absorbance change is not at a maximum. At 350 nm the extinction coefficient for the change is 800 l $mol^{-1}$ $cm^{-1}$.

### 2.10.5 Chromogenic reagents

When an enzyme-catalysed reaction creates a product with a reactive grouping not present in the substrate, it may be possible to include a reagent which will react directly with the product to form a chromophore. For a satisfactory continuous assay, the chromogenic reactant should not interfere with the enzymic reaction. The rate of the second reaction will be proportional to the concentration of the reagent and this must be present at a concentration high enough to avoid an unacceptable lag.

*i. Acetylcholinesterase*

Dithio-bis(2-nitrobenzoic acid) (DTNB, Ellman's reagent) reacts with free thiols in an exchange reaction that produces the yellow 4-nitrothiolate anion. Acetylcholinesterase is conveniently assayed by replacing the normal substrate with acetylthiocholine. Hydrolysis releases a thioester and the rate of release is continuously measured by a reaction with DTNB included in the assay solution. Conditions are 30°C; 0.1 M sodium phosphate, pH 7.5; 10 mM DTNB; 12.5 mM acetylthiocholine iodide.

### 2.10.6 Use of pH indicators in enzyme assays

Many reactions produce or consume protons and therefore are capable of altering the pH of a solution. The ionization of an appropriately chosen pH indicator may be exploited to convert the system into a spectrophotometric assay and in such systems the pH must be allowed to vary. However, so long as the pH change is small linear velocities are observed. A knowledge of the pH profile of the enzyme will help to decide the magnitude of the pH change that can be tolerated. The sensitivity of such a system decreases with increasing concentration of buffer and increases with increasing concentration of indicator. Linearity is best when indicator and buffer have the same pK (6). Both sensitivity and linearity are best at the pK of the indicator used. Clearly, if an enzyme is being assayed from different samples it is important that the sample does not contribute to the buffering capacity of the system or alter the pH. Quantification of rates of change must be accomplished by titration of the system using small volumes of standard acid or base and measuring the resulting absorbance changes. Amongst the enzymes that have been assayed in this way are the amino acid decarboxylases (6).

*i. Arginine decarboxylase*

This enzyme is very simply assayed in 0.05 M sodium acetate, pH 5.0, 0.025 M arginine, and including 10 μM bromocresol green as indicator. The rate of absorbance change is determined at 615 nm.

## 3. Turbidimetry

Enzymes that act upon insoluble polymers will frequently clarify a turbid solution and this property may be used to quantify the amount of enzyme present. The process involved is light scattering and not absorbance but it may be measured with a standard spectrophotometer. Such turbidimetric measurements are less easily standardized than absorbance measurements, partly because it is not easy to provide reproducible suspensions of insoluble polymeric substrate but also for reasons based upon instrumentation.

When an absorbance photometer is used to make measurements of turbidity, the reading that it gives is determined from the light transmitted by the solution in the same way that the instrument makes genuine absorbance measurements. In the case of turbidimetric measurements, however, some of the scattered light reaches the photodetector and the proportion depends on the distance of the detector from the cuvette (*Figure 12*). Clearly, an instrument (a) in which the cuvette is close to the detector, will receive more scattered light than one (b) in which detector and cuvette are further apart. For the same solution, instrument (a) will give a lower apparent absorbance reading than instrument (b). As an illustration, a turbidimetric assay for lysozyme is described.

The enzyme lysozyme has the function of hydrolysing bacterial cell walls and is conveniently assayed by observing the change in turbidity that occurs when it is added to a suspension of dried bacterial cells. This decrease in

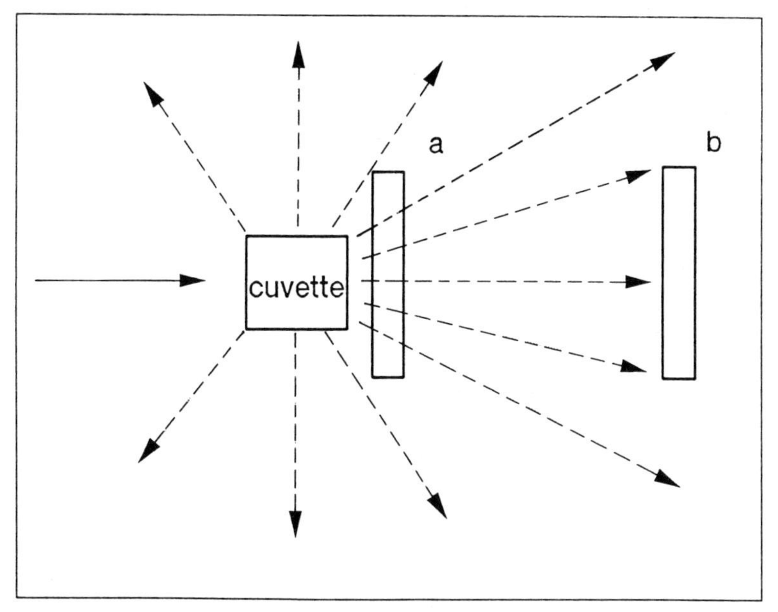


**Figure 12.** Use of a spectrophotometer for turbidimetric measurements. An instrument with photodetector in position **a** will detect more scattered light than one with the detector in position **b**.

turbidity is clearly the result of a complex process of progressive random hydrolysis and it is therefore not possible to express the rate in molar terms. The unit is defined in terms of the rate of decrease in turbidity. The wavelength chosen for these turbidity measurements is arbitrarily set at 450 nm and one unit of activity is defined as that which produces an initial rate of change in 'absorbance' of 0.001 per min when the volume in the cuvette is 2.6 ml (other conditions being pH 6.24 and 25 °C). It is important to note that in this case the volume in the cuvette must be specified. The same number of units added to a smaller volume of the same suspension would produce a proportionately higher change in turbidity.

## 4. Fluorescence

The phenomenon of fluorescence is, like that of absorption, the result of an electronic transition which converts the absorbing molecule to an excited state. Thus excitation and absorption are two words describing the same physical process. The difference between fluorescent and non-fluorescent compounds is determined by what happens when the excited state returns to the ground state. Whereas in non-fluorescent molecules the energy of the excited molecule is lost as heat, a fluorescent compound emits part of the energy as light. During the period ($\sim 10^{-9}$ s) between absorption and emission, the molecule loses some of its energy by vibrational relaxation so that the emitted light is of lower energy and consequently higher wavelength than the exciting light.

Fluorescence-based enzyme assays are potentially capable of much greater sensitivity than absorbance assays. This is because of a fundamental difference in the way that the measurements are made. Measurements of low values of absorbance are intrinsically difficult to make because they are based on the determination of two values of transmitted light that are high and nearly equal. The small amounts of light emitted by low concentrations of fluorescent material are intrinsically more readily measured because comparison is being made with the complete absence of emitted light when no fluor is present. It is by no means always true, however, that a fluorescence-based assay will be more sensitive than an absorbance-based assay based upon observation of the formation of the same compound.

## 4.1 The fluorimeter

*Figure 13* shows a schematic diagram of a fluorimeter. Up to the point of emergence of light from the cuvette the system is essentially the same as for an absorbance spectrophotometer. However, because fluorescence emission occurs in all directions, the detector is normally set to collect light emitted at 90° to the incident beam. A second monochromator is included for selecting the emission wavelength.

## 4.2 Quantitation of fluorescence

Although conditions may be arranged so that the relationship between measured fluorescence emission and concentration is virtually linear, the concentration of fluorophore in the sample cannot be calculated from the measured fluorescence emission by the application of a universal constant equivalent to an extinction coefficient.

A fluorescent compound emits a constant fraction of the light energy that it absorbs. This fraction is known as the quantum yield ($Q_f$). Strongly fluorescent materials have both high extinction coefficients and quantum yields close to 1. Although quantum yield is constant for a given set of experimental conditions, several features of the instrumentation prevent its use in the direct determination of concentration from fluorescence intensity measurements. The light incident on the cuvette varies from one instrument to another, and even with the same instrument from one occasion to another. Furthermore, although a given fraction of the absorbed light is emitted, only that part captured by the photodetector is registered, and this depends upon a variable feature of instrument design, namely the geometric relationship of cuvette to photodetector. Finally, although the photodetector response is proportional

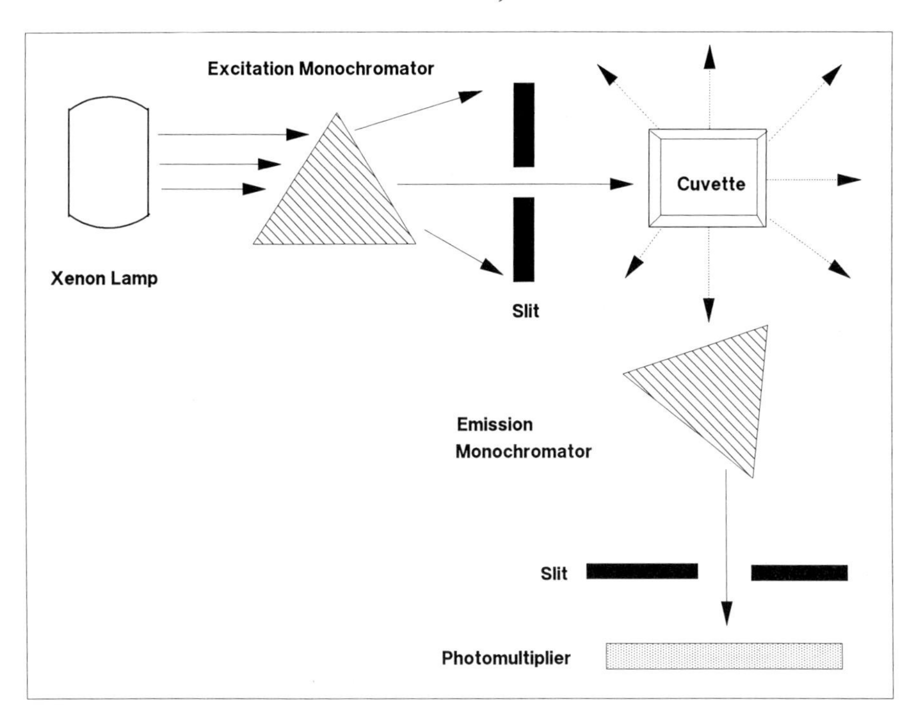


**Figure 13.** Essential features of a fluorimeter.

to the emitted light at a given wavelength, it does not provide an absolute measurement of light intensity, nor is its response the same at all wavelengths. Conversion of fluorescence emission values to units of concentration must therefore be achieved by the use of an appropriate standard.

The most straightforward standard to use is the same fluorescent compound that is generated by the assay. The concentration of this may be determined by weighing, or more simply and accurately by measuring its absorbance at the appropriate wavelength and calculating concentration using the extinction coefficient. When running an assay for the first time, it is essential to construct a calibration line relating fluorescence to concentration over the range that is expected from the assay. Thereafter, so long as a linear relationship exists, calculations may be based upon the measured fluorescence of a single sample of the standard at known concentration. It is essential that these calibration measurements be made using the standard sample of fluorophore under the same conditions as the fluorophore produced in the assay. In cases where the fluorescent product is not readily available an alternative secondary standard having fluorescence exitation and emission in the same region as the test material may be substituted. Alternatively, and very conveniently, manufacturers supply solid state fluors cast in the shape and size of a cuvette and having a wide range of fluorescent properties.

## 4.3 Causes of non-linearity—the inner filter effect

All fluorescent compounds absorb light at the excitation wavelength. This means that the intensity of incident light ($I_0$) falls exponentially as the beam progresses through a solution of a fluorescing compound. The relationship between fluorescence emission ($I_f$) and concentration of fluor ($c$) is therefore non-linear (Equation 6).

$$I_f = I_0 Q_f (1 - 10^{\varepsilon c l}) \quad (6)$$

In practice the relationship between *measured* fluorescence and concentration is more complex because of the way in which the instrument is constructed. When the absorbance of the solution at the exciting wavelength is high, the most intensely emitting part of the solution, that closest to the light source, is hidden behind the cuvette holder and the emitted light cannot be detected. A plot of measured fluorescence against concentration has the form of *Figure 14*, and it is quite possible therefore to conclude that a sample contains no fluorescent material when in fact it contains so much that the fluorescence cannot be detected. (An experimentalist beginning to use techniques based on fluorescence will gain a real understanding of the processes involved, together with much aesthetic pleasure from the colours produced, by observing the fluorescence in a cuvette directly with the lid of the fluorimeter open and adjusting the excitation wavelength, *taking care not to expose the eyes and skin to wavelengths shorter than 320 nm.*)

From a practical point of view a near-linear relationship between concentration and fluorescence emission is obtainable so long as absorbance values are sufficiently low (below 0.1).

## 4.4 Examples of fluorimetric enzyme assays

### 4.4.1 Direct observation of the natural reaction

Only a small proportion of the many naturally-occurring compounds that absorb UV and visible light are sufficiently fluorescent to provide useful enzyme assays.

*i. NAD(P)H dependent systems—glucose-6-phosphate dehydrogenase*

This enzyme is chosen as an example of the many dehydrogenases that may be assayed fluorimetrically by making use of the fluorescence of the reduced nicotinamide nucleotide coenzymes. The inner filter effect prevents the use of high concentrations of NAD(P)H and for this reason the systems most suited to fluorimetric assay are those run in the direction producing NAD(P)H. However, the very high sensitivity, often available from other fluorescence based assays, is not obtainable because the fluorescence of NADH is strongly quenched in aqueous solutions, the quantum yield being only approximately 0.03. The lower limits of detection of the two methods obviously depend upon the instruments used. Comparisons made by continuously recording over 3 min using two popular instruments (Beckman DU7 Spectrophotometer and

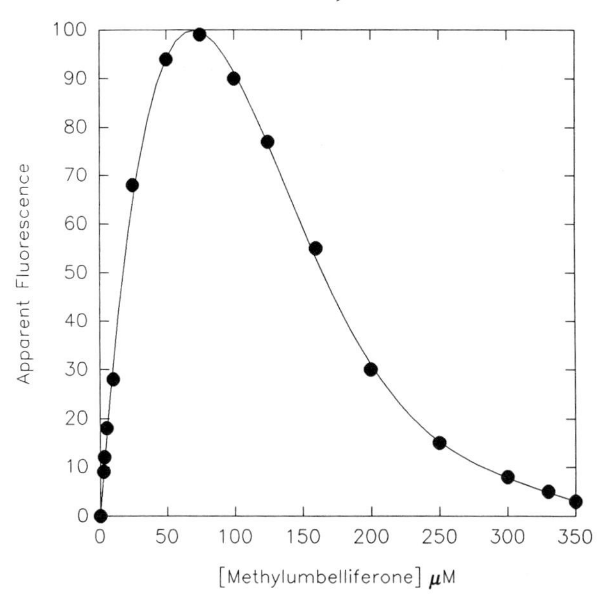


**Figure 14.** Relationship between measured fluorescence and concentration. The concentration dependence of the apparent fluorescence ($\lambda_{ex}$ = 355 nm, $\lambda_{em}$ = 460 nm) of methylumbelliferone was determined. The compound was dissolved in 0.1 M glycine, pH 10.0. At concentrations of fluor below about 10 μM the relationship is linear but becomes progressively more non-linear as light absorption becomes more significant. Eventually, at high concentrations, this inner-filter effect leads to a fall in measured fluorescence with increasing concentration.

Perkin Elmer LS 5 Luminescence Spectrometer) showed the lower limit of rate measurement by absorbance to be in the $10^{-7}$ M $min^{-1}$ range and that by fluorescence an order of magnitude more sensitive. At these levels of sensitivity both methods require the exclusion of dust from the solutions by filtration.

Glucose-6-phosphate dehydrogenase is assayed by adding 50 μl of enzyme sample to 3 ml of 0.1 M Tris–HCl, pH 7.8, containing 3 mM $MgCl_2$, 10 mM glucose 6-phosphate and 7 mM $NAD^+$ ($\lambda_{ex}$ = 340 nm, $\lambda_{em}$ = 465 nm).

*ii. Porphobilinogen deaminase*

Uroporphyrin I is a highly fluorescent compound formed by oxidation of uroporphyrinogen I, an intermediate that occurs in the later stages of the pathway leading to haem synthesis. Its determination by fluorimetry is used as the basis of an assay for porphobilinogen deaminase and for other enzymes in the same pathway (7).

---

**Protocol 2.** Fluorimetric assay of porphobilinogen deaminase

1. Mix 0.5 ml of Tris–HCl (0.1 M, pH 8.1) and 0.2 ml of 0.5 mM porphobilinogen.

**Protocol 2.** *Continued*

2. Start the reaction by adding 50 μl of enzyme sample.
3. Keep solution in the dark at 37°C for 30 min.
4. Stop the reaction by adding 0.25 ml of 50% (w/v) trichloroacetic acid.
5. Centrifuge to remove protein.
6. Expose to long-wavelength UV light to photo-oxidize the supernatant to uroporphyrin I.
7. Standardize using a solution of commercial uroporphyrin, the concentration of which has been determined by absorbance measurement and application of the extinction coefficient ($5.05 \times 10^5$ l $mol^{-1}$ $cm^{-1}$).

---

*iii. Anthranilate synthase*

Anthranilate, a key intermediate of aromatic amino acid metabolism, is fluorescent and its formation provides the basis of convenient linked assays of several enzymes as well as to anthranilate synthase itself (8).

$$\text{Chorismate} + \text{L-glutamine} \rightleftharpoons \text{anthranilate} + \text{pyruvate}$$

Conditions for assay are 20 mM L-glutamine, 10 mM $MgCl_2$, 0.1 mM chorismate, 25 mM mercaptoethanol in 50 mM potassium phosphate, pH 7.4. The reaction is followed at excitation wavelength 325 nm and emission wavelength 400 nm. Anthranilate is used as standard.

*iv. p-Aminobenzoate synthase*

In this assay *p*-aminobenzoate formed from the reaction of chorismate and glutamine is extracted into ethyl acetate and its concentration determined by comparison with standard PABA extracted in the same way (9).

The following reagents are mixed in the volumes indicated:

- Hepes 0.6 M, pH 7.8, 50 μl
- glutamine, 0.2 M, 100 μl
- guanosine 0.2 M, 50 μl
- chorismate 10 mM, 20 μl
- $MgCl_2$, 0.2 M, 50 μl
- solution containing 1.0 M EDTA, 60 mM mercaptoethanol, and 30% glycerol, 50 μl

The reaction is started by adding enzyme and water in a total volume of 50 μl. The reaction is stopped by adding 100 μl of 1 M HCl, the PABA formed is extracted into 1.5 ml of ethyl acetate and its fluorescence measured ($\lambda_{ex}$ = 290 nm, $\lambda_{em}$ = 340 nm). Concentrations are determined using standard PABA solution.

### 4.4.2 Synthetic fluorigenic substrates

Just as nitrophenol and nitroaniline have been used extensively in the preparation of artificial chromogenic substrates for esterases and peptidases, respectively, very many artificial substrates for hydrolases have been based upon the naturally occurring fluor umbelliferone (7-hydroxycoumarin) and its amino analogue 7-amino 4-methylcoumarin. A consensus of the many papers describing these methods is that they are potentially at least two orders of magnitude more sensitive than the corresponding assays in which nitrophenol or nitroaniline are released. Methylumbelliferone can be reliably measured at concentrations of $10^{-10}$ M. Unfortunately, the useful fluorescent properties ($\lambda_{ex} = 360$ nm, $\lambda_{em} = 455$ nm) of methylumbelliferone reside in the anion which is not formed at the acid pH values appropriate to the assay of many hydrolases. The enzyme-catalysed part of assays based on the release of this compound are therefore usually conducted at low pH. The reaction is then stopped and the fluorescent anion formed by the addition of a strong alkaline buffer. The pK for the conversion is 7.8 (10) so that this type of substrate can be used continuously for hydrolases such as alkaline phosphatase that act at alkaline pH values.

Many continuous fluorescent assays for peptidase substrates have been based on synthetic oligopeptides in which the scissile bond is an amide with 7-amino-4-methylcoumarin. There is sufficient difference between the fluorescence spectra of substrate and product that assays for peptidases can be made continuously over a range of pH values. However, there is significant overlap of the spectra and this means that the potential sensitivity of the fluorescence method is not fully realized. Despite conducting assays at excitation and emission wavelengths which are not maximal for the product, a significant fluorescence contribution from the substrate itself cannot be avoided. The assay of elastase with a fluorogenic peptide substrate of this type was found to be less sensitive than an absorbance assay based on thiopyridine (11). Attempts to overcome the overlap problem have been made by the introduction of fluors with different properties such as 6-aminoquinoline (12) and aminoaridone (13).

*i. β-Glucuronidase*

Assay of this enzyme using methylumbelliferyl glucuronide was first described in 1955 (14) and may be accomplished by adding the enzyme sample (10 μl) to 1 ml of 0.25 mM methylumbelliferyl glucuronide in 0.1 M sodium acetate, pH 4.6. After an appropriate period (1–30 min) the reaction is stopped by adding 4 ml of 0.1 M glycine/NaOH buffer, pH 10.4.

*ii. Lysozyme*

Assay of this enzyme provides an example of the use of a more complex fluorogenic substrate, namely methylumbelliferyl-β-D-N,N′,N″-triacetylchitotriose to make use of the enzyme's known substrate-binding properties. The enzyme may be assayed at 40°C using this compound (3 μM) dissolved in 20 mM sodium acetate, pH 5.0, containing 0.1 M NaCl (15). After time intervals appropriate to the amount of enzyme used, the reaction is stopped by adding an equal volume of 0.2 M glycine/NaOH at pH 10.4. Any precipitate that has formed should be removed by centrifugation and the fluorescence determined ($\lambda_{ex}$ = 360 nm, $\lambda_{em}$ = 450 nm).

*iii. Elastase*

The synthetic peptide substrate methoxysuccinyl–Ala–Ala–Pro–Val–aminomethylcoumarin provides a sensitive assay for elastase. By appropriate choice of excitation and emission wavelengths (370 nm and 460 nm) more than 70% of the maximum fluorescence of aminomethylcoumarin can be used while the contribution of the unhydrolysed substrate contributes only 0.5% of its maximum fluorescence. The system is sensitive enough to detect 10 nM product and is linear to about 5 μM. Elastase may be assayed at pH 7.6 and 25°C in 50 mM Tris–HCl, 0.5 M NaCl, 0.1 M CaCl2, (11).

### 4.4.3 Relief of quenching

An ingenious method for assaying hydrolases relies upon a process known as radiationless energy transfer in which the energy from a fluorescent group on one part of a polymeric substrate is transferred without emission of light to a chromophore nearby in the same molecule. Hydrolysis of the susceptible bond interrupts the process by separating the two interacting groups. In one adaptation of this system, energy absorbed as light by excitation of the donor fluorophore at its characteristic excitation wavelength is transferred directly to the acceptor fluorophore and then emitted as light at a wavelength characteristic of the acceptor. The rate of hydrolysis is monitored either by measuring the decrease in fluorescence of the acceptor or the increasing fluorescence of the donor.

*i. Carboxypeptidase A*

The fluorescence of tryptophan is quenched by the dansylation of the amino terminus of the carboxypeptidase substrate glycyl-tryptophan to give dansyl-L-glycyl-L-tryptophan. Quenching is relieved by hydrolysis of the peptide bond. Enzyme is added to the synthetic substrate dissolved in 50 mM Hepes buffer, pH 7.5, containing 1 M NaCl. Fluorescence excitation is at 290 nm and emission is measured at 350 nm. At the start of the reaction the fluorescence at this wavelength is only 1% of that which results upon complete hydrolysis of the substrate (16).

*ii. Aminopeptidase P*

This enzyme hydrolyses the amino-terminal residue from peptides in which the next residue is proline. In the fluorogenic substrate (17) the fluorescence

2-aminobenzoyl group is linked by ethylene diamine to the carboxy terminus of the tripeptide $NH_2$–Lys–Pro–Pro–COOH. The quenching dinitrophenyl group is linked to the ε-amino group of the N-terminal lysine. A 160-fold increase in fluorescence emission is observed when the substrate is cleaved at the DNP–Lys–Pro bond. In the assay, 20 μl of enzyme sample is added to 0.1 ml of 5 mM substrate in 0.2 M Tris–HCl, pH 8, containing 2.5 mM manganese sulphate and 10 mM trisodium citrate. The reaction is stopped after 20 min by adding 4.3 ml of a solution containing 1 mM dithiothreitol and 50 mM EDTA. Concentrations are determined by comparing increase in fluorescence ($\lambda_{ex}$ = 320 nm, $\lambda_{em}$ = 410 nm) with aminobenzoxyglycine standard.

*iii. Phospholipase*

Fluorescence is frequently quenched by a change in polarity of the environment of the fluorophore. The strong fluorescence of 6-carboxy fluorescein is quenched when it is incorporated into lecithin liposomes. A method for the assay of phospholipase is based on the disruption of the liposomes brought about by the hydrolysis of the phosphatidyl choline units from which the liposomes are composed (18).

### 4.4.4 Fluorogenic reagents

*i. Amine oxidases*

Enzymes generating hydrogen peroxide can be adapted to sensitive fluorometric assays using a continuous system in which homovanillic acid is converted to a fluorophore in the presence of horse radish peroxidase. As an example (19), diamine oxidase is assayed in 0.1 M phosphate, pH 7.8, containing 0.5 mM homovanillic acid and 10 μg/ml horse radish peroxidase. Enzyme in the form of tissue samples is first mixed and shaken at 37 °C for 10 min with buffer and peroxidase before starting the reaction by adding homovanillic acid and 0.1 mM putrescine as substrate. The reaction is followed continuously ($\lambda_{ex}$ = 315 nm, $\lambda_{em}$ = 425 nm).

## References


1. *Methods in Enzymology*. Academic Press, New York.
2. Bergmeyer, H. U. (1986). *Methods in Enzymatic Analysis*, Vol. 1–4. Verlag Chernie, Weinheim, Germany.
3. Burns, D. H. and Aberhart, D. J. (1988). *Anal. Biochem.*, **171,** 339.
4. Khalifah, R. G. (1971). *J. Biol. Chem.*, **246,** 2561.
5. Plummer, T. H. and Kimmel, M. T. (1980). *Anal. Biochem.*, **108,** 348.
6. Rosenberg, R. M., Herreid, R. M., Piazza, G. J., and O'Leary, M. H. (1989). *Anal. Biochem.*, **181,** 59.
7. Bishop, D. F. and Desnick, R. J. (1986). *Meth. Enzymol.*, **123,** 339.
8. Gozo, Y., Zalkin, H., Kein, P. S., and Heinrisburg, R. L. (1976). *J. Biol. Chem.*, **251,** 941.

9. Zalkin, H. (1985). *Meth. Enzymol.,* **113,** 293.
10. Yakatan, G. J., Juneau, R. J., and Schulman, S. G. (1972). *Anal. Chem.,* **44,** 1044.
11. Castillo, M. J., Kiichiro, N., Zimmerman, M., and Powers, J. C. (1979). *Anal. Biochem.,* **99,** 53.
12. Byrnes, P. J., Bevilaqua, P., and Green, A. (1981). *Anal. Biochem.,* **116,** 408.
13. Baustert, J. H., Wolfbeius, O. S., Moser, R., and Koller, E. (1988). *Anal. Biochem.,* **171,** 393.
14. Mead, J. A. R., Smith, J. N., and Williams, R. T. (1955). *Biochem. J.,* **61,** 569.
15. Yang, Y. and Hamaguchi, K. (1980). *J. Biochem.* (Tokyo), **87,** 1003.
16. S. Latt, S. A., Auld, D. S., and Vallee, B. L. (1972). *Anal. Biochem.,* **50,** 56.
17. Holtzman, F. Pittey, G., Rosenthal, T., and Vaner, A. (1987). *Anal. Biochem.,* **162,** 476.
18. Chen, R. F. (1977). *Anal. Lett.,* **10,** 787.
19. Snyder, S. H. and Hendley, E. D. (1971). *Meth. Enzymol.,* **17B,** 741.

# 3

# Radiometric assays

K. G. OLDHAM

## 1. Introduction

The accurate measurement of enzymatic activity in biological samples is important in many fields of biochemistry, not only in routine biochemistry and in fundamental research but also in clinical and pharmacological research and diagnosis. To be acceptable an assay should be specific, sensitive, quantitative, simple, and rapid. It should also be unaffected by side reactions and by the presence of drugs or anti-metabolites in samples, and it should allow the assay of enzymatic activity in both crude and purified enzyme samples.

Radiometric enzyme assays can fulfil all of these requirements and offer many advantages over non-radiometric techniques for nearly all enzyme assays, other than those which are carried out during enzyme-reaction mechanism studies which require the measurement of very rapid reaction rates.

First used in the 1950s, the application of radiometric enzyme assays was initially restricted by the limited commercial availability of suitable labelled substrates and counting equipment, and by a lack of simple and rapid techniques for the quantitative separation of substrates and products. The disappearance of these constraints in the sixties led to a rapid increase in the use of this technology, and publications involving its use are now numbered in thousands.

There are now a large number of suitable enzyme substrates commercially available, labelled with $^{14}C$, $^{3}H$, $^{32}P$ and, to a lesser extent, with $^{35}S$ and $^{125}I$. In most cases, the successful development of a radiometric assay for an enzyme will depend on the availability of a satisfactory quantitative method of separation of labelled substrate and product.

One problem, evident in the early days, still exists; because details of radiometric assays are invariably to be found only in the 'methods' sections of most published work, with no reference to their use appearing in either the title or abstract, it is difficult to find the necessary information without searching through a vast amount of published work. Another problem has been that, in much published work, the pressure was more on quick publication than on optimizing and simplifying the assays. As a result, many workers have continued to use complicated and tedious old, but tried, published methods when quicker and simpler methods are possible.

It is not possible in the limited space available here to give complete coverage of all the radiometric enzyme assays currently in use, nor even detailed protocols for sufficient assays to satisfy the majority of readers. Instead three important topics will be covered; these are:

(a) the major separation techniques available, paying particular attention to recent developments and improvements in their use
(b) experimental design to achieve optimum results
(c) problems and pitfalls, specific to individual techniques, which await the often unsuspecting users of labelled compounds

Avoidance of pitfalls is merely a matter of experience and common sense because they are simple to detect and overcome. Failure to recognize their presence may, at least, result in meaningless and wasted work, and at the worst in the gross misinterpretation of data.

Despite an increase in the use of non-radiometric techniques in many fields of biochemistry in the last few years, there has been a marked resurgence in the use of radiometric enzyme assays. This has been most evident in pharmacological and molecular biological research where there is a demand for large numbers of screening assays in drug-inhibition and gene-expression studies. The high sensitivity, specificity, and freedom from interference of radiometric assays make them highly suitable for screening studies. Cost- and time-efficiency are important in large-volume work. The last section of this chapter reviews recent developments in which the automation/semi-automation of assays has led to marked reductions in the 'hands on' time required, and where the miniaturization of assays has reduced the cost of reagents and consumables, and also minimized the use of flammable—and sometimes toxic—solvents and scintillants.

## 2. Techniques

The general principles of enzyme assay are covered in detail elsewhere in this book (Chapter 1). It is sufficient here to say that enzymatic activity should be measured under conditions where a linear relationship exists between initial rate and enzyme concentration. Methods of following enzymatic reactions are of two types; continuous and sampling. In both methods either product formation or substrate utilization may be measured but, unless the primary product is rapidly metabolized by other enzymes, it is normal practice to follow the reaction by estimating product formation, a procedure inherently more sensitive than the following of substrate utilization.

With but a few exceptions, radiometric enzyme assays are based on the conversion of radio-labelled substrates to labelled products and are carried out by sampling methods. The rate of reaction is determined by removing samples from the reaction mixture at appropriate time intervals and measuring the radioactivity of either product or residual substrate after their quantitative separation. Conversion of d.p.m. product formation to μmol product can then be calculated using the known substrate specific activity.

The two major requirements of a radiometric enzyme assay are the availability of a suitable labelled substrate of known specific activity, and of a simple and rapid method of quantitative separation of substrate and product. As there are now few enzymes for which suitably-labelled substrates are not readily available commercially, the most important requirement is the availability of a satisfactory separation technique.

Bibliographies are available (1, 2) giving references to several hundred radiometric assays in much more detail than is possible in this brief chapter. Much useful information is contained in two other publications (3, 4) which, though unfortunately out of print, may still be found in many departmental libraries.

People wishing to set up their own assays are advised not only to refer to these earlier publications, but also to carry out a computer search to identify recent publications on the enzyme of interest and to obtain details of any radiometric assays used by the authors (bearing in mind that the choice of assay method is often a personal choice with different authors using different techniques to assay similar enzymes). By using the guidelines in this article, readers may then develop and optimize the most appropriate assays for their desired application.

Unless otherwise stated, all the radioactive procedures discussed here employ liquid scintillation counting for the measurement of radioactivity.

## 2.1 Ion-exchange methods

Ion-exchange techniques are used extensively in preparative organic chemistry and it is not surprising that micro-scale ion-exchange separations have been used to separate substrates and products in a large number of radiometric enzyme assays. Both ion-exchange resin and ion-exchange paper (disc and sheet) methods are widely used.

### 2.1.1 Ion-exchange resin methods

In the simplest methods, small columns—commonly made from Pasteur pipettes plugged with glass wool—are used, one component being retained whilst the other passes straight through. If necessary the retained component can be removed with a suitable eluent.

Although convenient for many small-volume assays, this method has some disadvantages when large numbers of assays are needed; preparing and using the columns (application, washing, and elution) is time-consuming and wasteful of consumables, and reduced sensitivity can result if the product is eluted in a volume larger than is compatible with the scintillant so that only a fraction of the analyte sample can be counted.

Considerable reductions in 'hands on' time can be made by using commercially-available mini-columns such as Amersham's Amprep™ and Waters' Sep-Pak™ cartridges. These are now available containing a wide range of sorbents, together with detailed protocols covering a number of common separations. Vacuum manifolds, which permit the simultaneous and rapid processing of up to 40 samples at a time (5), are also available.

An added benefit of these mini-columns is that they usually give elution in small volumes, which allows the use of mini scintillation vials leading to considerable savings in scintillant usage. Production of columns on a commercial scale also avoids the variable reproducibility which often results when small numbers of columns are produced at intervals in-house.

Although these mass-produced columns are highly reproducible, a recent report (6), where Sep-Pak exchange cartridges were used to separate inositol phosphates, warned that changes in the manufacturing process had adversely affected the separation performance. People using such columns would, therefore, be well advised to carry out periodic controls to check that the separation characteristics remain unchanged.

In assays where it is necessary to count the activity not retained by the resin, it is simpler and quicker, instead of using a column, to carry out a batch-wise separation by adding the sample to an appropriate amount of resin in a centrifuge tube, mixing to equilibrate, and then counting the radioactivity in the supernatant after centrifugation (7, 8).

In another assay (9), of myristoyl CoA:protein *N*-myristoyl transferase, where both unreacted substrate (and decomposition products) and the reaction product required large-volume washes for satisfactory elution, the separation process was simplified by applying the assay sample to a small CM-Sepharose column and removing unreacted myristoyl-CoA and free myristic acid by washing seven times with Tris–HCl buffer, pH 7.8. The bottom of the column was then capped and the resin suspended in a 'stripping' eluent (Tris–HCl/NaCl) and then transferred to a scintillation vial using a wide-mouthed disposable plastic pipette.

When tritium-labelled substrates are used, advantage may be taken of the quenching, by the resin, of the soft β-emissions of adsorbed compounds. The reaction and ion-exchange separation are carried out in the scintillation vial; quenching by the resin reduces the counts due to the adsorbed substrate to a negligible level so that, on addition of the scintillant, only the activity of the unadsorbed product is counted.

Some workers, using this simple technique in a cyclic nucleotide phosphodiesterase assay (10), reported reduced sensitivity due to blank readings of 5–7%. Because tritiated purine nucleotides—particularly those labelled in the 8-position—can undergo a slow exchange of tritium with aqueous solvents on prolonged storage (11), the most probable cause of the reported high blanks was tritium exchange on storage. This problem is, however, easily overcome when high sensitivity is required, by lyophilizing bulk solutions before use to remove any tritiated water formed during prolonged storage (1–3, 12).

A generally-applicable assay for amino acid decarboxylases (13), which combines the advantages of ion-exchange with those of solvent extraction methods, used a liquid cation-exchanger, DEHPA (bisdiethylhexyl phosphoric acid), and $^{14}$C-labelled substrates. Amine products were extracted into a DEHPA-chloroform phase leaving the acidic substrates in the aqueous phase. The assay was easier and quicker than the usual $^{14}CO_2$-release assays and also easier to adapt to microscale. Assays capable of detecting less than $10^{-11}$ mole $^{14}$C-product could be performed in a 10 μl volume.

### 2.1.2 Ion-exchange paper methods

Using both conventional paper-chromatographic and paper-disc filtration methods, ion-exchange papers have been extensively used in radiometric assays of a wide range of enzymes (2). Disc-filtration methods are preferred because separation by filtration is rapid, and large numbers of samples can be processed simultaneously using commercially-available vacuum manifolds.

Non-specific binding of both the labelled substrate and trace impurities to the paper discs can contribute to blank readings. For maximum sensitivity in assays, such blanks can easily be reduced by methods described in references 1, 3, and 14.

Some workers (15) reported that a DEAE–cellulose disc assay of glucose-1-phosphate adenylyltransferase consistently gave lower results than three other assays. They attributed this to a reduction in product retention caused by the high $MgCl_2$ concentrations used in the assay. An alternative explanation (2) could be that high $Mg^{2+}$ concentrations resulted in the formation of a Mg–ADP–glucose complex, retained by the DEAE disc less effectively than free ADP–glucose. Whatever the cause, dilution of the samples before filtration overcame the problem. Researchers authenticating new ion-exchange paper disc assays should, therefore, avoid possible retention problems by investigating the effect of pre-filtration dilution of samples.

## 2.2 Precipitation of macromolecules

Many enzymes involved in the synthesis of macromolecules such as DNA, RNA, amino-acyltRNA, polypeptides and polysaccharides, and related kinases have been assayed using chiefly $^{14}$C- and $^{3}$H-labelled substrates and ATP–γ–$^{32}$P (for the kinases) and measuring the radioactivity incorporated into a form insoluble in acidic or organic solvents (1–4). Precipitation methods, using $^{14}$C-, $^{3}$H-, $^{35}$S-, $^{32}$P-, $^{125}$I-, and $^{131}$I-labelled substrates, have also proved useful in assays of a wide range of macromolecule-degrading enzymes (1–4).

### 2.2.1 Assays of synthetic reactions

In macromolecular synthesis studies, although samples can be processed simply by isolating the labelled product by centrifugation, most workers use modifications of Bollum's paper-disc method (16). In this method, which allows large numbers of samples to be processed simultaneously, samples are applied to discs or squares of filter paper or glass fibre; these are then dropped into a bath of the appropriate cold solvent which both stops the reaction and precipitates the macromolecule throughout the disc or square. Various ways of simplifying this process, and some problems and pitfalls specific to the technique, are described in detail elsewhere (1–4, 12, 14).

Two publications (12, 14) have drawn attention to the fact that radiation-decomposition products can spuriously be precipitated in many assays of macromolecular synthesis. Although only small amounts of these impurities are likely to be present in most labelled substrates, their effect will be magnified if all the trace impurity, but only a small proportion of the labelled substrate, is incorporated. A very simple way of removing such troublesome trace impurities has been described (12, 14).

Tritium's soft β-emission can result in large self-absorption losses when tritiated precipitates are counted in heterogeneous solution (Section 4.1). The following comments are specific to precipitation assays using tritiated substrates.

When macromolecular precipitation methods are used, the weight of precipitate, and hence the size of any self-absorption losses, will be dependent not only on the amount of acceptor present but also on the amount of protein or other macromolecule in the sample. This will vary both with the amount of enzyme extract and with its purity. A simple method of checking if self-absorption is a problem is to prepare duplicate or triplicate samples in which different weights of macromolecule are precipitated from different volumes of reaction mixture. If self-absorption losses are insignificant, there will be a linear relationship between the count rate and the volume of sample. If a non-linear plot is obtained, its shape will indicate the magnitude of the self-absorption losses. The possibility of self-absorption losses varying from sample to sample can be minimized, and reproducibility enhanced, by ensuring that the precipitate is distributed in a uniform layer over the filter.

Many workers have tried to correct for self-absorption losses by calibrating against a standard prepared by drying a known amount of the low-molecular-weight tritiated precursor on a paper disc or membrane filter. This procedure has led to up to four-fold overestimation of product formation (48, 51). This arose because the smaller molecules, which could penetrate filter fibres, suffered much more self-absorption losses than did the thin layers of macromolecules which could not penetrate the fibres.

In recent years there has been an increasing interest in the role of the post-translational modification of proteins in cellular regulation. Much attention has focussed on assays of two important protein kinases: tyrosine kinase (17) and protein kinase C (18), and an acyltransferase—protein, N-myristoyl transferase (9).

Assays for these enzymes have the two major requirements of sensitivity, because they normally occur at very low levels, and specificity, because the tissues in which they are found usually contain higher levels of similar but less specific enzymes. As an example, tyrosine kinase assays can suffer interference from the more abundant serine and threonine kinases, the calcium- and phospholipid-dependent protein kinase C from a variety of other protein kinases, and protein *N*-myristoyl transferase from the more common acyltransferases, which catalyse the palmitoylation of cysteine sulphydryl groups and serine and threonine hydroxyl groups.

In these assays, specificity was greatly increased by replacing naturally-occurring protein substrates with synthetic peptides which could not act as acceptors for the contaminating enzymes. A tyrosine-specific kinase assay (17), for example, used as acceptor the synthetic polymer (Glu:Tyr, 4:1). Protein *N*-myristoyl transferase specifically acylates $NH_2$-terminal glycine, enabling a specific assay (12, 19) to be developed using, as acceptor, the synthetic $NH_2$-terminal glycine peptide Gly–Asn–Ala–Ala–Ala–Arg–($NH_2$).

When synthetic peptides are used as substrates in these assays instead of the much higher molecular weight proteins, TCA precipitation methods are often substituted by methods in which the modified peptides bind tightly to ion-exchange paper discs. This is easily done and often results in lower blanks and higher sensitivity (18, 20).

### 2.2.2 Assays of degradation reactions

Precipitation methods may also be used for the rapid assay of macromolecule-degrading enzymes (1–3, 16). For the highest sensitivity and precision the best method is to precipitate the unreacted substrate and remove it by centrifugation before measuring the radioactivity of the product in the supernatant. Processing may easily be simplified and speeded up by using centrifugal micro-concentrators, commercially available from Amicon Ltd, Alltech Associates, and others. However, in screening assays, when large numbers of samples need processing, it may prove quicker, simpler, and cheaper, though less sensitive and precise, to use a paper-disc method and measure the radioactivity of the residual substrate.

Problems have been reported in assays of some macromolecule-degrading enzymes where products are large enough to be co-precipitated with residual substrate and so cause an underestimation of the enzymatic activity (1, 2, 16). Two methods have been used to overcome this problem. Instead of using straightforward centrifugation, samples can be filtered through cellulose–ester membrane filters with an appropriate molecular size cut-off, so that the whole macromolecular substrate but not the smaller, but still relatively large, product is retained (1, 2, 16). Commercially-available micro-concentration tubes, fitted with a range of molecular size cut-off membranes, may also be used.

Differential precipitation methods (21) can be used to ensure complete separation of product and residual substrate. Stetler-Stevenson *et al.* (22), for example, have recently described an assay of type IV collagenase in which the final percentage of TCA/tannic acid precipitant was carefully adjusted to ensure optimum separation of the tritiated substrate and product.

## 2.3 Solvent extraction methods

A simple, rapid method of enzyme assay is possible when substrate and product can be separated by solvent extraction (1–3). When substrate and product are not directly separable in this way, separation can often be achieved after forming a derivative of one component (2, 3); for example, inorganic phosphate can be converted to the butanol-soluble phosphomolybdate complex; acetylcholine forms a ketone-soluble complex with tetraphenylboron; keto acids can be converted into 2,4-dinitrophenylhydrazones which are soluble in ethyl acetate.

For maximum sensitivity, the conflicting demands of quantitative recovery and complete separation of substrate and product must be met (Section 3). Unfortunately, this is rarely achieved. In many assays a complete separation of substrate and product is not possible, although usually less than 1% of the substrate is extracted with the product. The resulting reduction in sensitivity is normally more than compensated for by the speed and simplicity of the method.

If a higher sensitivity is required, this can be achieved by modifying the extraction procedure by back-washing the organic solvent with water, using a different solvent or different substrate, or all three. As an example, in an assay of catechol-*O*-methyltransferase, a 1000-fold increase in sensitivity was attained by using dihydroxybenzoic acid as substrate instead of the more usual adrenaline, and by using ethyl acetate instead of toluene:isoamyl alcohol as the extractant (2, 23).

Expression of chloramphenicol acetyltransferase (CAT) is the most widely-applied measure of promoter strength in transfected cells. In the usual assay (24) residual substrate and product are first extracted into ethyl acetate before separation by thin-layer chromatography and counting. A simple, solvent-extraction assay, linear over two or three orders of magnitude, was developed by exploiting the enzyme's low specificity for the acyl donor, and using butyrylCoA instead of acetylCoA as acyl donor, and $^{14}C$- or $^{3}H$-labelled chloramphenicol as acceptor (25). The greater hydrophobicity of butyryl-chloramphenicol facilitated an effective separation of acylated chloramphenicol by solvent extraction into a 2:1 mixture of tetramethyl pentadecane (a well-defined non-volatile alkane) and xylene. Blanks, particularly with the tritiated substrate, were markedly reduced by pre-washing the labelled substrate with solvent before use in the assay.

Many solvent extraction assays have been simplified by extracting products into a solvent containing scintillant so that the extract can be counted directly (1, 2). Some workers have further speeded up these assays by freezing the lower aqueous layers and pouring the upper, organic phases directly into scintillation vials.

Tritium is a soft β-emitter with a very short (6 μm) range in water and therefore, when tritiated substrates are used, product activity extracted into scintillant-containing solvents can be counted by liquid scintillation without separating the aqueous phase. When the more energetic β-emitting $^{14}C$ label is used, high blanks, caused by a contribution from residual substrate in the aqueous phase, usually make it necessary to separate the two phases and count an aliquot of the organic phase. This technique has been used in routine assays of many proteolytic enzymes (1), and more recently in a CAT assay (26) which reported good results using $^{3}H$- or $^{14}C$-acetylCoA or $^{14}C$-butyrylCoA as substrate.

As with all assay techniques, a few pitfalls are possible in solvent extraction assays; for example, some organic solvents can cause quenching in scintillation counting. This problem may be overcome easily by selecting an alternative solvent which does not quench, by removing the solvent by evaporation (providing the product is not volatile) before adding scintillant, or by minimizing quenching by using a micro-technique so that only a small amount of sample is added to each vial.

Much pharmacological research involves studies of the effects of drugs, activators, inhibitors, and anti-metabolites on enzymatic activity. The presence of such additives in analytical samples can affect the solubility of substrate and/or product in the solvent. This could cause errors by influencing either the degree of separation or the recovery of products. In all assays, not only ones using solvent extraction methods, it is therefore important to check that such interference is absent.

Some solvent extraction assays (1, 27) have been complicated by the presence of volatile impurities present in the substrate or formed in side reactions and this has resulted in relatively high blank readings. Provided that the product itself is not volatile, such blanks may easily be reduced by drying to remove the solvent and volatile impurities before counting.

Irrespective of the separation technique used, resolution of substrate and product is enhanced by the addition of non-radioactive carriers (Section 4.2). This is particularly so in solvent extraction assays. Because many solvents can undergo decomposition on prolonged storage, forming potentially-interfering impurities, it is always advisable to use freshly-prepared solvents. The importance of this was shown in sensitive chloramphenicol acetyltransferase assays in which product and residual substrate are extracted into ethyl acetate before TLC analysis. It has been shown (28) that the solvent (or more likely acetic acid formed during prolonged storage) could acetylate chloramphenicol, forming traces of acetylchloramphenicol and thereby giving unacceptably high blanks in assays where very high sensitivity was required. This problem can easily be overcome by using fresh solvent containing chloramphenicol as carrier, and by avoiding storage of ethyl acetate extracts before TLC analysis.

## 2.4 Release or uptake of volatile radioactivity

### 2.4.1 Release of tritium as tritiated water

Tritium release, from specifically-labelled substrates, first used for the assay of the amino acid hydroxylases (29), has since been used as the basis of the rapid and simple assay of a large number of enzymes (1–3).

Some enzymes, which do not themselves catalyse tritium release from their substrates, can be assayed by coupling to another enzyme system capable of catalysing tritium release from the product. A recent example of this approach is a new assay for nucleoside phosphorylase, claimed to be better than previous radiometric assays relying on chromatographic separation of substrate and product (30). Using a [2-$^{3}$H]inosine substrate, the addition of xanthine oxidase to the assay reaction mixture catalysed tritium release from the primary product.

Tritiated water formed in assays may be recovered easily for counting by simple distillation, if the substrate is not volatile. This method of isolating the product is, however, too time-consuming when large numbers of samples need processing. It is more usual nowadays to use small columns of ion-exchange resin, charcoal, or alumina to retain residual substrate. Flow rates, particularly through charcoal and alumina, are often low, and assays may be much speeded up by centrifugation using commercially available disposable columns such as Amicon C-10 micro-concentrators (31).

Several pitfalls await the unwary user of tritium-release assays and these can result either in apparently low rates of reaction, which can be misinterpreted as kinetic isotope effects (32), or in high blank readings which reduce assay sensitivity (12). These pitfalls, and easy ways of detecting and overcoming them, have been dealt with in greater detail elsewhere (1–4, 12, 32), but are summarized briefly here.

Low rates of tritium-release will result if the position is not exactly as specified. The advent of Tritium NMR as an analytical tool to identify the exact position of labelling (33) and the demise of tritium gas-exchange reactions in the preparation of tritiated compounds mean that this problem is unlikely to be encountered nowadays.

Intramolecular migration of tritium—the so-called NIH shift—which causes tritium to be retained in the product, instead of being released, is believed to be a common feature of all enzymatic aromatic amino acid hydroxylations (32, 34), and may occur in other reactions (32). If such an effect occurs, it will result in an underestimation of enzymatic activity. Its occurrence, however, will be easy to detect as the product, which should be unlabelled (assuming the substrate was singly-labelled), will become labelled.

Such effects have been overcome easily by the chemical displacement of tritium from the labelled product (35).

High blank readings may be caused by chemical labilization of the tritium label (33). Many tritiated compounds undergo a very slow exchange of tritium with the solvent on storage, without necessarily undergoing significant radiolysis. If such solutions are used in tritium-release assays, the tritium released on storage may cause a significant blank reading. High blanks, caused in this way, can easily be reduced by lyophilizing the substrate solution, before use, to remove tritiated water formed by exchange (1–4, 36).

### 2.4.2 Release or uptake of carbon-14 dioxide

*i. Release of $^{14}CO_2$*

A large number of decarboxylating enzymes have been easily and accurately assayed by measuring $^{14}CO_2$ evolution from [$^{14}$C]-carboxyl-labelled substrates. Other enzymes, such as the carbamoyl transferases, arginosuccinate synthetase, and arginase, have been assayed by using [$^{14}$C]-carbamoyl-labelled substrates, and chemically or enzymatically decarboxylating the products. Several enzymes, such as collagenase, prolyl hydroxylase, dopamine-β-hydroxylase, and citrate synthase, which do not immediately suggest the possibility of $^{14}CO_2$-release assays, have also been assayed by this technique (1–3).

Although $^{14}CO_2$-release assays are simple and precise, they can be time-consuming when large numbers of assays are required. Also, many have previously been carried out in relatively large volume (~1 ml), probably as a result of earlier experience with conventional non-radioactive $CO_2$-release assays using the Warburg apparatus. Much attention has therefore been concentrated on attempts to increase sensitivity and to carry out assays in small volumes, both of which offer large cost savings, and to develop continuous and semi-automatic assay methods which would be less time-consuming.

Developments in these directions up to 1977 are described in detail elsewhere (1). Three recent developments, generally applicable for sensitive decarboxylase assays and highly suitable for carrying out large numbers of assays, use alternative separation methods to avoid time-consuming $^{14}CO_2$-trapping methods.

(a) A novel assay for glutamate decarboxylase has been carried out in a total volume of only 3 μl, and has permitted up to 180 assays a day to be carried out with a coefficient of variation of 2% (37). The reaction chamber was a piece of polyethylene tubing (40 mm long, 1 mm internal diameter). A 100 μl pipette tip was inserted into each end of this tubing which was then bent into a U-shape and introduced into a glass tube (100 × 15 mm) containing 0.5 ml of water. 2 μl enzyme solution and 1 μl [U-$^{14}$C]glutamate substrate solution in 2 μl and 1 μl microcapillaries (Drummond microcaps), respectively, were placed in the pipette tips at each end of the reaction chamber tubing. The reaction was initiated by centrifuging the glass tube for 5 s at 2000 *g*. After a 20-min incubation in a 37 °C shaking water bath, the reaction was stopped by placing the tubes in ice-cold water. Product and substrate were then separated by high-voltage paper electrophoresis (50 V/cm, 15 min) and quantified by liquid scintillation. Enzymatic activity was then calculated from the ratio of product counts to residual substrate counts, taking into consideration that [U-$^{14}$C] glutamate loses one fifth of its radioactivity on decarboxylation. This new micro-assay technique has great potential and could be used in the assay of any enzyme whose substrate and product can be separated by HVE or thin-layer chromatography.

(b) A second development—a general method for amino acid decarboxylases which is claimed to be simple, rapid, and more sensitive than $^{14}CO_2$-trapping methods—uses disposable SepPak Accell QMA and CM ion-exchange columns to isolate $^{14}$C-labelled amine products (38). Assays were carried out in 125 μl volume using ~0.05 μCi of [U-$^{14}$C]amino acid per assay. The claimed 3–5 fold increase in sensitivity was achieved by using uniformly-labelled substrates, available at higher specific activity than the carboxyl-$^{14}$C-labelled amino acids normally used in $^{14}CO_2$-trapping assays. The increase in sensitivity, therefore, related to an increase in gradient of the dose–response curve rather than to an improved detection limit (Section 3).

(c) An assay for phosphatidylserine decarboxylase, which combined solvent extraction and ion-exchange column chromatography, was reported to be simpler and faster than previously-used paper chromatographic, ion-exchange column, and $^{14}CO_2$-release assays (39). Using a phosphatidyl-[$^{3}$H]serine substrate, product and residual substrate were extracted with a chloroform:methanol solvent and applied to a small DEAE–cellulose column. Unreacted substrate was retained on the column and the product isolated by sequential solvent elution.

Radiometric assays of 2-oxoglutarate dehydrogenase complex, using ion-exchange column and $^{14}CO_2$-trapping methods, have suffered from high blanks due to the presence of trace impurities in the $^{14}$C-labelled substrate. A recent assay (40) based on the estimation of [1-$^{14}$C]succinate product, after removal of residual 2-oxo[5-$^{14}$C]glutarate substrate by precipitation as the 2,4-dinitrophenylhydrazone, gave lower blanks and higher sensitivity.

Several pitfalls are possible in $^{14}CO_2$-release assays and have been covered in some detail elsewhere (1). Most relate to problems in the trapping and subsequent counting of $^{14}CO_2$ and are summarized briefly here.

(a) Unsatisfactory results have sometimes followed inadequate acidification of reactants to ensure complete release of $^{14}CO_2$, and from insufficient time being allowed for its complete diffusion from solution and absorption by the alkaline trapping agent.

(b) Some workers have found that when $^{14}CO_2$ was trapped on filter papers soaked in hyamine, the hyamine bicarbonate was eluted only slowly by the scintillant, and it was necessary to allow papers to soak in the scintillant for 8 h to ensure complete elution and give reproducible results.

(c) Even though the amount of alkali in trapping solutions was much more than that theoretically necessary for complete absorption, some combinations of alkali and scintillant resulted in marked losses of $^{14}CO_2$. This problem, which was minimized by using KOH or NaOH solutions as trapping agents, can be overcome by using a scintillant cocktail containing added NaOH, or by adding ethanolamine:methyl cellosolve (1:2 v/v) to each scintillation vial.

The presence of volatile labelled compounds in reaction mixtures can cause trouble in $^{14}CO_2$-release assays, especially when $^{14}CO_2$-release is estimated after the mineral-acid acidification of reaction mixtures containing labelled short-chain fatty acids. This problem can be overcome easily by acidification with lactic acid, or by stopping the reaction with methanol and then acidifying with phosphate buffer (pH 6) to release $^{14}CO_2$. At this pH, $^{14}CO_2$ is volatile but fatty acids form non-volatile salts.

Many slightly-volatile trapping agents used in $^{14}CO_2$-release assays produce vapours which can diffuse into reaction mixtures and cause serious inhibition of the enzymes being assayed. In such circumstances, it is better to use non-volatile KOH or NaOH.

Many decarboxylase assays are subject to interference by concomitant non-enzymatic decarboxylation reactions which cause high blank readings. It has been suggested (12) that the use of phosphate buffers is a major cause of high blanks in these assays, and that their use in decarboxylase assays should be avoided where possible.

*ii. Uptake of $^{14}CO_2$*

Many carboxylases have been assayed using [$^{14}$C]bicarbonate as substrate and measuring the fixed activity after removal of residual substrate by acidification and drying (1–3). Unstable products may be stabilized chemically or enzymatically before acidification.

One pitfall is common in these assays, namely that the low cost of [$^{14}$C] bicarbonate has resulted in many workers using unnecessarily high substrate concentrations. This magnifies the effect of non-volatile trace impurities and causes relatively high blank readings. The problem is easily resolved by using the optimum concentration of [$^{14}$C]bicarbonate which has been prepared and stored at low specific activity to minimize the formation of non-volatile impurities. Alternatively, blanks can be reduced by preparing fresh substrate solutions by regenerating $^{14}CO_2$, from 'old' high blank [$^{14}$C]bicarbonate, and trapping it in dilute NaOH.

## 2.5 Paper and thin-layer chromatographic (TLC) methods

Paper and TLC methods of separating substrates and products have been used in many radiometric enzyme assays (1–3). The use of paper chromatographic methods was quickly superseded by TLC, and paper methods are very rare nowadays. The development and use of faster-running TLC systems, which permit adequate separation in a few hours or less, resulted in an increase in their use. For large-scale screening experiments, however, they were considered too time-consuming, but this objection has now been overcome by the commercial availability of Radioanalytic Scanning Systems (Section 5.3.2) which can, with high precision, sensitivity, and resolution, quantitatively scan and analyse TLC plates (19,41).

A major advantage of these chromatographic methods, particularly when working with crude tissue extracts, is that, by occasionally using autoradiography to locate the products, they allow a check to be kept on the stoichiometry of the reaction. This ensures easy detection of any degradation of substrate or product during the assay by other enzymes present in the extract or by chemical reaction.

Previous reviews (1–3) have discussed paper and TLC chromatographic techniques in some detail with particular reference to possible pitfalls. Most pitfalls relate to problems caused by self-absorption losses when soft β-emitting $^{3}H$- and $^{14}C$-labelled compounds are counted on solid support media (Section 4.1).

## 2.6 Electrophoretic methods

The advent of high-voltage electrophoresis considerably reduced the time taken for the separation of substrates and products. This led, in the late sixties and early seventies, to use of the technique in the radiometric assay of many enzymes (1).

Electrophoretic methods, however, failed to fulfil their early potential due to the limited availability of high-voltage electrophoretic equipment in many laboratories. The current demand for large-output cost-effective screening assays such as that cited earlier (37; Section 2.4.2 *i*), which allowed rapid, multiple 3 μl-reaction-volume assays, should stimulate the use of this versatile method.

## 2.7 Other methods

Several other separation methods have been used on occasions in the past (1, 2), but have been superseded by the methods described already. In exceptional circumstances, however, they might prove useful.

### 2.7.1 Adsorption on charcoal or alumina

The adsorption of either substrate or product by an insoluble adsorbent has formed the basis of many rapid and simple radiometric enzyme assays (1, 2). Charcoal and alumina have been used extensively as adsorbents in such assays.

The ability of charcoal selectively to adsorb nucleosides and nucleotides has frequently been used in assays. As a result, $^{32}$P-labelled substrates are preferred because the hard β-emission makes it possible to measure the activity of material adsorbed on the charcoal without self-absorption counting losses.

Although charcoal adsorption assays have been restricted to assays involving nucleosides and nucleotides, alumina adsorption methods have been used in a much wider range of assays. When alumina is used, separations are invariably carried out on columns rather than by batch-wise centrifugation as with charcoal. Alumina columns, however, frequently have slow flow rates and separation can be much speeded up by centrifugation.

Various problems and pitfalls, specific to adsorption assays are reviewed in more detail elsewhere (1–3).

### 2.7.2 Coupled enzyme assay methods

Many non-radiometric enzyme assays involve coupling to indicator enzymes such as the NAD or NADP oxido-reductases. Coupled enzyme reactions have also proved useful in many radiometric assays in four ways: to generate labelled substrates *in situ*, to form labelled derivatives of unlabelled reaction products, to give improved separations in assays, and to overcome interference from other enzymes or inhibitory reaction products. Typical examples are given below.

*In situ* generation of substrates has proved useful when the desired labelled substance is not commercially available (or perhaps too expensive for high-volume work) or is unstable. Oxaloacetate, for example, substrate of many transaminases, is highly unstable and is not available in a labelled form. It is, however, easily synthesised *in situ* from labelled aspartate and aspartate transaminase.

Coupled reactions have been used to stabilize unstable products, and for cost-efficiency in substrate specificity studies. As an example, many hydroxylation reactions result in the formation of unstable di-hydroxy compounds. By coupling the hydroxylation reactions with catechol-*O*-methyltransferase, and using methyl-labelled *S*-adenosylmethionine as the methyl donor, the unstable dihydroxy compounds can be converted into stable *O*-methyl derivatives which are easy to isolate by solvent extraction (42). This approach has proved particularly useful in studies of the substrate specificity of a number of hydroxylases, and has been carried out using just one labelled compound and one readily-available coupling enzyme.

Many enzymes use labelled ATP as substrate. These assays are frequently complicated by the presence of active ATPases which rapidly degrade the labelled substrate. This problem is routinely overcome by the addition of ATPase inhibitors or an ATP-regenerating system such as creatine phosphate and creatine kinase, or phosphoenolpyruvate and pyruvate kinase. This coupled enzyme approach cannot be used in assays of kinases which use γ-$^{32}$P-ATP or γ-$^{32}$P-GTP as the phosphoryl donor. An alternative approach, which has proved successful in assays of adenylate and guanylate cyclases, is to use as substrates analogues, such as the nucleotidylimidodiphosphates, which are substrates for the cyclases but not for the nucleoside triphosphate hydrolases (43).

Product inhibition of enzymes plays an important part in the regulation of many biological systems. It can also cause problems in enzyme assays. As an example, methyltransferase assays, using *S*-adenosyl-methionine as methyl donor, suffer very potent product inhibition by *S*-adenosyl homocysteine (SAH) formed during the reaction. This can be overcome easily by adding adenosine deaminase to reaction mixtures to destroy SAH and thus prevent inhibition (44).

# 3. Experimental design

Many assays, radiometric and otherwise, could be improved by considering the factors which influence enzyme rates and their measurement, and then paying attention to the design and optimization of the assay, a topic surprisingly absent from general textbooks on enzymology.

## 3.1 Sensitivity

When applied to enzyme assays, the term sensitivity can have two meanings: the detection limit, or the slope of the response curve. Here, unless specified otherwise, sensitivity will refer to the detection limit, i.e. the smallest amount of activity that can be measured. This has been defined (45) as being that amount of activity which gives rise to a signal which is twice that obtained from the blank.

Many factors must be considered when setting up an enzyme assay, and therefore it is impossible unequivocably to define optimum conditions applicable to all assays.

Many attempts to increase the sensitivity of radiometric assays, by using higher specific activity substrates, have failed because this is one of the few applications of labelled compounds where the use of higher specific activity does not automatically result in higher sensitivity. This anomaly, which has been discussed in more detail elsewhere (1–3, 12, 46), results largely because substrates at high specific activity are used (for cost reasons) at much lower concentrations than are lower specific activity substrates. As product formation falls with decreasing substrate concentration, this results in an often insignificant increase in the product radioactivity, coupled with poor kinetics and increased interference from endogenous inhibitors and contaminating enzymes. This frequently causes a reduction, rather than an increase, in sensitivity.

It is often relatively easy to increase sensitivity, and signal:noise ratios, by reducing blanks. Three major factors contribute to blank readings in both radiometric and non-radiometric assays.

(a) The first factor is impurities in the substrate which separate with the product. These can result from radiolysis of labelled substrates on storage (11, 47) or from chemical decomposition, particularly when bulk stocks of labile substrates are repeatedly thawed and re-frozen. To reduce radiolysis, labelled substrates should be stored at specific activities as low as are compatible with their intended use; labile substrates should be stored in small aliquots to avoid the need for repeated thawing and re-freezing. Potentially interfering impurities can also be removed by treating the substrate, before use, by the separation process to be used in the assay (1, 12, 14, 25, 46).

(b) Separation procedures that fail to give complete separation of substrate and product will contribute to blank readings. This effect, however, may often be reduced significantly by a slight modification of the separation procedure (e.g. Section 2.3).

(c) Concomitant non-enzymatic reactions can also contribute markedly to blanks in some assays, particularly those of the amino acid decarboxylases. Such contributions to blanks can often be reduced significantly by separately studying the non-enzymatic reaction and then selecting conditions where this reaction is minimal (4, 12). It should be remembered, too, that the usual zero-time blank will not give adequate correction for concomitant non-enzymatic reactions.

## 3.2 Selection of optimum assay conditions

Many factors must be considered when optimizing an assay. The choice of conditions will also depend on whether the aim is maximum simplicity, maximal sensitivity, maximum throughput, or minimum cost. It is impossible to define unequivocally conditions which will be optimal for all enzymes.

The following four factors, important in the choice of optimum conditions, have been reviewed in more detail elsewhere (1, 2, 12, 46).

### 3.2.1 Substrate concentration

It is commonly believed that enzyme assays should be carried out at high substrate concentrations ($\gg K_m$). This has many advantages, such as linear, near-maximum rates. There are also some disadvantages which are usually less well appreciated: only a small fraction of the substrate will be utilized (*Table 1*), and this is wasteful and costly when rare and expensive substrates are used; substrate inhibition can occur at high concentration; and blanks will be magnified at high concentrations because the three factors which cause blanks are directly proportional to substrate concentration, unlike the enzyme reaction which reaches a limiting value.

Maximum sensitivity will be attained when the signal:noise ratio is a maximum, i.e. when blanks are minimized and product formation is maximized. These are conflicting requirements, and the substrate concentration giving maximum sensitivity will depend on the extent to which each applies (1, 12). If equal weight is given to both factors, the rate $v_0$ (μmol product/min) and the percentage substrate conversion (proportional to $v_0/[S]$), conditions will be optimal when $v_0^2/[S]$ is a maximum. *Table 1* shows that this condition is met at a substrate concentration, [S], equal to $K_m$. There is, however, little variation in sensitivity at substrate concentrations in the range 0.5–2 $K_m$. This can be illustrated by taking a typical example of reducing the substrate concentration from 20 × $K_m$ to $K_m$; the enzyme rate will be reduced by less than 50% whereas the blank reading will be reduced 20-fold.

**Table 1.** Sensitivity of an enzymatic reaction as a function of the substrate concentration

| **Substrate concentration (mM) [S]** | **% $V_{max}$** | **Initial velocity ($v_0$) (μmol/min)** | **% reaction/min** | **$v_0^2/[S]$** |
|---|---|---|---|---|
| 1000 | 100 | 10.0 | 0.001 | 0.10 |
| 100 | 99.0 | 9.9 | 0.010 | 0.98 |
| 10 | 91.0 | 9.1 | 0.091 | 8.26 |
| 5 | 83.0 | 8.3 | 0.167 | 13.8 |
| 2 | 66.7 | 6.7 | 0.333 | 22.2 |
| 1 | 50.0 | 5.0 | 0.500 | 25.0 |
| 0.5 | 33.3 | 3.3 | 0.667 | 22.2 |
| 0.2 | 16.7 | 1.67 | 0.833 | 13.8 |
| 0.1 | 9.1 | 0.91 | 0.909 | 8.26 |
| 0.01 | 0.99 | 0.099 | 0.990 | 0.98 |
| 0.001 | 0.10 | 0.010 | 0.990 | 0.10 |

These results are calculated for a 'typical' example in which $K_m = 1$ mM and $V_{max} = 10$ μmol/min, in a volume of 1 litre.

There is a relationship between sensitivity, $K_m$, and $V_{max}$, and when an enzyme can react with several alternative substrates (Section 2.3) maximum sensitivity will normally be achieved by using the substrate which gives the highest ratio of $V_{max}/K_m$ (e.g. 52).

In assays where maximum sensitivity is not required, the main requirement is usually for good assays. This can be achieved at concentrations slightly higher than $K_m$, but precision will always be higher when blanks are low. The use of very high substrate concentrations should therefore be avoided.

It is sometimes supposed that carrying out assays at substrate concentrations close to $K_m$ will result in poor assays. However, it can be shown that the error introduced by carrying out an assay at an initial substrate concentration equal to $K_m$, allowing 20% reaction, and then calculating the initial rate by drawing a line between the zero-time and 20% reaction point, will give a value only ~5% lower than the true initial rate. This is well within normal assay precision.

### 3.2.2 Optimum reaction volume

As well as aiming to improve sensitivity and precision, there is much interest in making assays, which result in the incorporation of only a small amount of a sometimes expensive substrate, more cost- and time-effective. This can be done, in many assays, by selecting the optimum volume needed to maximize the percentage conversion (1, 2, 12, 46).

When an assay is carried out with a fixed amount of enzyme, under fixed conditions (e.g. pH, temperature, substrate concentration, and so on), the initial reaction rate will be independent of the reaction volume. As a result, the percentage reaction will decrease as the volume, and hence total amount of substrate, is increased. Thus, when only a limited amount of enzyme sample is available, maximum sensitivity will be achieved by carrying out the assay in the smallest possible volume. In the majority of assays, where maximum sensitivity is not the aim, carrying out assays in small volumes might increase the technical difficulties. Optimum conditions will then be obtained by carrying out reactions in the smallest volume most appropriate for the separation technique used.

### 3.2.3 How high should the specific radioactivity be?

When the optimum substrate concentration and assay volume have been selected, the substrate should be used at a specific activity high enough to ensure that the product gives satisfactory counting characteristics. It is unnecessarily wasteful and expensive to use higher specific activities.

Using $^{14}C$-, $^{35}S$-, or $^{32}P$-labelled substrates, 0.02 μCi of activity per assay sample will normally give ~400 c.p.m. for each 1% reaction. With $^{3}H$, 0.04 μCi will be needed to give a similar count rate. The specific activity necessary to give this activity depends on the sample volume and on the substrate concentration. *Table 2* shows the results of some typical calculations. High specific activity substrates are essential only in assays of enzymes with low $K_m$ values ($<10^{-5}$ M), in assays carried out in very small volumes, and in assays which result in very low conversion to product.

### 3.2.4 Purity of the substrate

The only reasonable interpretation of purity is that a compound is pure if it is pure enough for its intended application (47). A consequence of this is that a compound which is sufficiently pure for use in one particular assay, or with one particular separation technique, may be considered grossly impure in other assays or with other separation techniques. Unless the supplier has

**Table 2.** Specific activity (mCi/mmol) of substrate needed to give 0.02 μCi per assay sample

| Volume of assay sample (μl) | Substrate concentration | | | | |
|---|---|---|---|---|---|
| | **10 mM** | **1mM** | **100 μM** | **10 μM** | **1 μM** |
| 5 | 0.4 | 4.0 | 40 | 400 | 4000 |
| 10 | 0.2 | 2.0 | 20 | 200 | 2000 |
| 100 | 0.02 | 0.2 | 2.0 | 20 | 200 |
| 500 | 0.004 | 0.04 | 0.4 | 4.0 | 40 |

SI units of radioactivity: 1 millicurie = $3.7 \times 10^7$ becquerels

provided a labelled compound specifically for use in a particular assay, individual research workers will usually be the best judges of a substrate's suitability.

Impurities may be divided into two classes: those which interfere in the assay and those which do not. Those which interfere usually do so by being separated with the product and causing high blank readings. Very small amounts of impurity can be serious if all the impurity separates with the product, but only a small proportion of substrate is converted to product (Section 3.2.1; 1, 12, 14, 46).

Impurities that do not interfere are much less important because their effect is usually to make the substrate concentration lower than calculated, and most assays are carried out under conditions where the rate is not sensitive to variation in substrate concentration. However, if an impure labelled substrate is diluted with non-radioactive carrier without allowing for the presence of labelled impurity, then the calculated specific activity will be wrong. This can result in an error in the calculated rate, which might be misinterpreted as an isotope effect (32).

# 4. Problems and pitfalls

Problems and pitfalls specific to individual separation techniques have already been considered in Section 2. The aim of this section is to alert newcomers to some more general problems and pitfalls which await them and, in fact, all users of labelled compounds. Experienced workers will already have learned most of this, perhaps by hard and painful experience!

## 4.1 Special effects with tritiated substrates

Tritium-labelled substrates are often used in enzyme assays. Due to certain special effects, which occur only with tritiated compounds, spurious isotope effects have been inferred from rate measurements made using these as substrates (32). Most of these special effects, and ways of detecting them, have been mentioned in Section 2; counting problems have not, and these are discussed below.

The low energy of tritium β-particles can result in the possibility of large self-absorption losses when tritiated compounds are counted in heterogeneous solution (1–4). It should not be imagined that instrumental counting-efficiency determinations compensate for such self-absorption losses; they do not. It is safer to count tritiated samples in homogeneous solution, but this is often tedious and most workers use heterogeneous methods to measure tritiated products present as precipitates, or adsorbed onto various support media (e.g. TLC plates).

When tritiated compounds absorbed on inert media are counted by liquid scintillation, they should be counted in a scintillant which either does not elute any of the active compound, or else elutes it completely. Failure to do so will result in a continuous drift in counts as the active material is gradually eluted and counted at higher efficiency. It is best first to elute the active material with a suitable aqueous extractant in a counting vial, before adding a water-compatible scintillant and counting (1, 4, 49).

Self-absorption problems can also arise when tritiated compounds are counted on paper chromatograms, electrophoretograms, or TLC plates. Unequal drying can lead to the unequal distribution of solutes throughout the support media (1, 3, 50). Although semi-quantitative results can be obtained by heterogeneous counting, fully quantitative results can be obtained only by elution or combustion, and then counting in homogeneous solution.

Both of these problems can occur, but to a much smaller extent (~25%) when the more energetic β-emitter $^{14}C$ is counted on inert media (3).

## 4.2 Problems associated with the handling of small masses and very dilute solutions

The high sensitivity of radiometric assays often results in small amounts of labelled compounds being handled in very dilute solution. This can cause easily-avoided problems both during separation and counting.

During separation by paper chromatography, electrophoresis, and TLC, a minute amount of the compound can be irreversibly adsorbed by the support medium thereby causing a spot on the origin and a streak behind the main spot. The amount of material irreversibly adsorbed in this way is extremely small, but it can represent a significant proportion of the total weight and activity applied. This problem can easily be avoided by diluting the sample with a small amount of non-radioactive carrier compound when stopping the enzyme reaction, so that the small weight irreversibly bound represents only a negligibly small proportion of the active material applied. Adding carrier in this way can also improve resolution in ion-exchange and solvent-extraction assays.

It has also been shown that, during the scintillation counting of very small amounts of labelled compounds, irreversible adsorption on to the walls of glass vials can cause low count rates (1, 53). As the counting geometry changes from $4\pi$ (homogeneous solution) to essentially $2\pi$ (flat surface), counting losses of up to 50% can occur. This effect can also be prevented by adding carrier.

### 4.3 Polynucleotide degradation by 'finger' nuclease

Unexpected losses of activity during the isolation of tRNA can be caused by a very potent nuclease present in a water wash of fingers (4, 54). This nuclease was inactivated only slowly by heating at 100°C and, because of its extreme stability, coupled with the fact that serious contamination could result from the contact of glassware with fingers, great care was recommended to avoid the possibility of such degradation of primers, templates, or labelled polynucleotide products in studies of polynucleotide synthesis.

### 4.4 Inhibition of RNA polymerases by traces of ethanol

An RNA polymerase has been shown to be markedly inhibited by traces of ethanol (4, 55). Ethanol concentrations as low as 0.5% (v/v) in the reaction mixture caused 25% inhibition, and 2% (v/v) ethanol caused 80% inhibition. The authors therefore warned of the dangers of using $^{3}H$- and $^{32}P$-labelled nucleotides, supplied in 50% (v/v) ethanol, without prior removal of the ethanol.

## 5. Automation of assays

The need to carry out large numbers of enzyme assays can be expensive in terms of both reagent costs and time. As assays and the necessary instrumentation have become more sophisticated, high-skilled technicians have been obliged to perform time-consuming repetitive tasks rather than delegate them to less-skilled assistants and devote their own time to more challenging work. It is not surprising, therefore, that in recent years much laboratory attention has focussed on attempts to improve productivity and cost-efficiency by automating, or semi-automating, radiometric assays. Nowadays automation is more a necessity for all laboratories rather than just a luxury for large and wealthy ones.

Most developments in this area resulted from the large interest, over the past two decades, in the automation of immunoassays (RIA and ELISA) and, more recently, receptor binding studies. Many of these automated and semi-automated methods are relatively easy to adapt for use in radiometric enzyme assays.

Improvements in this direction can be considered under four headings: sample preparation, sample handling, measurement, and full automation—

robotics. Space limitations here prevent a full discussion of these developments. The manufacturers of the instruments referred to briefly below do, however, provide a range of helpful technical literature and skilled technical advice.

## 5.1 Sample preparation

Manual sample-preparation is a weak link in all analytical methods. Being subject to human variation, manual methods can be a major source of errors and, being time-consuming, they increase costs and slow-down sample turn-round.

Sample preparation involves three steps: preparation of solutions, dilution, and dispensing. Several options are available, mostly developed for use with 96-well microtitre plates (Section 5.2). These range from fully-automated systems to manual equipment such as multi-channel pipettes, available in 6, 8, or 12 channel versions, and hand-held semi-automated diluting systems, which greatly speed up the more time-consuming parts of the sample preparation process.

Choice of a system should be based on the number of samples usually processed at one time, as a larger system than that routinely required would increase costs both in terms of initial expense and wasted reagents. It is usually only a minor inconvenience to use a part-automated piece of equipment twice on occasions when a larger throughput is required.

## 5.2 Sample processing

One of the earliest attempts to automate radiometric enzyme assays (1, 56) focused on the automation of filter disc procedures and allowed automation to the stage at which an aliquot of the reaction mixture is applied to a disc. Followed by manual processing, this allowed 40 samples to be processed an hour.

Two important developments which encouraged progress in the automation of assays were the advent of cell harvesters and the introduction of multi-well plate technology.

### 5.2.1 Cell harvesters

The development, in the late seventies, of multiple cell harvesters for the automated processing of large numbers of cell cultures was a major step forward towards the automation of radiometric enzyme assays.

Cell harvesters, now available from several suppliers, including Brandel (European distributor, Semat UK), Dynatech, Ilacon, Skatron, Tecan, and Zymark, were first used for cell-proliferation studies that measured the incorporation of labelled precursors into precipitable macromolecules. Their potential in radiometric assays for a large number of macromolecule synthesizing and degrading enzymes, and others capable of assay by ion-exchange paper methods, was soon recognized.

Recent publications give examples of the application of this technology to assays of two important enzymes: reverse transcriptase (57, 58), and protein kinase C (59, 60).

Several accessories, which greatly speed up the subsequent handling and radioactive measurement, are available. For assays in which filtrates rather than precipitates are needed for counting, harvesters can collect supernatants separately into micro-tubes or directly into scintillation vials. Scintillation cocktail can then be dispensed automatically, and simultaneously, into up to 96 vials held in counter-compatible racks, which can then be loaded directly into the scintillation counter.

For assays in which the precipitate is counted, harvesters can automatically score the glass-fibre or paper-filter mats for transfer, automatically if required, into scintillation vials. Alternatively, the whole sheet can be transferred to a direct radiation reader such as the Ambis or Berthold radioanalytic imaging systems (Sections 2.5 and 5.3).

When tritiated substrates are used in precipitation assays, it is essential to ensure even precipitation over the whole filter disc to avoid self-absorption counting problems (Section 2.2.1). For this reason the Skatron cell harvester, which has a special cone-shaped harvesting block and, unlike most other cell harvesters, delivers samples from underneath the filter mat, may have some advantages.

### 5.2.2 Micro-well plate technology

Micro-well plates, containing 12 rows of 8 wells, available with U- or V-shaped or flat-bottom wells, have proved useful in the semi-automation of many radiometric enzyme assays, particularly in combination with cell harvesters. Several applications and technical hints are worthy of mention.

Kany *et al.* (61) developed a simple microplate assay for glycosidases, using a [$^{3}$H]-labelled oligosaccharide substrate, in which residual substrate was removed by precipitation with conconavalin A-Sepharose followed by centrifugation of the plate at 500 *g* for 5 min.

In a protease assay, Ruckledge and Milne (62) avoided the need of precipitation and centrifugation, by immobilizing the substrate on the microcell walls by reaction with glutaraldehyde. The authors claimed that the method was generally applicable to inhibitor/activator screening assays for many proteases, and also that bulk preparation of the substrate-coated micro-well plates reduced inter-assay variation.

A novel and generally-applicable protein kinase assay (17) used micro-well plates made from polyacrylamide gel containing a synthetic peptide acceptor. After reaction, gels were washed with water and then electrophoresed to remove residual traces of substrate ATP-$\gamma$-$^{32}$P, thereby giving much reduced blanks and increased sensitivity.

Aftab and Hait (60), in describing their protein kinase C assay, recommended the addition of 0.005% Triton X-100 as a wetting agent to ensure even distribution of reagents across the bottom of the wells when carrying out reactions in small volumes. They also cut small notches out of opposite sides of the plate to ensure adequate water flow around the well bottoms during water bath incubation.

Anyone familiar with the pipetting of small volumes of colourless solutions into micro-well plates will know the difficulty of telling which row of wells has just been filled. This need no longer be a problem. A new device, the Sure-Vue™ microtiter plate visualization aid (63), uses the liquid lens created by adding liquid to a multititer well to act as a visual indicator by distorting the grid pattern on a backing placed under the plate.

## 5.3 Radioactive measurement

Three recent developments have had a great impact on radioactive measurement; simplifying counting, facilitating large sample throughput, and reducing both 'hands on' time and reagent costs. The methodologies encompassing these developments are described briefly.

### 5.3.1 Scintillation proximity assays (SPA)

This novel technique, exclusive to Amersham International, has so far been used only in ligand-binding assays, but it obviously has tremendous potential for use in radiometric enzyme assays and is a major breakthrough in assay technology (64–67).

SPA technology involves the use of fluomicrospheres containing scintillant molecules. The principle on which it depends is that weak β-emitters, such as $^{3}H$ and $^{125}I$ (Auger electrons), need to be close to scintillant molecules in order to produce light. Only labelled ligand bound to the fluomicrospheres causes light to be emitted. This light can be directly measured using a standard β-scintillation counter without physical separation of the microspheres from the reaction mixture (66).

SPA offers many advantages: simple, easily automatable steps; no centrifugation or sample aliquotting is required; convenient one-tube assays; cost efficiency, as no scintillant, vials, or scintillant disposal are needed; safety, as no toxic or flammable solvents are involved.

Although SPA technology has so far been used only in ligand-binding assays, the potential is enormous particularly if it is possible to attach acceptors, such as oligonucleotides and peptides, to fluomicrospheres.

### 5.3.2 Radioanalytic imaging systems

Radioactive β-scanning systems are increasingly being used in many laboratories routinely carrying out large numbers of sensitive radiometric enzyme assays.

The Ambis Radioanalytic Imaging System, for example, provides high-resolution imaging and simultaneous quantitation of any 2-D surfaces such as TLC plates, gel- and paper-electrophoretograms, ion-exchange paper, glass-fibre and paper filter mats, and microwell plates (41, 68). Suitable for $^{14}C$,

$^{32}P$, $^{35}S$, and $^{125}I$, but not $^{3}H$ (on account of large self-absorption losses), the system has a maximum resolution of 0.4 mm and a range spanning $10^5$ levels of activity. Typically, it has a sensitivity of at least 50 d.p.m./2 $mm^2$ spot in 60 min.

Berthold supply two radioactive β-scanning systems which can be used to quantify $^{14}C$, $^{3}H$, $^{32}P$, $^{35}S$, and $^{125}I$ on membranes and TLC plates. The Digital Autoradiograph LB 286 allows the automatic measurement of any planar 2-D matrix. Measurement of filter samples from cell harvesters is an important application of the Multi-Sample Counter MSC 2000 which has proved particularly useful in multiple micro-assays of reverse transcriptase (19) where it permits the assay of almost 2000 samples, corresponding to 20 microtiter plates, in one run.

### 5.3.3 An alternative method of liquid scintillation counting

Many assays involve the harvesting of radio-labelled samples onto filter mats. Pharmacia's new LKB Wallac Betaplate™ liquid scintillation counter now provides a way of reducing 'hands on' time in such assays (69, 70). With 6 detectors counting simultaneously, and a total loading capacity of 1920 samples, the Betaplate counter also achieves an extremely rapid throughput. Large cost reductions are possible because the system does not require the use of vials and much less scintillant is used, thereby greatly reducing the bulk of waste for disposal.

Up to 96 samples at a time are harvested onto a filter mat. This is then enclosed in a small plastic bag with a small amount (4–10 mls) of a specially-formulated biodegradable, non-toxic scintillation cocktail, placed in a cassette and loaded directly into the counter. Using specially-designed 96-well plastic T-trays, the system can also be used (without adding scintillant) in SPA assays.

## 5.4 Robotics

Robotics provide complete automation of assays and are the ultimate goal in many laboratories. Several suppliers, including Perkin-Elmer, Tecan, and Zymark, are active in this field. For further information, see (71).

# 6. The advantages of radiometric methods of enzyme assay

Rapid, simple, specific and sensitive radiometric assays are available for a wide range of enzymes. These assays have many advantages over non-radiometric assays, particularly in assays of low levels of enzymatic activity in crude extracts.

Radiometric assays, which measure only the radioactive products, are, unlike most conventional assays, unaffected by the presence of endogenous or added inactive product. They are, therefore, ideally suited for the measurement of enzyme activity in extracts containing high concentrations of the product, or in studies of product inhibition where enzyme activity must be measured in the presence of high concentrations of product.

Often an enzyme reaction may be catalysed by two enzymes, one of high and one of low specificity. To assay the high specificity enzyme, the non-specific enzyme must be inhibited, usually by adding a high concentration of one of its substrates which does not interfere in the assay of the high specificity enzyme. This is easy to do in radiometric assays where the non-specific enzyme can be blocked with an excess of non-radioactive substrate.

Many non-radiometric assays involve coupling to indicator enzymes. This makes them sensitive to interference from coloured, UV-absorbing, or fluorescent impurities in biological extracts. Conventional coupled assays cannot easily be used for studying the effects of activators/inhibitors which might effect the coupling enzyme as well as the enzyme under study. Radiometric assays do not suffer from interference of this sort.

The sensitivity of radiometric assays allows them to be used at very low substrate concentrations and with very little dilution of the enzyme. This is very important in the *in vitro* determination of the *in vivo* activity of enzymes inhibited by drugs that act competitively. Conventional assays, in which the enzyme is appreciably diluted and high substrate concentrations are used, can cause the *in vivo* inhibition to disappear and falsely high rates to be observed (72, 73).

The sensitivity and versatility of radiometric enzyme assays allows the use of a wider range of substrate concentrations than is possible with most other assays, making them ideal for screening assays for drugs which are normally carried out at low substrate concentrations.

Radiometric enzyme assays have contributed much to developments in many fields of biochemistry. The increasing use of existing assays, the development of new assays, and advances in automation will undoubtedly prove to be increasingly useful in the future.

## Acknowledgements

During my thirty years with Amersham, I had much contact with leading workers in the radiometric enzyme assay field. I thank them, without naming them individually for they are too numerous to mention here, for giving so freely of their time to discuss their enzyme assay techniques and problems with me. I hope also that this chapter will allow me to show my appreciation of their help by passing on their knowledge to a new generation of enzyme assayists so that they can avoid the pitfalls and mistakes we all made in the past.

## References

1. Oldham, K. G. (1977). In *Radiotracer Techniques and Applications*, Vol. 2, (ed. E. A. Evans and M. Muramatsu), pp. 823–91. M. Dekker, New York.
2. Oldham, K. G. (1973). *Methods Biochem. Anal.,* **21,** 191–286.
3. Oldham, K. G. (1968). *Radiochemical Methods of Enzyme Assay*, (Review No. 9). Amersham International, Amersham, Bucks, UK.
4. Monks, R., Oldham, K. G., and Tovey, K. C. (1971). *Labelled Nucleotides in Biochemistry*, (Review No. 12). Amersham International, Amersham, Bucks, UK.
5. Prusiner, S. and Milner, L. (1970). *Analyt. Biochem.,* **37,** 429.
6. Wreggett, K. A. and Irvine, R. F. (1987). *Biochem. J.,* **245,** 655.
7. Reed, D. J., Goto, K., and Wang, C. H. (1966). *Analyt. Biochem.,* **16,** 59.
8. Hölipert, M. and Cooper, T. G. (1990). *Analyt. Biochem.,* **188,** 168.
9. Paige, L. A., Chafin, D. R., Cassady, J. M., and Geahlen, R. L. (1989). *Analyt. Biochem.,* **181,** 254.
10. Jost, J-P., Hsie, A., Hughes, S. D., and Ryan, L. (1970). *J. Biol. Chem.,* **245,** 351.
11. Evans, E. A. (1977). In: *Radiotracer Techniques and Applications*, Vol. 1, (ed. E. A. Evans and M. Muramatsu), pp. 237–338. M. Dekker, New York.
12. Oldham, K. G. (1970). *Int. J. Applied Rad. Isotop.,* **21,** 421.
13. McCaman, W. W., McCaman, R. E., and Lees, G. J. (1972). *Analyt. Biochem.,* **45,** 242.
14. Oldham, K. G. (1971). *Analyt. Biochem.,* **44,** 143.
15. Roberts, R. M. and Tovey, K. C. (1970). *Analyt. Biochem.,* **34,** 582.
16. Bollum, F. J. (1966). In: *Procedures in Nucleic Acid Research*, (ed. G. L. Cantoni and D. R. Davies), pp. 296–300. Harper and Row, New York.
17. Sahal, D. and Fujita-Yamaguchi, Y. (1989). *Analyt. Biochem.*, **182,** 37.
18. Pack leaflet for Protein Kinase C enzyme assay system, code RPN 77, 1989. Amersham International, Amersham, Bucks, UK.
19. Faff, O., Filthuth, H., and Berthold, A. (1990). *Biotech. Forum. Europe,* **4(90),** 317.
20. Sahal, D. and Fujita-Yamaguchi, Y. (1987). *Analyt. Biochem.,* **167,** 23.
21. Liotta, L. A., Tryggvason, K., Garbisa, S., Robey, P. G., and Abe, S. (1981). *Biochemistry,* **20,** 100.
22. Stetler-Stevenson, W. G., Kreck, M. L., and Talano, J-A. (1990). *Du Pont Biotech Update,* **5(3),** 12.
23. McCaman, R. E. (1968). In: *Advances in Tracer Methodology*, Vol. 4, (ed. S. Rothchild), pp. 187–202. Plenum, New York.
24. Gorman, C. M., Moffat, L. F., and Howard, B. H. (1982). *Mol. Cell Biol.,* **2,** 1044.
25. Seed, B. and Sheen, J-Y. (1988). *Gene,* **67,** 271.
26. Novel Fluor Diffusion CAT Assay, (1990). *Du Pont Biotech Update,* **5(2),** 12.
27. Deguchi, T. and Axelrod, J. (1972). *Analyt. Biochem.,* **50,** 174.
28. $^{14}$C-Chloramphenicol, CFA 754, Application Notes, 1989. Amersham International, Amersham, Bucks, UK.
29. Nagatsu, T., Levitt, M., and Udenfriend, S. (1964). *Analyt. Biochem.,* **9,** 122.

30. Chang, C-H., Bennett, L. L., and Brockman, R. W. (1989). *Analyt. Biochem.*, **183,** 279.
31. Shackleton, D. R. and Hulmes, D. J. S. (1990). *Analyt. Biochem.*, **185,** 359.
32. Oldham, K. G. (1968). *J. Labelled Compounds,* **4,** 127.
33. Evans, E. A. (1974). *Tritium and its Compounds*, (2nd edn), Butterworths, London.
34. Guroff, G., Daly, J. W., Jerina, D. M., Renson, J., Witkop, B., and Udenfriend, S. (1967). *Science,* **157,** 1524.
35. Lovenberg, W., Bensinger, R. E., Jackson, R. L., and Daley, J. W. (1971). *Analyt. Biochem.,* **43,** 269.
36. Beaven, M. A. and Jacobsen, S. (1971). *J. Pharmacol. Exp. Ther.,* **176,** 52.
37. Hagel, C., Fleissner, A., and Seifert, R. (1989). *Analyt. Biochem.,* **182,** 64.
38. Heerze, L. D., Kang, Y. J., and Palcic, M. M. (1990). *Analyt. Biochem.,* **185,** 201.
39. Overmeyer, J. H. and Waechter, C. J. (1989). *Analyt. Biochem.,* **182,** 452.
40. Kaule, G. and Günzler, V. (1990). *Analyt. Biochem.,* **184,** 291.
41. Smith, I. (1988). *The AMBIS two-dimensional beta scanner. Proceedings of the 1987 Conference on Thin Layer Chromatography,* (ed. F. A. A. Dallas, H. Reid, R. J. Ruanc and I. D. Wilson). Plenum Press, New York.
42. Daly, J. W. and Manian, A. A. (1967). *Biochem. Pharmacol.,* **16,** 2131.
43. Maguire, M. E. and Gilman, A. G. (1974). *Biochim. Biophys. Acta,* **358,** 154.
44. Coward, J. K., Slisz, E. P., and Wu, S. Y. H, (1973). *Biochemistry,* **12,** 2291.
45. McCaman, R. E. (1968). *Adv. Tracer Methodol.,* **4,** 187.
46. Oldham, K. G. (1990). In: *Receptor-Effector Coupling: A Practical Approach*, (ed. E. C. Hulme) pp. 99–116. Oxford University Press, Oxford.
47. Sheppard, G. and Thomson, R. (1977). In *Radiotracer Techniques and Applications*, Vol. 1. (ed. E. A. Evans and M. Muramatsu), pp. 171–235. M. Dekker, New York.
48. Meyer, R. R. and Keller, S. J. (1972). *Analyt. Biochem.,* **46,** 332.
49. Phillips, R. F. and Waterfield, W. R. (1969). *J. Chromat.,* **40,** 309.
50. Ives, D. H., Durham, J. P., and Tucker, V. S. (1969). *Analyt. Biochem.,* **28,** 192.
51. Wenzl, M. and Stohr, W. (1970). *Analyt. Biochem.,* **37,** 282.
52. Ryan, J. W., Chung, A., Martin, L. C., and Ryan, U. S. (1978). *Tissue and Cell,* **10,** 555.
53. Davidson, J. D. and Oliverio, V. T. (1965). In: *Isotopes in Experimental Pharmacology,* (ed. J. L. Roth), p. 345. Univ. Chicago Press, Chicago.
54. Holley, R. W., Apgar, J., Doctor, B. P., Farrow, J., Marini, M. A., and Merrill, S. H. (1961). *J. Biol. Chem.,* **236,** PC42.
55. Straat, P. A., T'so, P. O. P., and Bollam, F. J. (1968). *J. Biol. Chem.,* **243,** 5000.
56. Goldstein, G., Maddox, W. L., and Rubin, I. B. (1968). In: *Technicon Symposia,* **1,** 47.
57. Spira, T. J., Bozeman, L. H., Holman, R. C., Warfield, D. T., Phillips, S. K., and Feorino, P. M. (1987). *J. Clin. Microbiol.,* **25,** 97.
58. Lee, M. H. (1988). *J. Clin. Microbiol.,* **26,** 371.
59. Parant, M. R. and Vial, H. J. (1990). *Analyt. Biochem.,* **184,** 283.
60. Aftab, D. A. and Hait, W. N. (1990). *Analyt. Biochem.,* **187,** 84.
61. Kang, M. S., Zwolshen, J. H., Harry, B. S., and Sunkara, P. S. (1989). *Analyt. Biochem.,* **181,** 109.

62. Ruckledge, G. J. and Milne, G. (1990). *Analyt. Biochem.*, **185,** 268.
63. Litt, G. J. (1990). *Du Pont Biotech. Update,* **5(3),** 8.
64. Udenfriend, S., Gerber, L. D., Brink, L., and Spector, S. (1985). *Proc. Nat. Acad. Sci.,* **82,** 8672.
65. Udenfriend, S., Gerber, L., and Nelson, N. (1987). *Analyt. Biochem.,* **161,** 494.
66. *Scintillation Proximity Assay: Principles and Practice* (1989). Amersham International, Amersham, Bucks, UK.
67. Bosworth, N. and Towers, P. (1989). *Nature,* **341,** 167.
68. Smith, I. (1985). *BioEssays,* **3,** 225.
69. Potter, C. G., Warner, G. T., Yrjönen, T., and Soinen, E. (1986). *Phys. Med. Biol.,* **31,** 361.
70. Hyypiä, T. E., Auvinen, E., Kovanen, S., and Ståhlberg, T. H. (1990). *J. Clin. Microbiol.,* **28,** 159.
71. *Laboratory Robotics Handbook* (1988). Zymark Corporation, Hopkinton, MA, USA.
72. Disney, R. W. (1965). *Biochem. Pharmacol.,* **15,** 361.
73. Oldham, K. G. (1968). *Biochem. Pharmacol.,* **17,** 1107.

# 4

# High performance liquid chromatographic assays

SHABIH E. H. SYED

## 1. Introduction

Liquid chromatography is a separation process in which the mixture is separated into its individual components followed by their detection with a suitable monitor. Optimization of the resolution of separated components, as well as speed of separation, has been a major interest of many researchers for a very long time. It is well known that the above parameters are directly affected by the size and nature of stationary phase particles. In conventional liquid chromatography, where gravity was usually used to pull the solvent or mobile phase through a column packed with a stationary phase, a lower limit on the size of particles was eventually reached beyond which flow under gravity completely diminished.

This raised the need for the development of pumps capable of generating high pressures. However, at the pressures generally operated in high pressure liquid chromatography (HPLC) (up to 5000 p.s.i.) conventional soft matrices, e.g. Sephadex and Sepharose, will collapse. These days, silica-based matrices are the most commonly used in HPLC; they are available in particle sizes as low as 3 μm and can additionally withstand high pressures. A typical HPLC separation may take between 5 and 30 min compared to several hours in the case of conventional liquid chromatography.

A whole range of stationary phases and instrumentation (types of detection, solvent delivery systems, and data processing) have now become available for enhancement of resolution, sensitivity, and rapid data analysis. Advancement in the geometry and volume of flow cells has considerably increased the sensitivity of detection.

It is the aim of this chapter to describe briefly the principles of the technique of HPLC, to give a detailed practical discussion of the instrumentation available, the various stationary phases, and the limitations of the technique. A major part of the chapter is then devoted to the application of HPLC specifically to enzymatic analysis. In this respect, examples for each class of enzymes will be given with complete practical details for carrying out the desired separation. It is hoped that this will provide a sufficient background for an intending user of HPLC to adapt the procedure to his/her particular enzymatic reaction.

# 2. Theory of HPLC

## 2.1 Introduction

The retention of a mixture of solutes to a stationary phase occurs as a result of different mechanisms, and thus it is a complex process. The elementary forces acting on the molecules are as follows:

- Van der Waals forces operate between molecules
- dipole interactions arise in molecules and result in electrostatic attraction
- hydrogen bonding interactions
- dielectric interactions resulting from electrostatic attraction between solute molecules and a solvent of high dielectric constant
- electrostatic interactions

## 2.2 Chromatographic parameters

### 2.2.1 Retention

Before considering the theory in more detail, the reader should become familiar with the basic parameters related to a chromatogram. *Figure 1* shows a typical HPLC chromatogram in which detector response versus time or elution volume is plotted. According to *Figure 1*:

- $t_0$ —time taken for unretarded solvent front or any components of the mixture to elute from the column
- $V_0$—void volume which represents the sum of the interstitial volume between particles and the accessible volume within the particle pores
- $t_R$ —retention time
- $V_R$—retention volume, i.e. volume passed during $t_R$

$V_R$ and $t_R$ can be related by the following equation:

$$V_R = t_R \times F \tag{1}$$

where $F$ is the flow rate of mobile phase. The retention volume is directly related to the distribution or partition coefficient of solute ($k$) between stationary and mobile phases:

$$V_R = V_m + kV_s \tag{2}$$

where $V_m$ is volume of mobile phase and $V_s$ is the volume of stationary phase. In the process of moving through the column the molecules continually

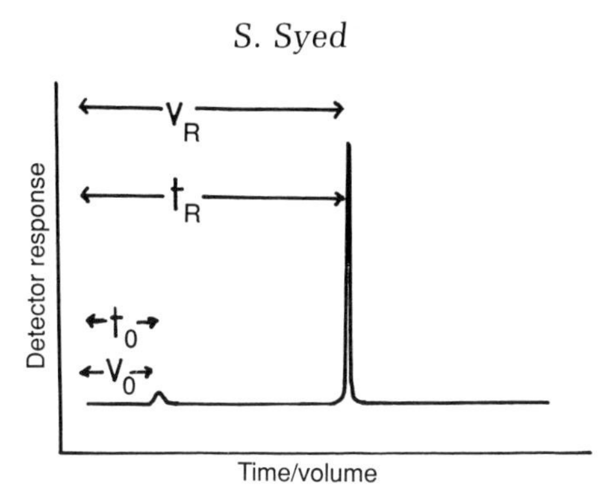


**Figure 1.** A typical HPLC profile showing various parameters defined in the text.

fluctuate between the mobile phase and stationary phase. $t_R$ can therefore be divided into the time the molecules spend in the mobile phase ($t_m$) and in the stationary phase ($t_s$):

$$t_R = t_m + t_s \tag{3}$$

The capacity factor ($k'$), which is a common measure of the degree of retention, is given by the equation:

$$k' = (t_s/t_m) = (t_R - t_m)/t_m = (V_R - V_m)/V_m \tag{4}$$

The capacity factor is related to the distribution coefficient by the following equation:

$$K = k'(V_m/V_s) \tag{5}$$

$$\text{and } K = (C_s/C_m) \tag{6}$$

where $C_s$ and $C_m$ are the concentrations of solute in the stationary and mobile phases respectively. The capacity factor is therefore the ratio of mass of solute in stationary phase to the mass of solute in the mobile phase.

The ratio of capacity factors of two solutes 1 and 2 is called the separation factor, α, or sometimes the selectivity.

$$\alpha = (k'_2/k'_1) \text{ where } (k'_2 > k'_1) \tag{7}$$

### 2.2.2 Characterisation of band broadening

On migration through a column, the individual zones of solute undergo dispersion or broadening as shown in *Figure 2*, which is a typical HPLC peak assumed to be Gaussian in shape.

$W_2$ is the width of peak at its base, and $W_1$ is the width at half height. For any Gaussian peak,

$$W_2 = 2 \times W_1 = 4\sigma \tag{8}$$

where σ is the standard deviation of the HPLC peak. The efficiency of the chromatographic column can be determined using the theoretical plate by

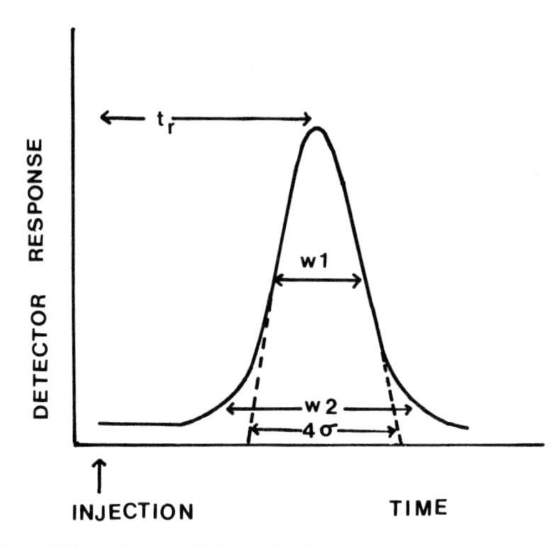


**Figure 2.** Terms describing the peak broadening.

subdividing the column into many theoretical plates. In each plate, the solute molecules achieve an equilibrium distribution between the mobile phase and the stationary phase. The number of theoretical plates is both a measure of column efficiency and band broadening. $N$ can be obtained from the equation:

$$N = (t_R/\sigma)^2 \text{ i.e. } N = (4t_R/W_2)^2 \tag{9}$$

The plate number can be easily derived from an HPLC chromatogram by measuring $t_R$ and $\sigma$, as shown in *Figure 2*. The former is obtained with relative ease and good accuracy in the case of most modern machines. A second method, perhaps more accurate, involves the measurement of peak width at half height, $W_1$, given by:

$$N = 5.54 \times (t_R/W_1)^2 \tag{10}$$

The larger the value of $N$, the better the separation. $N$ increases with smaller size of particles, low flow rates, higher temperatures, less viscous solvents, small solute molecules, and good column packing. Since $N$ is directly proportional to column length ($L$) it is more useful to measure height equivalent to a theoretical plate, $H$.

$$H = L/N \tag{11}$$

$H$ is a measure of band broadening over a certain distance. It should be the aim of the user to use a large number of plates for a given length and, therefore, a shorter plate height. HPLC columns are specified with the number of theoretical plates under the conditions of a given solute, temperature, flow rate, eluants, and so on. The column performance should be checked periodically by using the recommended solute or your own standard. The four main factors contributing to $H$, and therefore to band broadening, are eddy diffusion, longitudinal diffusion, variations in mass transfer, and excessive dead volume of the chromatographic system.

### 2.2.3 Resolution

The optimization process of chromatography involves balancing the maximization of separation with the minimization of band broadening. The ratio of the two effects is expressed as resolution, $R_s$, and is given by Equation 12:

$$R_s = \frac{2(t_R^B - t_R^A)}{W_2^A + W_2^B} \quad (12)$$

Resolution can also be expressed in terms of the fundamental chromatographic factors, namely the selectivity, $\alpha$, the capacity factor, $k'$, and the plate number $N$. To obtain the best resolution, these three factors can be optimized separately. It turns out that the selectivity factor is the most important in terms of resolution and can be modified by altering temperature or composition of the mobile phase or stationary phase. Change in the mobile phase can be most readily implemented in comparison to temperature and stationary phase.

# 3. Retention mechanism

## 3.1 Characteristics of silica

Nowadays, nearly all HPLC separations are carried out on chemically-bonded phases where alkyl groups are covalently linked to the surface of silica, thereby overcoming the problem of phase stripping which occurs in the case of non-covalently-bonded phases. However, the presence of silica cannot be ignored. Commercially-supplied silicas have differing physical properties including specific surface area, average pore diameter, specific pore volume, and particle shape. *Table 1* lists these properties for some commonly used silicas.

Silicas used in HPLC have an average surface area of around 300 $m^2/g$, an average pore diameter of 10 nm, and therefore a specific pore volume of $\approx$1 ml/g. These properties allow the separation of solutes of molecular weight less than 5000 daltons. Generally, the silicas in aqueous suspension are acidic, having a pH value in the range of pH 4–5. The differences in surface pH of silicas (*Table 1*) influence the selectivity of the bonded phase in polar solvents and, in the case of non-polar solvents, the solute retention and column efficiency. As an example, the separation of aromatic amines is significantly affected on using pH 7.2 and pH 9.0 silica with the same pore volume (1). Most silicas are stable in the pH range 2–8 only.

Silica on its own has been commonly used in normal phase chromatography where the stationary phase is either solid silica or a polar liquid phase coated

**Table 1.** Properties of some commercial silicas

| | Surface area ($m^2$/g) | Pore diameter (nm) | Pore volume (ml/g) | Total porosity | Packing density (g/ml) | pH* | Particle shape† |
|---|---|---|---|---|---|---|---|
| Eurosorb | 400 | 9 | 1.0 | 0.84 | 0.35 | 6.7 | I |
| Hypersil | 180 | 11 | 1.1 | 0.78 | 0.53 | 9.0 | S |
| LiChrosorb Si 100 | 320 | 10 | 1.0 | 0.83 | 0.34 | 7.0 | I |
| LiChrospher Si 100 | 250 | 10 | 1.2 | 0.84 | 0.35 | 5.5 | S |
| LiChrospher Si 500 | 60 | 27 | 0.8 | 0.85 | 0.38 | 9.9 | S |
| Nucleosil 100 | 300 | 10 | 1.0 | 0.82 | 0.34 | 5.7 | S |
| Partisil | 400 | 9 | 0.9 | 0.84 | 0.35 | 7.5 | I |
| Polygosil 60 | 500 | 6 | 0.8 | 0.83 | 0.37 | 8.0 | I |
| Porasil | 350 | 10 | 1.1 | 0.86 | 0.37 | 7.2 | I |
| Servachrom Si 100 | 300 | 10 | 1.2 | 0.84 | 0.34 | 6.7 | I |
| Spherisorb SW | 220 | 8 | 0.6 | 0.77 | 0.61 | 9.5 | S |
| Zorbax BP-Sil | 300 | 6 | 0.5 | 0.71 | 0.60 | 3.9 | S |

* in 5% aqueous suspension
† I, irregular; S, spherical

on to silica particles. However, most biochemical analyses have not used this type of chromatography and, in any case, it suffers from the serious problem of phase stripping during gradient elution.

## 3.2 Reverse phase chromatography (RPC)

In contrast to normal phase chromatography, in the case of reverse phase chromatography the stationary phase is non-polar while the mobile phase is polar and contains one or more non-polar organic modifiers. One of the huge advantages of RPC is the relative inertness of the silica beads where only the interactions between the solute and the covalently-linked hydrophobic alkyl chain are possible. This allows the exploitation of a wide range of solvent effects through addition of salts or organic modifiers to the mobile phase, temperature, and pH.

### 3.2.1 Preparation of reverse phases

The alkyl groups are bonded by chemical reaction of alkylsilanes containing reactive mono-, di-, tri-chloro, or alkoxy groups with silanols on the surface of silica to give a new siloxane bond as shown on the opposite page.

The unreacted silane should be removed by reaction with trimethylchlorosilane after hydrolysis of the chloro group. This is called end-capping and is essential for good and reproducible separations. The surface coverage of silica can be calculated from the carbon content of the bonded phase determined by organic analysis (2, 3). Due to the colloidal properties of silica and the processes involved in its production, the concentration of unreacted silanols

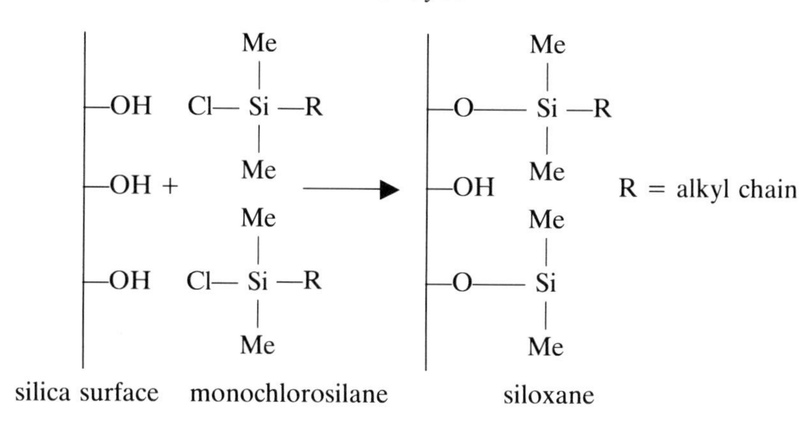


varies from batch to batch making an additional contribution to retention. This makes it difficult to compare separations carried out on columns purchased from different manufacturers.

In general, the $k'$-values of solutes increase with increasing carbon content/unit volume of column and with increasing chain length, provided the surface coverage is identical. Thus, a non-polar solute is retained to a greater extent on a C-18 than on a C-8 reverse phase. The former has a high selectivity for structurally homologous compounds, e.g. ATP, ADP, CMP, and so on, while the separation of highly lipophilic compounds, e.g. large peptides, is preferably carried out on C-8 reverse phase. *Table 2* lists some of the commonly used packings. Phenyl and alkyl-nitrile groups are also included as reverse phases. However, it must be said that C-18 is by far the most commonly used reverse phase and the reader is referred to (4–6) for a survey of its versatile applicability to separations of amino acids, peptides, proteins, vitamins, fats, steroids, antibiotics, nucleotides, and sugars.

**Table 2.** Other bonded phases available commercially. Hypersil, Zorbax, Nucleosil, and Partisil also offer the range of matrices given in this table.

| Matrix | Properties |
|---|---|
| LiChrosorb RP8 | C8 chains attached to silica, suitable for medium polarity compounds |
| LiChrosorb RP1 | C1 attached to silica. Highly lipophilic compounds. |
| LiChrosorb CN | Weak polarity. Suitable for lipids, steroids, and amino acids. |
| LiChrosorb Diol | Medium polarity column, separation of peptides and proteins. |
| Anagel TSK | Phenyl groups attached to G5000 PW. |
| 5PW | Large pore size. Used for hydrophobic chromatography of protein and peptide. |

### 3.2.2 Theory of reverse phase chromatography

Recent years have seen a rapid expansion in literature of separation mechanisms in RPC. The available experimental evidence suggests that the mechanism is not simple. However, it is out of the scope of this Chapter to discuss these mechanisms in detail and the reader is referred to (7–12) and references therein.

## 3.3 Influence of composition of mobile phase

In general, highest retention is obtained in pure water while water-miscible solvents such as methanol, acetonitrile, higher alcohols, dioxane, or tetrahydrofuran (THF) are used to enhance elution of a particular solute. The elution power increases in the order given due to decreasing polarity of the solvent and hence its increased retention on the reverse phase. The organic solvents have been arranged in order of increasing elution power in the so-called 'eluotropic series' given in *Table 3*. However, it must be pointed out that not all the solvents in *Table 3* are compatible with HPLC detectors.

**Table 3.** Physicochemical properties of various solvents

| **Solvent** | **Density** | **BP** | **RI ($n_D^{20}$)** | **Viscosity (cP, 20 °C)** | **Polarity [$\varepsilon°$ ($Al_2O_3$)]** | **UV cut-off (nm)** |
|---|---|---|---|---|---|---|
| Fluoroalkanes | | | | | −0.25 | |
| *n*-Pentane | 0.626 | 36.0 | 1.358 | 0.23 | 0.00 | 210 |
| 2,2,4-Trimethylpentane | 0.692 | 98.5 | 1.392 | | 0.01 | 210 |
| Hexane | 0.659 | 86.2 | 1.375 | 0.33 | 0.01 | 200 |
| Cyclohexane | 0.779 | 81.4 | 1.427 | 1.00 | 0.04 | 210 |
| Carbon tetrachloride | 1.590 | 76.8 | 1.466 | 0.97 | 0.18 | 265 |
| Toluene | 0.867 | 110.6 | 1.497 | 0.59 | 0.29 | 285 |
| Benzene | 0.874 | 80.0 | 1.501 | 0.65 | 0.32 | 280 |
| Diethyl ether | 0.713 | 34.6 | 1.353 | 0.23 | 0.38 | 220 |
| Chloroform | 1.500 | 61.2 | 1.443 | 0.57 | 0.40 | 245 |
| Methylene chloride | 1.336 | 40.1 | 1.424 | 0.44 | 0.42 | 240 |
| Tetrahydrofuran | 0.880 | 66.0 | 1.408 | 0.55 | 0.45 | 215 |
| Methylethylketone | 0.805 | 80.0 | 1.378 | | 0.51 | 330 |
| Acetone | 0.818 | 56.5 | 1.359 | 0.32 | 0.56 | 330 |
| Dioxane | 1.033 | 101.3 | 1.422 | 1.54 | 0.56 | 220 |
| Ethylacetate | 0.901 | 77.2 | 1.370 | 0.46 | 0.58 | 260 |
| Triethylamine | 0.728 | 89.5 | 1.401 | 0.38 | 0.63 | |
| Acetonitrile | 0.782 | 82.0 | 1.344 | 0.37 | 0.65 | 200 |
| Pyridine | 0.978 | 115.0 | 1.510 | 0.94 | 0.71 | 305 |
| *n*-Propanol | 0.804 | 97.0 | 1.380 | 2.30 | 0.82 | 210 |
| *iso*-Propanol | 0.785 | 82.4 | 1.377 | 2.30 | 0.82 | 210 |
| Ethanol | 0.789 | 78.5 | 1.361 | 1.20 | 0.88 | 210 |
| Methanol | 0.796 | 64.7 | 1.329 | 0.60 | 0.95 | 205 |
| Water | 1.000 | 100.0 | 1.330 | 1.00 | High | |
| Acetic acid | 1.049 | 117.9 | 1.372 | 1.26 | High | |

*n*-Alkanes are generally used as mixtures in the more polar solvents such as methanol and acetonitrile. For water–methanol and water–acetonitrile mixtures, a linear decrease of log $k'$ versus % of organic modifier is obtained.

As slight changes in eluent composition affect $k'$-values significantly, caution should be exercised when preparing eluent mixtures. The composition of the mobile phase changes during the de-gassing procedure or on storage due to selective evaporation of volatile components. To exemplify the use of such mixtures to separate and elute solutes, *Figure 3* shows the separation of a mixture containing acetyl-CoA and CoA, using a water–potassium phosphate–methanol mixture.

## 3.4 Effect of pH and salts

In the case of compounds which can undergo ionization, a change in pH can markedly affect their retention and selectivity. As an example, acidic solutes will be eluted before inert solutes at pH's near their pK values. The reasons for this behaviour are the presence of unreacted but dissociated silanols within the pores or the surface which can cause solute exclusion, and an

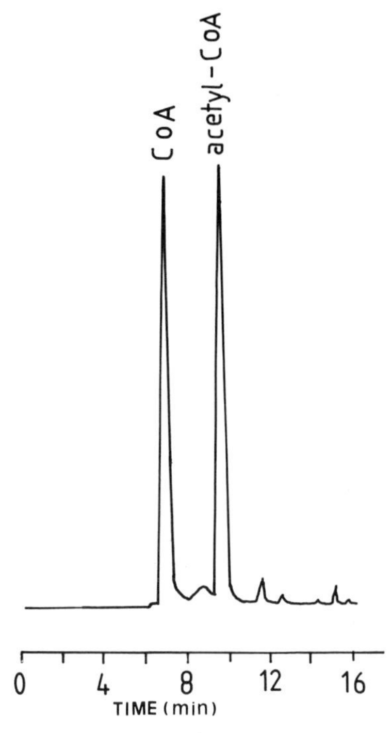


**Figure 3.** Separation of a mixture of acetyl-CoA and CoA on a Spherisorb C18, 5 μm column (1.5 × 250 mm). The column was eluted with a 0–20% methanol gradient in water containing 10 mM potassium phosphate, pH 6.7, at a flow rate of 1.5 ml/min. The detection wavelength was 254 nm.

increase in the electrostatic interactions with aqueous solvents. Both these effects can be minimized by lowering the pH by addition of acetic acid. The silanols can also interact with basic substances in an ion-exchange mechanism. If this extra mechanism proves to be undesirable, it can be suppressed by the addition of a base or an acid to form the salt. Of course, an appropriate concentration of NaCl could also be added to compete for ion-exchange sites. The addition of the latter also increases the polarity of solvent thereby causing the $k'$-value of non-polar solutes to increase and the resulting difference in selectivity between solutes can only lead to their enhanced separation on the chromatographic column.

## 3.5 Influence of temperature

The speed can usually be enhanced by increasing column temperature ($T$) and a number of different column heaters are now commercially available. *Figure 4* shows a Van't Hoff plot of $\ln k'$ versus $1/T$ for a series of catecholamines. In

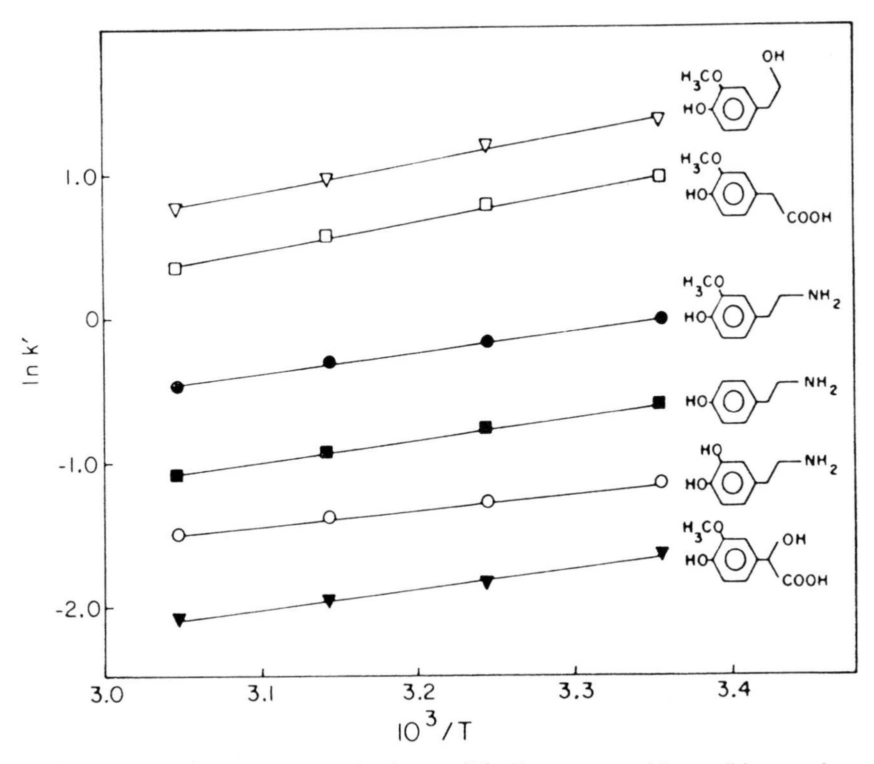


**Figure 4.** Van't Hoff plots of capacity factors ($k'$). Chromatographic conditions: column, Partisil 1025 ODS (250 mm × 4.6 mm ID); mobile phase, 0.05 M $KH_2PO_4$; flow rate, 1 ml/min; detection, UV at 254 nm. $T$ indicates the column temperature (°K). Reproduced from (10) with permission.

general, a 10°C increase in temperature reduces the capacity factor by two-fold. However, in some cases, the increase in temperature leads to increased retention, e.g. molecules which adopt a compact, near-spherical configuration are retained for a longer period. Furthermore, changes in temperature may also affect the ionization of either the buffers or the solute and this may result in an altered $k'$-value. It should also be noted that the increased temperature will lead to a reduction in viscosity of the mobile phase especially in the case of methanol–water mixtures, thereby allowing increased flow rates and a reduced separation time. However, caution must be exercised in that the decreased retention time must not be compensated by loss in selectivity which may lead to loss in resolution.

## 3.6 Ion-pair chromatography

This technique, which may be used to separate both ionic and non-ionized solutes, is similar to RPC with similar factors affecting separation. Thus, in addition to a hydrophobic reverse phase such as C-18 and an aqueous mobile phase containing an organic modifier, a counter ion is also added to form an ion-pair with the charged solute molecules. The retention properties of the solute are altered as a result of this. Generally, for separation of acidic solutes a hydrophobic organic base is added, while in the case of basic solutes a hydrophobic organic acid is used.

As with RPC, there is some controversy over the mechanism of separation. Five basic models have been proposed to explain the influence of ion-pairing agents on separation of solutes. For a detailed discussion of retention mechanisms, the reader should consult the review in (13). The mechanisms of ion-pair chromatography can basically be described by two extremes. Firstly, the counter ions and solute ions form dissociated ion-pairs in the mobile phase and are selectively retarded by the stationary phase and separated. Secondly, the counter ions are first adsorbed by the stationary phase and then interact with the solute ions. The extent to which a particular mechanism operates depends upon the counter ions and their hydrophobicity, amount of organic modifier in mobile phase, and the nature of solute.

The following equation represents the ion-pair formation:

$$\underset{\text{(less retarded)}}{\underset{\text{charged solute}}{RCOO^-_{(aq)}}} + \underset{\text{hydroxide})^+_{(aq)}}{\underset{\text{(tertiarybutylammonium}}{TBA}} \rightleftharpoons \underset{\text{stationary phase}}{\underset{\text{retarded by the}}{(RCOO^-\ TBA^+)_{org.}}}$$

Generally, a tertiary amine such as trioctylamine or a quaternary amine such as TBA is a suitable choice initially. For basic compounds, an alkyl sulphonate (e.g. heptane sulphonate) or an alkyl sulphate (e.g. sodium lauryl sulphate) are often used. It has been further shown that the $k'$ factor increases in a non-linear fashion with increasing chain length and, therefore, the hydrophobicity of the alkyl sulphate (14) (*Figure 5*).

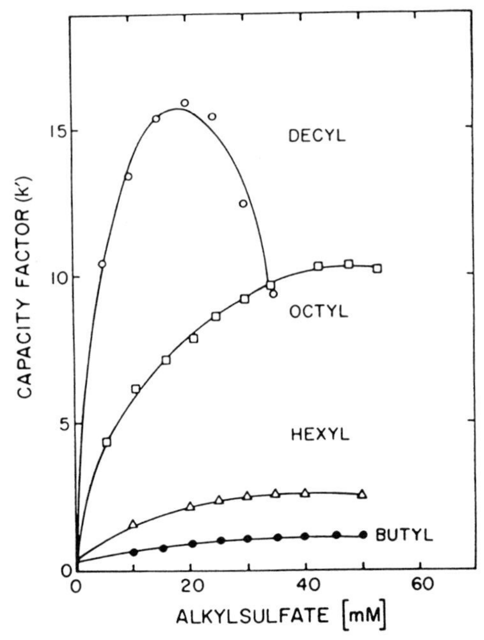


**Figure 5.** Plots of capacity factor ($k'$) of adrenaline versus the counter-ion concentration for various *n*-alkylsulphates. Chromatographic conditions: column, Partisil ODS (10 μm); mobile phase, 500 mM potassium phosphate buffer (pH 2.55) containing various concentrations of counter-ion; flow rate, 2 ml/min; temperature, 40°C; inlet pressure, 400 p.s.i.; detection, UV at 254 nm. Reproduced from (14) with permission.

## 3.7 Ion-exchange resins

*Table 4* shows a list of commercially-available ion-exchange stationary phases together with their properties.

The mobile phase usually consists of buffered solutions to which an organic modifier, such as methanol, acetonitrile, or dioxane, may be added for greater selectivity. The pH of the mobile phase is an important parameter in affecting retention in ion-exchange as this would affect the ionization of the ion-exchange group as well as the molecules to be separated. The ionic strength of the mobile phase may also be changed to alter $k'$-values through competition for exchange sites. It can be shown that $k'$ is inversely proportional to ionic strength in the case of monovalent ions. For divalent ions, $k' \propto 1/(\text{ionic strength})^2$. This is the case when only the ion-exchange mechanism is operating. However, in the case of hydrophobic interaction, two mechanisms operate; the ion-exchange mechanism which causes the $k'$-value to decrease with increasing ionic strength, and the salting-out mechanism which leads to increased $k'$-values due to increased hydrophobic interaction of solute with the organic stationary phase.

**Table 4.** Commercially-available ion-exchange resins

| Matrix | Type | Particle size (μm) | Pore size (Å) | Capacity | Comments |
|---|---|---|---|---|---|
| Anagel TSK | | | | | |
| DEAE 5PW | weak anion | 10 | — | 0.09 meq/ml | groups chemically attached |
| CM 5PW | weak cation | 10 | — | 0.09 meq/ml | to GP 500. PW hydrophilic |
| SP 5PW | strong cation | 10 | — | 0.13 meq/ml | support. |
| Hypersil APS and APS2 | weak anion | 3, 5, and 10 | 90–150 | — | 2% carbon coverage, amino-bonded |
| LiChrosorb AN | strong cation | 10 | porous | — | large surface area, separates sugars |
| LiChrosorb NH | weak anion | 10 | porous | — | and vitamins |
| Nucleosil $NH_2$ | weak anion | 7 | — | — | separates carbohydrates |
| Nucleosil SB ($NR_4$) | strong anion | 10 | — | 1 mM/g | ——— |
| Partisil ($NH_2$) PAC | weak anion | 5, 10 | 85 | — | separation of carbohydrates |
| Partisil SAX ($NR_4$) | strong anion | 10 | — | — | separation of nucleotides |
| Partisil SCX ($SO_3H$) | strong cation | 10 | — | — | separation of nucleic acids, amino acids, polyamines, and so on |
| Spherisorb SAX ($NR_4$) | strong anion | 10 | 80 | — | 2% carbon coverage |
| Spherisorb SCX ($SO_3H$) | strong cation | 10 | 80 | — | |
| Zorbax-$NH_2$ | weak anion | 6 | 70 | — | sugars and nucleotides |
| Zorbax SAX | strong anion | — | 300 | — | water soluble bases, proteins, and, |
| and SCX 300 | strong cation | — | 300 | — | nucleic acids |
| Hydropore AX (polyethyleneimine) | weak anion | — | — | 10 mg–1.6 g protein | ——— |

### 3.8 Size exclusion chromatography (SEC)

Soft dextrans and agaroses, which lack mechanical stability, are not suitable for HPLC. A list of commercially-available stationary phases is given in *Table 5*. The two most commonly used for aqueous SEC are cross-linked polyether or polyester and silica-based phases. Each has hydroxyl groups covalently attached to the surface. For aqueous SEC, TSK PW, RSK SW, and Zorbax GF series are quite suitable for proteins, peptides, and nucleic acids over very wide molecular weight ranges (e.g. TSK SW 30 000 → 500 000). For non-aqueous chromatography TSK HXL series can be used. The disadvantage of SEC over other forms of column chromatography is its low capacity since total volumes must not exceed 1–2% of total column volume. Similar factors to those involved in conventional gel filtration also affect the separations in the case of HPLC matrices, e.g. pore size distribution, pore volume, and pH.

## 4. Instrumentation

### 4.1 Essential components of an HPLC system

*Figure 6* is a schematic representation of a typical HPLC system. The various components shown will be discussed briefly in turn in this section.

**Table 5.** Resins used for size exclusion HPLC

| Matrix | Particle size (μm) | Pore size (Å) | Comments |
|---|---|---|---|
| Zorbax Bio Series GF 250 | 4 | 300 | Silica based, separation range of 4000–400 000 daltons. |
| Anagel TSK | | | |
| G 2000 SW | 10 | 125 | Separation range for proteins of <30 000 |
| G 3000 SW | 10 | 250 | 30 000–500 000, |
| G 4000 SW | 13 | 500 | >500 000. All are rigid spherical hydrophilic gels. Capacities of 5–200 mg. |
| $SW_{XL}$ | 5 and 7 | — | Separation range 5000–1 × $10^6$, stable over pH 2–12. Silica with a bonded hydrophilic layer with reduced pore volume. Run times halved compared to SW series. |
| HXL series 1000–7000 | — | — | Mol. wt. range <500–>1 × $10^6$, made from styrene/divinyl/benzene suitable for non-aqueous gel permeation chromatography, e.g. THF, toluene, and so on. Capacity <5 mg. |
| PW and PWXL series | 11–17 | — | Mol. wt. range 4000–800 000. Aqueous |
| G 2000 PW–G 6000 PW | 6–13 | — | gel filtration. |

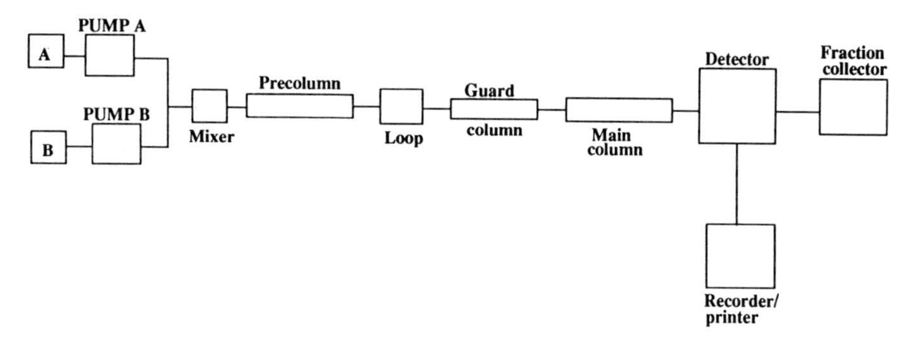


**Figure 6.** Essential components of an HPLC system.

## 4.2 Pumps

The pump is the most important feature of an HPLC system (apart from the column). Over the past decade or so, instrument manufacturers have sought to develop various operating principles with the main aim of producing pumps which could deliver pulseless, constant, and reproducible flow of solvent over short and long periods.

Amongst the early designs were gas pressure-driven and high pressure single- and twin-headed syringe pumps. These have largely been superseded by reciprocating single, dual, or triple pistons with or without diaphragm. The characteristics of each type of pump have been discussed by Snyder and Kirkland (15). Most common types of reciprocating pumps consist of a piston which displaces a volume of liquid (*via* a cam drive). The direction of flow is controlled by check valves. This design suffers from a major drawback in that, as the piston moves through the whole length of the chamber, the pulse of solvent being generated leads to uneven flow. Some pump designs have incorporated a hydro-pneumatic dampener consisting of a length of flattened tubing between pump and column leading to unwanted extra dead-volume. A number of recent pump designs have instead incorporated a fast refill cycle in the piston in conjunction with a dampener, thereby reducing pulsation considerably and leading to relatively smooth solvent delivery, e.g. Altex 100 model, Beckman System Gold models 116 and 126. The twin-headed pistons operating 180° out-of-phase (e.g. Waters M45 type) and triple-headed pumps where the pistons operate 120° out-of-phase (e.g. Du Pont) deliver essentially pulseless flow since the refill stroke of one piston is compensated for by the fill stroke of the other. Recently, a pump with reciprocating piston speeds of 23 Hz has been introduced (e.g. ACS model 400) where the stroke rate is too fast to be recorded by detectors leading to pulse-free flow. More sophisticated pumps have, in addition, incorporated mechanisms to compensate for the compressibility of different solvents. The delivery flow rate decreases when the pressure increases due to liquid compressibility. Common solutions to this problem include the use of:

(a) 'roll-off' (decrease in flow-rate) compensation circuitry which corrects the pump drive based on measurement of pump outlet pressure
(b) flow feed-back control where a flow sensor, operating independently from the pump, measures flow rate and, via feed-back circuity, corrects pump drive when measured flow-rate deviates from the setting
(c) dual-stage pumps or metering pumps (e.g. Milton Roy constametric pumps)

It may also be noted that some recent designs have employed the use of floating pistons to minimize the seal wear (e.g. Beckman System Gold 116 and 126).

The usual flow rates attainable with a ±0.1% error are in the range 0.1–10 ml/min for analytical applications (e.g. LDC/Milton Roy). However, Beckman claim to have reproducible flow rates (±0.1% in the range 0.001–10 ml/min, Gold System 116 and 126 module and 0.01–10 ml/min for the 110B model. However, caution must be exercised in accepting such figures.

## 4.3 Gradient modules

In most HPLC systems, the use of binary gradients is made where two different solvents are separately pumped in a single-headed, two-pump system and then mixed in a low-volume mixing chamber (a high pressure dynamic mixing using a stirring magnet with an internal filter) on the high-pressure side of the pump. Reciprocating pumps are less precise at low flow rates (0.1 ml/min) and therefore the initial and final stages of the gradient are less accurate. Beckman System Gold models 116 and 126 and LDC/Milton Roy constametric pumps have such a dynamic mixing system. The multi-headed pumps also have the option of using low-pressure mixing, and operate by controlling the solvent through switching valves which are, in turn, controlled by solenoids. This allows the delivery of more than two solvents. In the Beckman Gold System 126 model, with a dual-piston system, up to eight solvents can be delivered (high-pressure mixing).

## 4.4 Sample injection

A precise, quantitative result of HPLC analysis requires injection of well-defined sample volumes in a highly reproducible manner. Commercially-available injector types allow the variation of volume in the range 0.5–20 μl for analytical separations and 0.1–10 ml in the case of preparative runs. Early designs were influenced by those used in gas chromatography. These have now been largely superseded by the high pressure injection valves (the loop type, *Figure 7*).

Beckman model 210A injection valve (four-ports), for example, is able to accommodate 5–2000 μl sample volumes and can withstand pressures of up to 10 000 p.s.i. To reduce sample dilution, the diameter of the ports has been

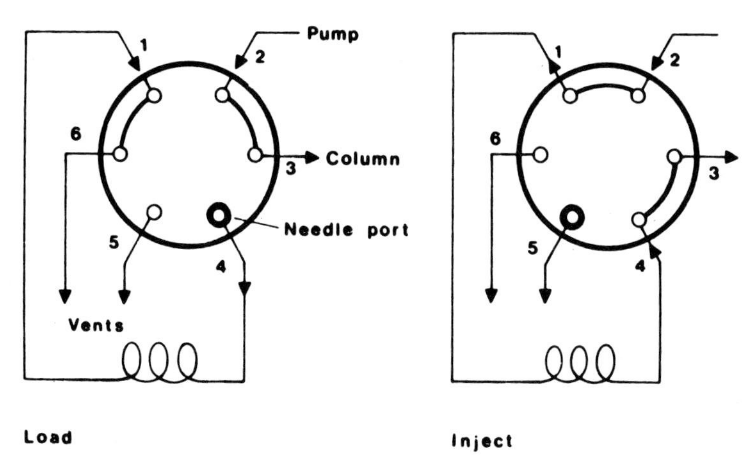


**Figure 7.** Illustration of the arrangement of a high pressure injection valve. In the load position, the sample is introduced via the sample loop (positions 1,4). Upon injection the solvent flow from the pump (2) is directed through the sample loop to the column (3).

reduced to only 0.25 mm. Other types of valve include the recycling valve used for cycling effluent from the column, back into the pump and out through the column again (e.g. Beckman Derlin). Alternative methods of, for example, stopped-flow septum injection, have been found to be inferior due to generation of particulate matter from the septum after repeated injections. The injector type shown above can also be operated automatically when a large number of samples are to be analysed. Such a system would be run overnight, for example.

## 5. Detectors

Much progress has been made in the design of specific (depends on some property of solute) and non-specific (depends on bulk property of mobile phase) detectors with regard to addressing the need for achieving high sensitivity and satisfactory lower limit of detection, in addition to obtaining low levels of noise and drift. It is outside the scope of this Chapter to discuss each type of detector in detail and the reader is referred to (3) and (4) as well as the literature produced by the various manufacturers for their respective detection systems. A brief survey of the most commonly used systems is therefore given below.

### 5.1 UV/visible detectors

The UV/visible type detectors are the most widely-used detectors due to their versatility, high sensitivity, wide dynamic range, and relative insensitivity to temperature and flow variations. The detectors can be classified as follows.

### 5.1.1 Fixed-wavelength detectors

Fixed-wavelength detectors utilize lamps which emit light at discrete wavelengths (pressure mercury lamp is in common use emitting at wavelengths of 254, 313, 365, 405, 436, 546, and 578 nm). By use of suitable cut-off filters, a particular wavelength can be selected. Light sources are also available for monitoring at lower wavelengths, e.g. zinc at 214 nm and cadmium at 229 and 326 nm. The advantages of using fixed-wavelength detectors include low noise level, low operating cost, and simplicity of operation. It should be noted that the majority of HPLC detection is carried out at 254 nm. The Gilson model 112, for example, offers detection at two standard wavelengths (254 and 280 nm), as well as the possibility of using other wavelengths in the range 214–640 nm by selecting the appropriate pre-focused lamp/filter assembly. The sensitivity is in the range 0.001–1.0 absorbance units (AU) with a noise level of less than $4 \times 10^{-5}$ AU at 254 nm. The detectors manufactured by Hewlett-Packard and Bio-Rad (model 1740) are reported to give similar sensitivity range and noise levels of less than $2 \times 10^{-5}$ AU.

### 5.1.2 Variable wavelength detectors

Variable wavelength detectors use light sources which give a continuous emission spectrum in the range 190–600 nm. A continuously-adjustable monochromator is used for wavelength selection. The Beckman System Gold model 166 is a programmable detector module. The light path travels from the lamp (deuterium or optional tungsten) to focusing and steering mirrors, then to the diffraction grating, through the flow cell and, finally, to the photodiodes. The entrance slit, mirrors, diffraction grating, and the grating drive mechanisms are all part of the monochromator assembly. On selecting a new wavelength, the grating is moved by a stepper motor at a speed of 70 nm/sec. This type of mechanism is also operated by detectors from Bio-Rad (model 1706). More recent models, e.g. Beckman model 166, Gilson models 115 and 116, and Hewlett-Packard utilize a monochromator assembly in which the light is diffracted by a holographic grating suspended in a magnetic field. The wavelength of light reaching the exit aperture end-beam splitter depends on the angular position of the grating about its vertical axis. Multi-wavelength monitoring is achieved by rapidly oscillating the grating between any two selected wavelengths. Rapid scanning of entire spectral ranges at speeds of up to 100 nm/sec means that spectra of peaks can be monitored as they are eluted. These data can be useful in peak identification.

The latest development in this field is the Diode Array Detector (DAD). The light diffracted by a holographic grating is detected by an array of photodiodes applied to a silicon substrate, 2 cm in length. The Beckman model 168, for example, has an array of 512 diodes with a spectroscopic resolution of 1 nm/diode over the range of 190–600 nm. The image of the absorbance spectrum of the eluting substance is formed on the diode array.

Records of spectra at different positions of a single peak can indicate the purity of the eluting peak. Two major advantages of the DAD over motor-driven monochromators are that they possess fewer moving parts and that they are less prone to peak skewing at high flow rates.

The sensitivity of UV detectors is generally in the low nanogram range. It is also affected by flow cell geometry and volume. A well-designed cell should minimize the refractive index effects, contain no unswept areas, and be able to withstand high pressures. Most cells have volumes of 8–10 μl and an optical pathlength of 10 mm. Cells with lower volumes are also available, e.g. Gilson offer a range of 0.3–11 μl with variable path lengths. The choice of the cell is dependent upon the particular application, e.g. analytical versus preparative.

## 5.2 Fluorescence detectors

Fluorometric detection possesses inherently higher sensitivity than absorption since the intensity of emitted light is directly proportional to power of exciting radiation. This method can be used to detect any compound which cannot be conveniently detected by any other method (e.g. UV/refractive index) and has intrinsic fluorescence or can be derivatized by reacting with a fluorescent compound, e.g. fluorescamine, dansyl chloride. This higher sensitivity and selectivity have been extensively applied in biochemical systems since many biologically-important compounds strongly fluoresce, such as porphyrins, riboflavins, vitamins, certain drugs, nucleotides, and so on.

The conventional fluorescence detector consists of:

(a) a light source (xenon, mercury arc, quartz halogen)

(b) a wavelength reflector using any monochromator assembly suitable for a UV-vis spectrophotometer. Thus, excitation spectra can be obtained for the compound of interest. Several filters are also available between 254–520 nm (excitation) and 420–680 nm (emission). Bio-Rad model 1700, for example, provides a series of lamps and filters (excitation and emission) to optimize detection. High signal throughput is achieved by using a phosphur-coated mercury lamp. Light is emitted in every direction, but is usually monitored at right angles to the excitation beam. A parabolic reflector is inserted inside the cuvette chamber to reflect more of the light towards the photomultiplier.

(c) single and dual flow cells are available commercially to take account of the fluorescence of the mobile phase. The flow cells vary in size in the range 1–40 μl, e.g. Bio-Rad provide two flow cells for their model 1700 of volumes 6.5 μl and 19 μl. The stated maximum sensitivity is 0.05 p.p.b. quinine sulphate in 0.1 N $H_2SO_4$ or, more generally, picogram amounts can be detected.

Further improvements have been introduced recently, e.g. the use of a laser light source which is more monochromatic than conventional light sources.

Increased sensitivity can introduce problems of contaminating fluorescent material in the mobile phase. It may also be noted that by proper choice of excitation conditions, the quantum efficiency of fluorescence of the sample can be optimized. The concentration of solute should not be so high as to cause quenching and thus reduce sensitivity or overload. It should also be mentioned that the photomultiplier or amplifier fluorescence can also be quenched by impurities and dissolved oxygen.

Compounds that do not fluoresce can be detected by carrying out pre- or post-column derivatization with fluorescamine, dansyl chloride, or other fluorophores (16).

## 5.3 Refractive index (RI) detectors

This type of detector measures the change in the refractive indices of the liquid in the reference and sample cells and, due to its non-specificity, is a universal detector. However, the difference in refractive indices between different solutes is very small. The sensitivity of detection is therefore much lower than the fluorescence and UV/visible detectors and is in the μg range. The lower limit of detection (LLD) of a modern RI detector is in the order of $1 \times 10^{-8}$ $\Delta$RI units. The LLD is dependent upon temperature ($2–4 \times 10^{-4}$ $\Delta$RI units/°C) and pressure fluctuations. A temperature control of ±0.001°C is required if a noise level of less than $1 \times 10^{-7}$ RI units is to be maintained. Additionally, the detector response is affected by mobile phase composition and gradient elution is not possible with the commonly-used RI detectors. The three most widely used RI detectors are the deflection, fresnel, and interferometric type detectors.

In the deflection type instrument, a collinated light beam passes two parallel chambers in a glass prism acting as reference and sample cells. If the refractive index of solution ($\eta_s$) is equal to the refractive index of reference ($\eta_r$) the beam is slightly shifted parallel to the incident beam. If the $\eta_s$ changes, then the beam is deflected and then measured by a differential photodiode. The output signal can either be positive or negative depending upon $\eta_s$ being larger or smaller than $\eta_r$. Instruments of this type are supplied by Waters (R400) and typical noise levels of $5 \times 10^{-9}$ RI units are obtained.

In the fresnel type detector, a light beam is split into a reflected beam ($\alpha_2$, *Figure 8*) and a penetrating beam ($\alpha_1$) at the boundary surface between glass and liquid. If $\alpha_1$ is close to 90° (i.e. almost total reflection) a small change in refractive index of the liquid will result in a significant change in intensity of penetrating beams. Two beams are used, one for reference and the other for sample flow cells. Small differences in the RI of the two cells will cause a difference in intensity of light back-scattered from a metal surface at the back wall of the two cells. Noise level is comparable with the deflection-type detector. Such an instrument is supplied by Laboratory Data Control. A point of caution should be made in that, although these cells contain no unswept areas (typical volume 3 μl) they are susceptible to formation of solid

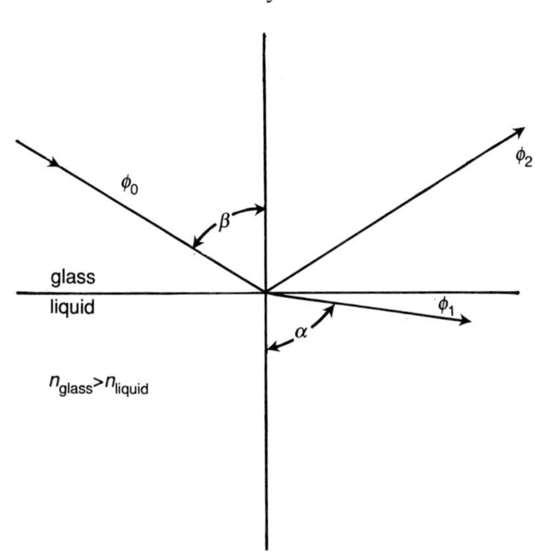


**Figure 8.** Measurement principle of a fresnel type RI-detector: $\phi_0$ intensity of incident light; $\phi_1$ intensity of light penetrating boundary surface between glass and liquid (into liquid); $\phi_2$ intensity of reflected light. A small change of $n_{liquid}$ changes $\alpha$ and therefore $\phi_1/\phi_0$.

films which lead to cell imbalance. Careful cleaning is thus required. For preparative work larger cells are also available.

The interferometric detector measures the difference in the speed and hence the phases of the two beams passing through reference and sample cells. If the $\eta_s = \eta_r$ then the beams after passage through cells interfere constructively, i.e. in phase. On change in $\eta_s$, destructive interference occurs and leads to decreased level of photomultiplier output (e.g. Optilabs). Recently, LKB have produced a new instrument which incorporates features of both the fresnel optics and deflection refractometer.

## 5.4 Electrochemical detectors

The development of electrochemical detectors has been prompted by the need for the quantitation of trace quantities of biologically important compounds, in particular the biogenic amines. These detectors offer a higher sensitivity and selectivity than the methods already discussed. Picogram and femtogram levels of electroactive compounds have been reported. The detailed theory of such instruments is outside the scope of this Chapter; the reader is referred to (17, 18).

Briefly, these detectors are of two types.

(a) Bulk property detectors, of which the most commonly used are conductivity detectors which measure the change in cell resistance.

(b) Solute property detectors which monitor the change in potential or current as the solute passes through the flow cell. The more popular detectors are either amperometric or coulometric. Solute is passed over an electrode held at a constant voltage which is sufficiently high to cause either reduction or oxidation; then the current produced is proportional to solute concentration. The concentration of solute does not change significantly (~5% conversion). In coulometric detectors, the solute is almost totally converted (~95%). The signal given by the latter is greater but also leads to increased background signal leading to reduced signal-to-noise ratio.

Usually a three-electrode system is used. The working electrode used in the oxidative mode is usually carbon paste (a mixture of graphite and either paraffin oil, wax, or grease) or glassy carbon. Both give low background currents and are reproducible. However, electrochemical detection is limited to phases containing <25% (v/v) methanol or 5% (v/v) acetonitrile. In the reductive mode, a mercury electrode (mercury deposited as a thin film on a gold substrate) is usually used. In the older designs, a dropping mercury electrode, usually known as polarography, was used.

It should be emphasized that the constituents of the mobile phase must be of the highest purities to minimize background currents. The voltage range within which detection of solutes can take place depends on the electrode material. For glassy carbon it is −1.0–+1.3 V and for mercury −2–+0.3 V with reference to a saturated calomel electrode. The thin film detector is the most commonly used, e.g. Bio-Rad model 1340 uses a glassy carbon electrode with a voltage range of ±2 V and can detect 20 pg amounts/injection. Bioanalytical systems provide a more popular design also using thin film where the mobile phase flows parallel to the electrode surface and a planar auxillary electrode is placed opposite with the thin-layer channel acting as a salt bridge to the reference electrode. Briefly, the classes of compounds that have been investigated include aromatic amines, phenols, thiols, nitro-compounds, quinones, purines, ascorbate, and uric acid.

## 5.5 Radioactivity monitors

The technique of scintillation counting has been adapted over the last decade to on-line detection of HPLC effluents. Most radioactivity detectors use flow cells of different types positioned between two photomultiplier tubes which usually possess reflectors to ensure high counting efficiencies. The flow cells are of three types:

(a) Homogeneous: the column effluent is mixed with liquid scintillant before passing through flow cell and is sent to waste.

(b) Heterogeneous: the eluant passes through scintillant granules inside the flow cell.

(c) High-energy: the scintillator material (liquid or solid) surrounds the flow cell through which the eluate passes.

A major limitation of these detectors is their inability to compensate for changing counting efficiency in gradient HPLC. Some detectors, however, use a standard curve to take account of this. Several detectors are now linked to microprocessors allowing detailed data collection and processing. The counting efficiencies and sensitivity of detection depend on the type of detector, type of flow cell, and the isotope, e.g. sensitivity of detection for $^{14}C$ is 80–100 d.p.m. while for $^{3}H$ it is 250–400 d.p.m. in the case of heterogeneous and homogeneous detectors.

### 5.6 Chemical reaction detectors

Where the solute possesses no chromophore, electrophore, or bulk property, the detection can be carried out by chemically introducing electrophores or chromophores using either pre- or post-column derivatization. The derivatized substance can then be detected in the normal manner either spectrophotometrically or fluorometrically. A number of reactor types are available, namely, packed bed reactors (19, 20), tubular detectors (20, 21), and segmented flow reactors (22–24).

## 6. Practical considerations

### 6.1 Selection of a chromatographic mode

Generally speaking, the particular separation of a compound depends upon molecular weight, range of solubility, and its chemical structure. *Table 6* summarizes the category of compounds and the appropriate modes of separation.

**Table 6.** Mode selection

| Sample | | | Mode |
|---|---|---|---|
| MW <2000 | Non-ionic | High polarity | Normal phase |
| | | Low polarity | Reversed phase |
| | Ionic | Acidic | Anion-exchange |
| | | | Reversed phase ion-pair |
| | | Basic | Cation-exchange |
| | | | Reversed phase ion-pair |
| MW >2000 | Water soluble | Ionic | Ion-exchange |
| | | Polar | Normal phase |
| | | Non-ionic | Reversed phase |
| | | | Affinity |
| | | | Ligand exchange |
| | | | Normal phase |
| | | | Size exclusion with aqueous mobile phase |
| | Water insoluble | | Size exclusion with organic mobile phase |
| | | | Normal phase in organic solvents |

## 6.2 Solvent selection

The criteria for selection of a suitable solvent to cause elution depend upon the physico-chemical properties of the sample and the solvent being considered. When preparing the sample for a separation, dissolve the sample in the same mobile phase as that being used to effect elution. In cases of low solubility, attempt to use a larger injection volume. If the sample is dissolved in a different solvent (e.g. a stronger eluting solvent) from the mobile phase then resolution may be impaired. The choice of the eluting solvent can be facilitated by the use of the eluotropic series shown earlier in *Table 3*. However, from a practical point of view, avoid the use of solvents of high viscosity which can lead to high column back pressures, and solvents of high volatility since it may prove difficult to maintain a fixed composition of mobile phase for any significant length of time. As UV is the most common method of detection, do not use solvents with a high UV cut-off point. These factors preclude the use of a number of solvents in the eluotropic series. When halide salts are included in the mobile phase, flush the whole system with plenty of water to prevent corrosion of stainless-steel tubing, column, and pumps, and remember that salt crystals can also form over long periods of storage.

Solvent purity can have a major effect on the column performance. The impurities in solvents can be lower or higher homologues, acids and bases, UV-absorbing compounds, and water. They can adsorb to the surface of the stationary phase and cause a change in selectivity of a column, increase in back pressures, and peak distortion. However, a large number of purified solvents, including HPLC-grade water, are now commercially available. Normal deionized water is unsuitable for use in HPLC.

## 6.3 De-gassing and filtration of solvents

The solvents and the sample may contain particulate matter which can impair the sealing of pump valves and clog the column frits or the surface of the column bed. Therefore, filter all solvents and samples through a 5 μm solvent-compatible filter (e.g. Millipore). It is also advisable to attach additional filters (sintered) to the ends of tubing placed inside the solvent reservoir.

Dissolved gases can become problematic especially in the case of gradient systems where the gas is less soluble in a mixture than in the single solvent (e.g. a water/methanol or water/acetonitrile gradient) due to the exothermic mixing process. Various ways of de-gassing include boiling, vacuum de-gassing, agitation by sonication, and sparging with helium, argon, or nitrogen. Amongst these, sonication and sparging with helium or argon are the most commonly used, although the latter are expensive but very effective. As de-gassing itself will lead to a change in composition of the mobile phase, filter and de-gass the solvents separately and then mix them. If helium or argon is used, then seal the reservoir after de-gassing. Some people continue bubbling helium at a very small flow rate throughout the HPLC separation.

## 6.4 Sample preparation

As crude biological samples may contain particulate matter and protein, make sure that the sample has been homogenized, centrifuged or filtered (5 μm Millipore), and deproteinized. The protein can be removed by perchloric or trichloroacetic acid. Alternatively, use methanol (1:1 ratio) and leave sample for 5 min. In each case, remove the protein precipitate by centrifugation or filtration through a semi-permeable membrane (e.g. Amicon) with the appropriate cut-off point. If the compound is in a dilute solution, then attempt concentration of the sample by evaporation, for example, using either nitrogen, vacuum, or lyophilization.

## 6.5 Column packing

Considerable saving in cost can be made by using 'in house-packed' columns instead of commercially packed ones. Columns can be efficiently packed by using a slurry of the stationary phase. Use a high capacity, pneumatic amplifier pump (25) as HPLC pumps do not give sufficiently high impact velocities. First remove fines by suspending the support in an organic solvent, leave to stand for 15 min, and then decant the supernatant. Repeat this several times. Suspend the material in a high-density solvent and pour into a reservoir. Pump the HPLC column at about 3000 p.s.i. Fit the column with end pieces and test for proper performance. A poorly-packed column can lead to column voids and compression during operation causing loss in resolution via peak splitting or peak distortion and broadening. For alternative procedures consult the review by Kaminski *et al.* (26).

## 6.6 Column protection

Protection of the column from particulate material by the use of on-line filters and pre-column filtration has already been mentioned. It is also important to protect the column from components of the sample which may bind to the column irreversibly and, in time, cause build up on top of the column leading to high back pressures and loss in performance. Therefore, attach a guard column between the injector and the main column. The guard column usually contains the same stationary phase as the main column. Any components with high $k'$-value will be removed at this stage. The guard column can be replaced at a small cost by repacking it with new stationary phase. In the author's experience, adding small portions of dry matrix with a spatula and gently tapping, while rotating the column on a flat bench, proves quite satisfactory. The column can then be packed by pumping water through it at a high flow rate (a flow rate of 9 ml/min will give ~1000 p.s.i. for a 4.6 mm internal diameter column with 5 μm particles). The guard column, however, can contribute to extra-column band-broadening effects.

The use of aqueous buffers also causes loss of column material to give void at the top. This is especially the case with silica-based matrices at pH values above pH 7. This leads to loss in peak resolution and peak splitting. Again, place a pre-column between the pumps and the injector to pre-saturate the solvents with silica, thereby preventing the main column from dissolution.

Never allow the column to go dry as dry stationary phase will shrink, leading to voids. Wash the column with a pure solvent such as methanol at the end of the day and cap it securely at both ends. If buffer solutions or salts have been used, wash the column thoroughly with water before storage to avoid corrosion of stainless steel. Always mark the top and bottom of the column so that the same direction of flow is always used. Only one end of the column is then contaminated by impurities. It is then easier to replace the top portion of the column stationary phase with new material.

### 6.7 Tubing

Use minimal lengths of stainless steel tubing interconnecting the various parts of the HPLC, as it contributes to extra-column band-broadening. Generally, 1/16″ steel capillary tubing is used with an internal diameter of either 0.5 or 0.3 mm.

## 7. Application of HPLC to enzymatic analysis

### 7.1 Hydrolases

#### 7.1.1 Dihydroorotase from rat liver (EC 3.5.2.3)

In mammals, the first three reactions of pyrimidine nucleotide biosynthesis are carried out by a trifunctional enzyme, the protein Pyr 1–3. Of the three activities, the dihydroorotase catalyses the reversible cyclization of *N*-carbamyl-L-aspartate to L-dihydroorotic acid (27). At neutral pH, the equilibrium lies towards the acyclic molecule. The assay of dihydroorotase activity is based on the separation of radio-labelled substrate and product by ion-pair reverse phase HPLC.

---

**Protocol 1.** Assay of dihydroorotase

1. Carry out the assay in a total volume of 100 μl containing 50 mM Hepes, pH 7.4, and L-[6-$^{14}$C] dihydroorotate.
2. Add 5–20 μl of enzyme to give the required percentage conversion.
3. Incubate the reaction mixture for 20–30 min at 37°C and then terminate by adding 100 μl of 1% (w/v) SDS.
4. After a brief incubation, add 200 μl of elution buffer and centrifuge to remove precipitate.
5. Use the supernatant directly for HPLC. Carry out the separation of

substrate and product on a Waters C-18 column (Nova-Pak C-18 cartridge, 0.5 × 10 cm, 5 μm particles). Use a radioactive flow detector (Flo-One/Beta system with 2.5 ml flow cell) for continuous monitoring of radioactivity. The elution buffer should be 3.5 mM tetrabutylammonium phosphate, pH 7.0, and acetonitrile (85:15).

**6.** Elute the column with buffer at 1.5 ml/min, while the flow scintillant (Flo-Scint 3, Radiomatic Instruments) should be pumped at 5 ml/min. The elution profile shown in *Figure 9* consists of c.p.m. versus time. The observed percentage conversion can be obtained from the chromatogram as c.p.m. carbamyl aspartate/total c.p.m. For preparation of rat liver dihydroorotase, consult the procedures described elsewhere (28).

### 7.1.2 Angiotensin-converting enzyme (EC 3.4.15.1 dipeptidyldipeptidase)

Angiotensin-converting enzyme converts angiotensin (I) to angiotensin (II) and the dipeptide, histidyl–leucine (His–Leu). It also degrades the vaso-depressor peptide, bradykinin (29). The enzyme is found in lung, kidney, serum, brain, and testicles (29). The assay of its activity relies upon determination of hippuric acid liberated from Hip–His–Leu (31).

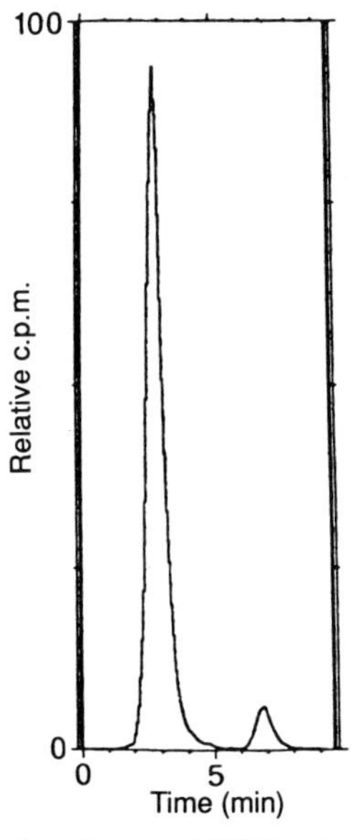


**Figure 9.** HPLC chromatogram of a mixture of [$^{14}$C]carbamyl aspartate and [$^{14}$C]dihydroorotate in the dihydroorotase reaction. Conditions were as described in the text. (Full scale = 3200 c.p.m.; total c.p.m. = 23 728; observed % conversion = 6.8%.) Reproduced from (27) with permission.

**Protocol 2.** Assay of angiotensin-converting enzyme

1. Incubate the enzyme for 30 min at 37 °C with 5 mM Hip–His–Leu in Tris–HCl, pH 7.8, containing 300 mM NaCl, 5 mM magnesium acetate, 0.25 M sucrose, and Nonidet-P40.
2. Terminate the reaction by adding 3% (w/v) metaphosphoric acid.
3. Centrifuge for 5 min before injecting 20 μl of supernatant onto the HPLC column.
4. For the purpose of separation of hippuric acid from Hip–His–Leu, use a 25 × 0.4 cm ID Nucleosil 7 C-18 column with 7.5 μm particle (Macherey-Nugel and Company).
5. The mobile phase should be methanol: 10 mM potassium dihydrogen orthophosphate (1:1) and adjusted to pH 3.0 with phosphoric acid.
6. Use a flow rate of 1.0 ml/min for eluting the column.
7. Detection can be carried out by spectrophotometric means at 228 nm. Peak height can be used for the purpose of quantitation.

*Figure 10a* shows the chromatographic separation of a standard mixture containing hippuric acid, Hip–His–Leu and His–Leu. The upper diagrams (*Figure 10b–e*) show the formation of hippuric acid after incubation of various biological samples with Hip–His–Leu. No endogenous interfering substances are detected even when using Nonidet-P40, a detergent used for solubilizing tissue. The angiotensin-converting enzyme can be obtained from blood collected from the abdominal aorta of rat. Lung and kidney are removed immediately after sacrifice, gently rinsed with saline, chopped into small pieces, and then homogenized in Tris–HCl, pH 7.8, containing 30 mM KCl, 5 mM magnesium acetate, 0.25 M sucrose, and Nonidet-P40. The supernatant, after centrifugation at 20 000 *g*, acts as the enzyme preparation.

## 7.2 Isomerases

### 7.2.1 Diaminopimelate epimerase (EC 5.1.1.7)

LL-2,6-diaminopimelate 2-epimerase (DAP epimerase) catalyses the interconversion of the LL- and *meso*-isomers of DAP. The *meso*–DAP is then decarboxylated to yield L-lysine as the final step in the lysine biosynthetic pathway in bacteria (32, 33). Both enzymes are of interest with regard to regulation of the lysine pathway.

$$\text{LL-DAP} \underset{\textit{DAP-epimerase}}{\rightleftharpoons} \textit{meso}\text{-DAP} \xrightarrow[\textit{DAP decarboxylase (EC 4.1.1.20)}]{CO_2} \text{L-Lysine}$$

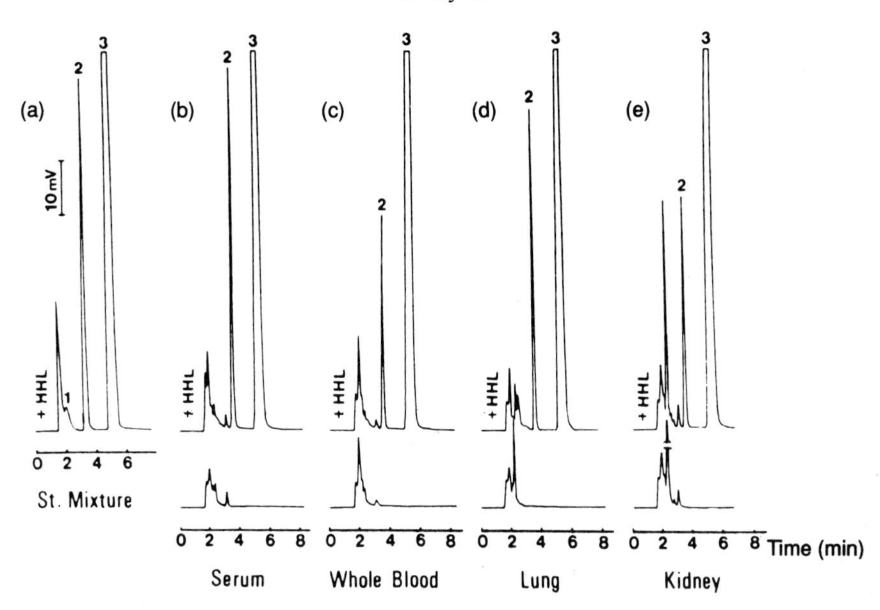


**Figure 10.** Chromatograms obtained from various samples incubated with (upper diagram) or without (lower diagram) Hip–His–Leu (HHL). (a) Standard mixture of 2.7 nmol His–Leu, 2.7 nmol hippuric acid and 100 nmol Hip–His–Leu. (b) A 50-µl aliquot of serum or (c) whole blood was incubated with (upper diagram) or without (lower diagram) 5 mM Hip–His–Leu according to the assay method described in the text. After 30 min, 0.75 ml of 3% (w/v) metaphosphoric acid was added and centrifuged. (d) Lung or (e) kidney was homogenized in 5 volumes of chilled Tris–HCl buffer containing 0.5% Nonidet-P40, and centrifuged. The supernatant was incubated with (upper diagram) or without (lower diagram) 5 mM Hip–His–Leu. In the case of lung, the supernatant was diluted 20 times with the buffer prior to incubation with Hip–His–Leu. Analytical conditions as described in the text. Peaks: 1, His–Leu; 2, hippuric acid; 3, Hip–His–Leu. Reproduced from (31) with permission.

---

**Protocol 3.** Assay of diaminopimelate epimerase

1. Prepare a 1.5 ml assay mixture containing 15 µmol recrystallized *meso*–DAP, 0.1 µmol pyridoxal 5-phosphate, 3.75 µmol norvaline (internal standard), and about 0.5 mg protein as de-salted cell-free extract in 0.1 M potassium phosphate, pH 7.0.
2. Carry out the reaction at 37°C for approximately 15 min.
3. Take 0.5 ml before and after incubation for HPLC analysis.
4. Terminate the reaction by quickly transferring samples to 10 ml Pyrex tubes each containing 1.0 ml of 1.2 M sodium-acetate buffer, pH 5.2
5. Boil the tubes for 5 min and centrifuge to remove denatured protein.
6. Dilute the resulting supernatant 20-fold in water for subsequent HPLC analysis (34).

---

The assay relies on the separation of *O*-phthaldehyde (OPA) derivatives of LL–DAP and *meso*–DAP diastereomers.

---

**Protocol 4.** HPLC analysis of LL–DAP and *meso*–DAP

1. Prepare the OPA/2-mercaptoethanol derivatizing solution according to Unnithan *et al.* (35).
2. Mix 200 μl aliquot of the deproteinized sample above with 200 μl of a 2% (w/v) solution of SDS in 400 mM sodium borate, pH 9.5, and 400 μl of the derivatizing reagent at room temperature. SDS improves the stability of OPA–lysine derivative.
3. Inject 20 μl of this mixture after exactly 1 min.
4. Separate the derivatives on a Spherisorb C-18 ODS reverse phase column (250 ×4.5 mm ID, 5 μm particles).
5. Elute with a linear gradient from 100% solvent A (30% methanol, 70% 50 mM sodium acetate, pH 5.9) to 30% solvent A and 70% methanol over a period of 35 min. The OAP derivatives can be detected by fluorescence. The excitation and emission wavelengths are 340 and 455 nm, respectively.

---

*Figure 11* is a typical separation of OAP derivatives of LL–DAP, *meso*–DAP, norvaline, and lysine. As can be seen, by using this HPLC method both the decarboxylase and the epimerase activities can be simultaneously assayed.

## 7.3 Lyases

### 7.3.1 $C_{17-20}$ Lyase (Cytochrome P-$450_{21SCC}$)

Human cytochrome P-$450_{21SCC}$ has two activities as shown below:

$$\text{Pregnenolone} \xrightarrow[\text{adrenal and gonads}\ (1)]{O_2,\ \text{NADPH}} 17\alpha\text{-hydroxy pregnenolone} \xrightarrow[\text{gonads}\ (2)]{O_2,\ \text{NADPH}}$$

$$17\alpha\text{-hydroxy pregnenolone} \xrightarrow{\text{adrenal}} \text{Cortisol}$$

$$\rightarrow \text{Dehydroepiandrosterone} \xrightarrow[\text{testes, ovaries}]{} \text{Androstenedione testosterone} \xrightarrow[\text{ovaries}]{} \text{Oestrogens}$$

where activity (1) is that of 17α-hydroxylase and activity (2) of $C_{17-20}$ lyase. The lyase reaction is predominant in gonads (36). The assay described here uses normal phase chromatography for the separation of steroids and on-line measurements of radioactivity (37).

---

**Protocol 5.** Assay for lyase

1. Prepare an assay mixture containing 0.8 μM [7-$^3$H]-17 α-hydroxypreg-

nenolone as the substrate, 1 mM NADPH, 5 mM glucose-6-phosphate, 1 IU/ml of glucose-6-phosphate dehydrogenase and 0.02 mg microsomal protein. The total volume of assay should be 100 μl in 50 mM phosphate buffer, pH 7.4, while the total $^3H$ per assay is 0.2 μCi.

2. Carry out the incubation at 34°C for 6 min.
3. Terminate the reaction by adding 5 ml of a 2:1 mixture of $CHCl_3$: methanol, and 0.9 ml water.
4. Shake the tube for 5 min and centrifuge to separate phases.
5. Discard the upper aqueous phase and wash the interface with $CHCl_3$–MeOH–$H_2O$ (4:48:47).
6. After discarding the wash, use a stream of nitrogen to evaporate the lower phase to dryness in a water bath at 40°C.
7. Rinse the sides of the tube with a little chloroform and repeat evaporation.
8. Dissolve the residue in THF in hexane prior to HPLC.
9. Perform the separation of steroids on a silica gel column (LiChrosorb Si-60, 5 μm, 250 × 4 mm) by eluting with a THF-hexane gradient at a flow rate of 1 ml/min. Use a silica pre-column to saturate buffers with silica. The gradient conditions are 18–22% THF in hexane over 30 min and isocratic at 22% THF for 8 min. A Flo-One model HS radioactivity detector, for example, can be used for on-line detection.

---

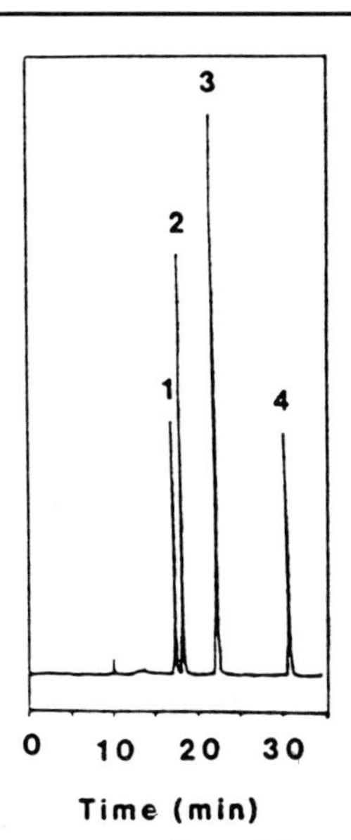


**Figure 11.** HPLC analysis of an enzyme reaction mixture after 30 min incubation (see text for experimental details). The peaks are *O*-phthaldialdehyde derivatives of LL-DAP (1); *meso*-DAP (2); norvaline (3); lysine (4). Reproduced from (34) with permission.

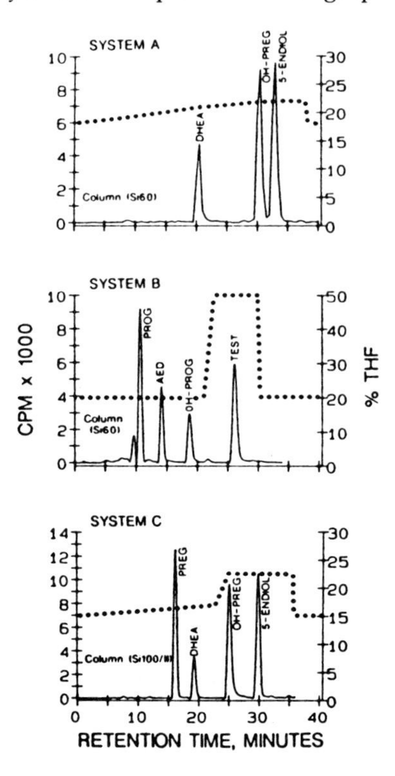


**Figure 12.** Chromatograms of steroid mixtures. The flow detector was in the scaler mode and reported a count every 6 s. The dotted line represents the mobile phase gradient, as percentage THF in hexane-THF. The mixture contained the following steroids: pregnenolone (PREG); dehydroepiandrosterone (DHEA); androst-5-ene-3β:17β-diol (5-ENDIOL); progesterone (PROG); androst-4-ene-3,17-dione (AED); testosterone (TEST). Reproduced from (37) with permission.

*Figure 12* shows the separation of 17 α-hydroxy pregnenolone and its products, dihydroepiandrosterone (i.e. lyase) and androstenedione [see (37) for the preparation of lyase from human testes].

### 7.3.2 Uroporphyrinogen decarboxylase (EC 4.1.1.37) from mouse liver

Uroporphyrinogen decarboxylase is an enzyme of heme biosynthesis and converts uroporphyrinogen, which contains eight carboxyl groups, to coproporphyrinogen which is a tetracarboxy product (38). The reaction proceeds via the hepta-, hexa-, and pentacarboxy intermediates. The enzyme from mouse liver is assayed for its activity using reverse phase HPLC (39). A similar system has been described in the case of the enzyme from chicken erythrocytes (40). Uroporphyrinogens (I) and (III) as well as intermediates can be used as substrates for the enzyme.

## **Protocol 6.** Assay of uroporphyrinogen decarboxylase

*Preparation*

1. Reduce solutions of porphyrins (0.5–0.7 mmol/μl in a total volume of 100 μl) to porphyrinogens with 5% sodium amalgam in 5 mM NaOH. Carry out the reduction under $N_2$ until no fluorescence is observed under UV.
2. Remove the porphyrinogen solution from the mercury with a Hamilton syringe and neutralize with 0.5–2 μl of 4M $H_3PO_4$.
3. Dilute the solution with an equal volume of incubation buffer containing 0.1 mM EDTA/0.1 M sodium mercapto-acetate.

*Assay*

1. Mix 0.2 ml of the enzyme extract with 0.78 ml of 0.1 M sodium phosphate buffer, pH 6.8, containing 0.1 mM EDTA/3 mM mercaptoacetate and keep at 4°C.
2. After saturating the mixture with a stream of $N_2$ for 20 s, add 20–50 μl of the porphyrinogen and then stopper the incubation under $N_2$.
3. Rapidly shake at 37°C in the dark.
4. Terminate the incubations with uroporphyrinogens after 30 min and those with pentacarboxyporphyrinogen after 10 min by rapid cooling in ice and then mixing with 50 μl of 6 M HCl.
5. Leave the mixtures in the light for at least 30 min to oxidize porphyrinogens to porphyrins and then add a further 50 μl of 6 M HCl.
6. Centrifuge the tubes and analyse the supernatants by HPLC.

*HPLC analysis*

1. Carry out the HPLC on a Spherisorb ODS column (25 cm × 4.6 mm ID, 5 μm particles). Use a CO:PELL ODS as a pre-column.
2. Elute the porphyrin standards as well as those formed from porphyrinogens with a linear gradient of 65–95% methanol over 20 min in 50 mM lithium citrate, pH 3.
3. Continue washing with 95% methanol for a further 5–10 min. Use a flow rate of 1 ml/min.
4. The detection can be carried out by absorption at 400 nm.

*Figure 13a* and *b* shows separation of a mixture of standard porphyrins (A) and the products of the decarboxylase reaction (B). The identity of HPLC peaks is given in the legend. Peak 8 corresponds to the substrate while peaks 4, 5, 6, and 7 are the products.

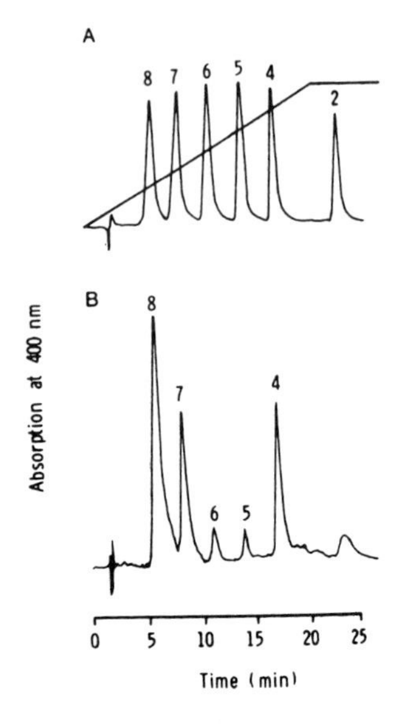


**Figure 13.** Separation of free prophyrins by HPLC using a Spherisorb ODS 1 5-μm column (25 cm × 4.6 mm ID) with a linear gradient system of methanol/50 mM lithium citrate (pH 3) from 65–95% methanol at 1 ml/min, described in the text. (A) A scintillant mixture of the I isomers of uroporphyrin (8), heptacarboxyporphyrin (7), hexacarboxyporphyrin (6), pentacarboxyporphyrin (5), and coproporphyrin (4), together with mesoporphyrin IX (2). The slope of the solvent gradient system is also shown. (B) Products of an incubation of mouse liver supernatant with uroporphyrinogen (10 nmol) for 60 min as described in the text. Reproduced from (39) with permission.

## 7.4 Ligases

### 7.4.1 Glutaminyl cyclase from bovine pituitary homogenate

Glutaminyl cyclase catalyses the conversion of the peptide, Gln–His–Pro–$NH_2$ to thyrotopin releasing hormone (TRH) (41, 42). The method, involving dansylation of the N-terminus (43) followed by detection of the fluorescent derivative, is precluded in the present case due to the N-terminal location of glutamine cyclization. The model peptide Gln–Leu–Tyr–Glu–Asn–Lys–OH can be used instead since it possesses a C-terminal lysine which could provide the E-$NH_2$ group for dansylation (44). The dansylated derivative acts as the substrate.

The substrate and the cyclized product can be synthesized by conventional solid-phase peptide synthesis.

---

**Protocol 7.** Assay of glutaminyl cyclase

1. Carry out an incubation of glutaminyl cyclase at 37°C by adding up to 0.5 mg/ml of bovine pituitary extract (prepared according to reference 45)

to the dansylated peptide substrate present at concentrations of 1–98 μM in 50 mM Tris–HCl, pH 8.0, or Mops, pH 7.0.

2. Remove 10–20 μl aliquots at various times and stop the reaction by the addition of 10 μl of 10 mM phenanthroline.
3. Inject 10 μl for HPLC analysis.
4. Carry out the separation at 50°C on a thermostated Hypersil ODS column (4.6 mm ID × 100 mm, 5 μm particles). Use a Brownlee RP-18 guard column (3.2 mm ID × 15 mm) as added protection for the main column.
5. Elute the substrate and cyclized peptide product by isocratically washing the column with 24% (v/v) acetonitrile in 0.1 M sodium acetate, pH 6.5, at a flow rate of 1.2 ml/min. Detection can be accomplished by the use of a fluorescence detector with an excitation filter of 352–360 nm and an emission cut-off filter of 482 nm.

---

*Figure 14a* clearly shows the separation of substrate and product peptides, while *Figure 14b* is a time-course of enzymatic conversion of the dansylated peptide substrate.

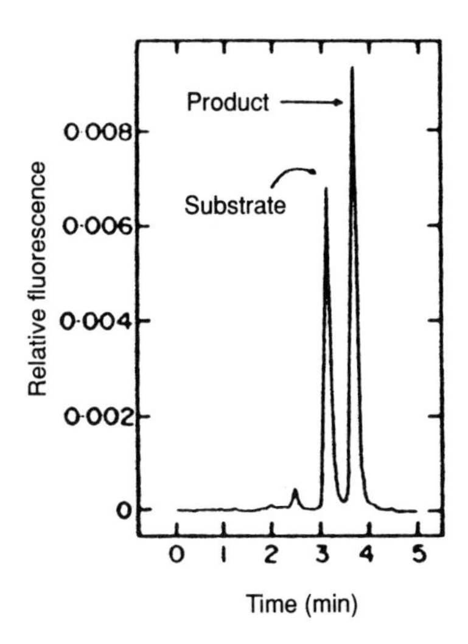


**Figure 14a.** HPLC separation of a reaction mixture containing the substrate Gln–Leu–Tyr–Glu–Asn–Lys–ε–(Dns)–OH and the enzymatically generated product <Glu–Leu–Tyr–Glu–Asn–Lys–ε–(Dns)–OH). Partial conversion of the substrate to product was achieved by incubation of Gln–Leu–Tyr–Glu–Asn–Lys–ε–(Dns)–OH (43 μM) with crude bovine pituitary homogenate (0.5 mg $ml^{-1}$) in 50 mM Tris–HCl, pH 8.0, for 10 min. HPLC conditions are as described in the text. Peak heights correspond to the injection of 80 pmol substrate and 130 pmol product. Reproduced from (45) with permission.

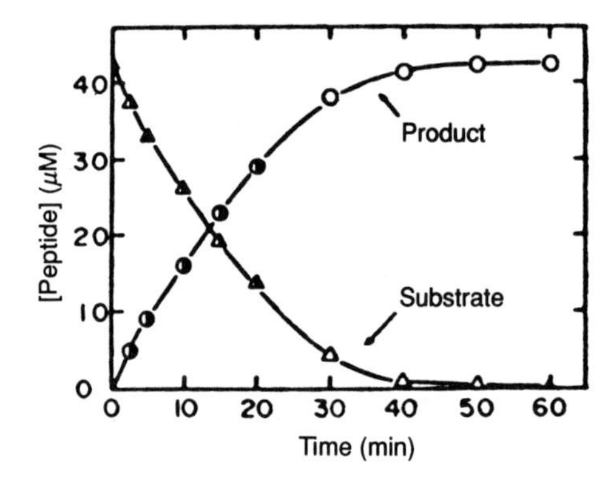


**Figure 14b.** Time course of enzymatic conversion of Gln–Leu–Tyr–Glu–Asn–Lys–ε–(Dns)–OH to <Glu–Leu–Tyr–Glu–Asn–Lys–ε–(Dns)–OH. Reactions were initiated by the addition of enzyme preparation to achieve S-300 purified glutaminyl cyclase Peak II (160 μg $ml^{-1}$) and peptide substrate (43 μM), in 50 mM Tris–HCl, pH 8.0. At the indicated times, aliquots (10 μl) of reaction mixture were removed, mixed with 10 mM phenanthroline (10 μl), and subjected to HPLC analysis. The time course of product formation (○) and substrate depletion (△) were obtained from HPLC analysis of samples either immediately following collection (open symbols) or after a 3 h delay (filled symbols). Reproduced from (45) with permission.

### 7.4.2 δ-(L-α-aminoadipyl)-L-cysteinyl-D-valine (ACV) synthetase

ACV synthetase acts on L-α-aminoadipic acid, L-cysteine, and L-valine to form ACV which is the precursor to the β-lactam ring in fungi and streptomyces. The assay of this enzyme depends on the separation and detection of derivatives of *O*-phthaldehyde (OPA) with the ACV and the unreacted amino acids by reverse-phase HPLC (46).

---

**Protocol 8.** Assay of ACV synthetase

*Assay*

1. Mix 0.5 ml of a crude or fractionated cell extract of *Streptomyces clavuligurus* with 0.5 ml of a solution containing 150 mM KCl, 45 mM ATP, 45 mM $MgCl_2$, 15 mM EDTA and 3 mM chloramphenicol in 0.3 M Mops buffer, pH 7.2.
2. Sparge the mixture with $N_2$ at 0°C for 10 min and seal the tube with a rubber septum.
3. Initiate the reaction with injection of 0.5 ml of a de-gassed solution containing 15 mM L-α-aminoadipic acid, 15 mM L-cysteine hydrochloride and 15 mM L-valine adjusted to pH 7.2.
4. Incubate the reaction mixture at 27°C with gentle agitation.

**Protocol 8.** *Continued*

5. Terminate the reaction by injecting 0.4 ml of 20% (w/v) trichloroacetic acid and clarify the suspension by centrifugation.
6. Store at −20°C until analysis by HPLC.

*HPLC analysis*

1. Oxidize a 100 μl aliquot of the sample with an equal volume of performic acid for 2.5 h at 0°C, add 2 ml water and lyophilize before re-dissolving in 100 μl water.
2. Prepare fluorescent isoindole derivative (OPA-derivative) by mixing 20 μl of above sample with 40 μl Fluo-R reagent. 20 μl homoserine solution can also be added to act as an internal standard.
3. Quench the reaction with 120 μl 0.1 M sodium acetate buffer, pH 6.25, and use 20 μl aliquots for HPLC analysis.
4. Separate the isoindole derivatives on a 45 × 4.6 mm reverse phase Ultrasphere ODS column containing 5 μm particles.
5. Develop a binary gradient between solvent A (0.1 M sodium acetate, pH 6.25-methanol-THF, 90:9.5:0.5) and solvent B (methanol).
6. Carry out the detection with a fluorescence detector with excitation at 300–395 nm and emission at 420–650 nm.

---

*Figure 15* is a chromatogram of an oxidized sample from a 1 h incubation mixture containing extract of mycelium. This simple and rapid assay has also been used for monitoring purification of ACV synthetase (46).

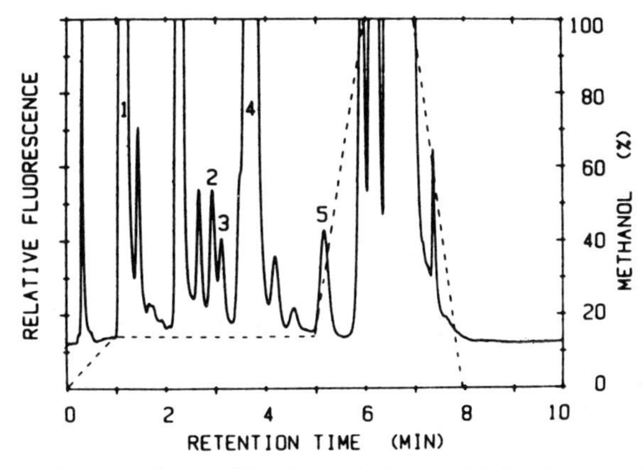


**Figure 15.** Chromatogram of an oxidized sample from a 1 h incubation mixture containing the crude extract from 24 h mycelium. The concentration of methanol in the gradient is shown by the dashed line. The peaks are: 1, cysteic acid; 2, ACV sulfonate; 3, unidentified but associated with cysteine; 4, α-aminoadipic acid; 5, homoserine, the internal standard. Reproduced from (46) with permission.

## 7.5 Oxidoreductases

### 7.5.1 Glutamate synthase (EC 1.4.7.1) from cyanobacteria

The assimilation of ammonia occurs mainly via the glutamine synthetase (EC 6.3.1.2) and glutamate synthase (EC 1.4.7.1) pathway and in some micro-organisms via glutamate dehydrogenase (47). These enzymes form a key role in bridging nitrogen and carbon metabolism. Both the synthetase and synthase activities can be monitored by detecting glutamine and/or glutamate. Both metabolites can be separated by reverse phase HPLC (48).

---

**Protocol 9.** Assay of glutamate synthase

1. Prepare an assay mixture for glutamate synthase with a total volume of 0.9 ml and containing 45 μmol of potassium phosphate, pH 7.0, 5 μmol L-glutamine, 1 μmol 2-oxo-glutarate, 5 μmol aminooxyacetate, 10 nmol *Synechococcus* ferredoxin and an aliquot of enzyme.
2. Start the reaction with a 0.1 mol solution containing 0.8 mg sodium dithionite in 0.12 M $NaHCO_3$ and incubate at 30°C for 15 min. Terminate the reaction with 0.6 ml of 1 M HCl.
4. Centrifuge 0.4 ml of the sample at 12 000 *g* for 4 min and dilute 25-fold with 50 mM potassium phosphate, pH 7.5.
5. Mix 50 μl of this with 150 μl of a derivatizing solution of OPA according to (49).
6. Inject 20 μl into the injector loop after 90 s.
7. Carry out the HPLC analysis on a μ Bondapak C18 or a Novapak C18 (3.9 mm × 4 cm) column thermostatted at 45°C.
8. Elute the column with 20 mM sodium phosphate, pH 6.5, containing 22% (v/v) methanol and 2% (v/v) THF at a flow rate of 1–1.5 ml/min.

---

*Figure 16* shows separation of glutamate and glutamine for the *in situ* assay. The amounts of glutamate or glutamine can be obtained from a calibration curve prepared with standards of these amino acids. Use amounts in the range 0–3 nmole/injection.

## 7.6 Transferases

### 7.6.1 Aryl alkylamine (serotonin) *N*-acetyltransferase (EC 2.3.1.87)

Serotonin *N*-acetyltransferase (NAT) catalyses the *N*-acetylation of serotonin to *N*-acetyl serotonin and is a key regulatory enzyme in the melatonin (5-methoxy-*N*-acetyltryptamine) pathway (50). The present assay for the NAT activity is based upon ion-pair HPLC using either fluorescence or electrochemical detection of *N*-acetyltryptamine (51). In the present case, only the former detection method will be described.

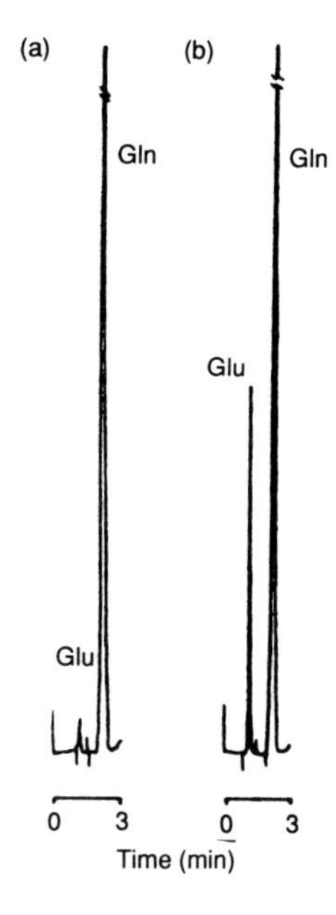


**Figure 16.** Chromatograms corresponding to a glutamate synthase *in situ* assay. (a) Sample taken at zero time; (b) sample taken after completion of the assay. Reproduced from (48) with permission.

---

**Protocol 10.** Assay of NAT

1. Prepare a homogenate of retinal or pineal gland tissue by sonicating in different volumes of ice-cold 0.25 M potassium phosphate buffer, pH 6.5, containing 1.4 mM acetyl-CoA to give the desired protein concentration.
2. Centrifuge at 28 000 *g* for 10 min at 4°C.
3. Mix 75 μl aliquots of the resulting supernatant or whole homogenate (cultured retinal cells) with 25 μl of 8 mM tryptamine in the same buffer and incubate at 37°C for 15 min.
4. Stop the enzyme reaction by the addition of 20 μl of 6 M perchloric acid and centrifuge at 28 000 *g* for 10 min at 4°C.
5. Use 10 μl aliquots for HPLC analysis.
6. Carry out the separation on a Partisphere C18 (5 μm particles, 110 × 4.7 mm) reverse phase column. Wash the column with 50 mM phosphoric acid containing 33% (v/v) methanol and 0.65 mM sodium octylsulphate, adjusted to pH 3.5 at a flow rate of 1.5 ml/min.
7. Set the excitation and emission wavelengths at 285 and 360 nm respectively for detection.

---

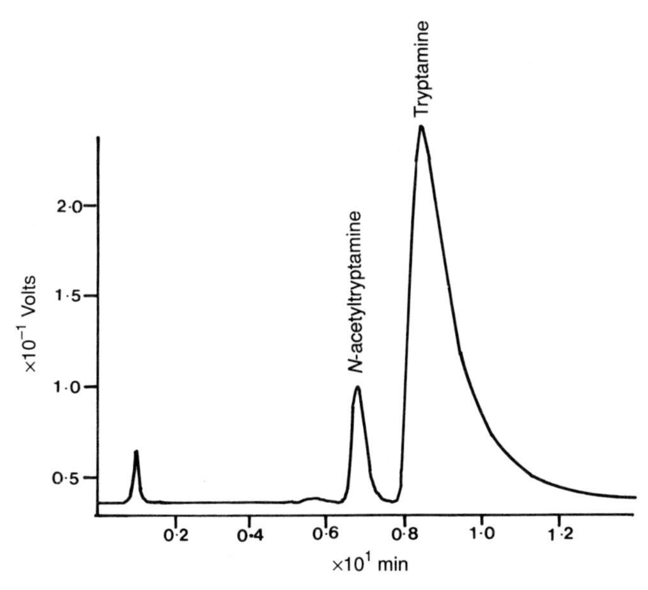


**Figure 17.** Representative HPLC-fluorescence chromatogram of a sample prepared from chicken pineal gland. Pineal glands isolated in the middle (6th h) of the dark phase of the light–dark cycle were assayed for NAT activity. The enzymatic reaction was stopped by addition of 6 M perchloric acid and samples were further processed for HPLC analysis. The *N*-acetyltryptamine peak represents 448 pmol injected in a volume of 10 μl. For methodological details see text. Reproduced from (51) with permission.

The separation of *N*-acetyltryptamine from the substrate tryptamine is shown in *Figure 17*. Sodium octylsulphate acts as an ion-pairing reagent and increases the retention time of tryptamine which elutes as a broad peak. The synthesis of N-acetyltryptamine standard and procedures for the preparation of tissues and cell cultures are fully described in (51).

### 7.6.2 Ornithine aminotransferase (OAT) (EC 2.6.1.13) from rat liver

Ornithine aminotransferase is a mitochondrial matrix enzyme catalysing the conversion of L-ornithine and 2-oxoglutarate to form glutamic-γ-semialdehyde and glutamate. The semi-aldehyde spontaneously cyclises to give $\Delta^1$-pyrroline-5-carboxylic acid (P5C). The assay for the aminotransferase activity in the present case depends on the reaction of P5C with *O*-aminobenzaldehyde (OAB) and separation of the resulting dihydroquinozolinium (DHQ) by reverse phase HPLC (52).

---

**Protocol 11.** Assay of ornithine aminotransferase

1. Homogenize the liver of a female rat in a 20% (w/v) solution containing 0.1 M potassium phosphate, pH 7.4, with 0.2 M sucrose and 4 μg pyridoxal 5′-phosphate/ml at 4°C.

2. Centrifuge at 14 000 *g* for 15 min to remove cell debris and particulate matter and use it for subsequent assays.
3. Prepare a fresh assay mixture containing 35 mM L-ornithine, 3.7 mM 2-oxoglutarate and 4 μg pyridoxal-phosphate/ml in 50 mM potassium phosphate, pH 7.4, in a total volume of 2 ml.
4. Incubate the enzyme at 37°C for 0–60 min withdrawing samples periodically and terminating the reaction with 1 ml of 3 M HCl containing 7.5 mg OAB/ml.
5. Centrifuge samples to remove precipitated protein. Monitor the formation of DHQ at 440 nm using a UV/visible detector.
6. Calculate its concentration by using an extinction coefficient of 2.59 $mM^{-1}$ $cm^{-1}$.
7. Elute the DHQ and OAB from a LiChrosorb $C_{18}$ (4.6 × 250 mm, 10 μm particles) column by isocratically washing with methanol/$H_2O$ (1:2) mixture, with detection at 254 nm. Use a flow rate of 1.5 ml/min.

---

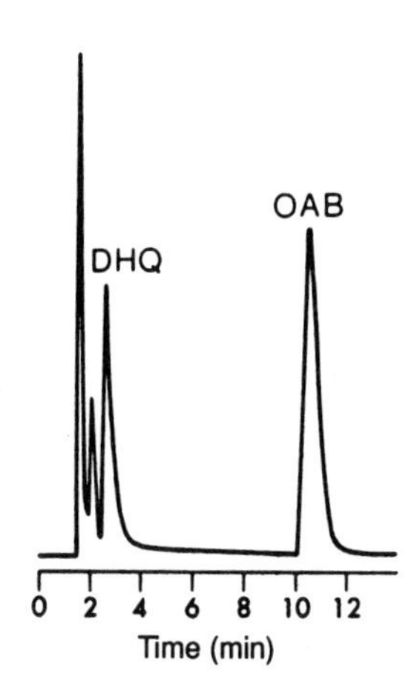


**Figure 18a.** Isocratic reverse-phase HPLC showing separation and detection of dihydroquinozolinium (DHQ) and *O*-aminobenzaldehyde (OAB). The column was LiChrosorb $C_{18}$, 10 μm and 4.6 × 250 mm, and the solvent system was 1 part methanol:2 parts $H_2O$, pumped at a flow rate of 1.5 ml/min. Reproduced from (52) with permission.

*Figure 18a* shows the separation of DHQ and OAB standards while *Figure 18b* is a plot of OAT activity versus time. The OAT activity is represented as μmoles P5C which can be calculated from a standard curve.

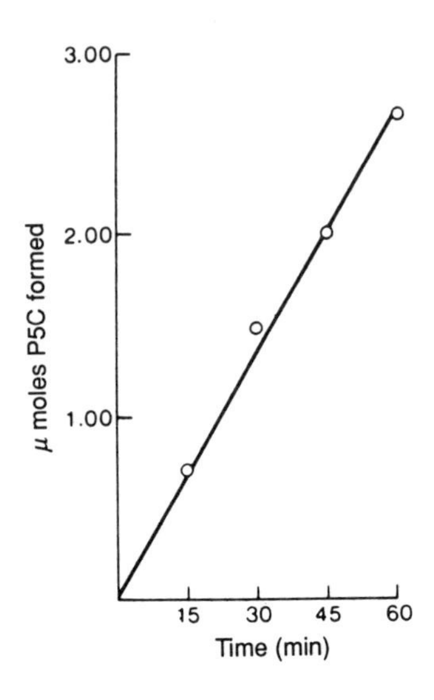


**Figure 18b.** OTA activity as determined by HPLC (see text) plotted against time. The reaction is linear for 60 min.

# References

1. Engelhardt, H. and Muller, H. (1981). *J. Chromatogr.,* **218,** 395.
2. Unger, K. K., Becker, N., and Roumeliotis, P. (1976). *J. Chromatogr.,* **125,** 115.
3. Engelhardt, H. and Ahr, G. (1981). *Chromatographia,* **14,** 227. *High-Performance Liquid Chromatography* (1985). (ed. A. Hanschen, K. P. Hupe, F. Lottspeich, and W. Voelter). VCH Verlagsgesellschaft, Germany.
4. Krstulovic, A. M. and Brown, P. R. (1982). *Reversed-Phase High-Performance Liquid Chromatography—Theory, Practice and Biomedical Applications.* John Wiley and Sons, New York.
5. Horvath, C. (ed.) (1983). *High-Performance Liquid Chromatography (Advances and Perspectives),* Vol. 3. Academic Press, New York.
6. Hearn, M. T. W. (ed.) (1985). *Ion-Pair Chromatography (Theory and Biological and Pharmaceutical Applications).* Marcel Dekker, Inc., New York.
7. Knox, J. H. and Pryde, A. (1975). *J. Chromatogr.,* **112,** 171.
8. Karch, K., Sebastian, I., and Halasz, I. (1976). *J. Chromatogr.,* **122,** 3.
9. Frank, H. S. and Evans, M. W. (1945). *J. Chem. Phys.,* **13,** 507.
10. Horvath, C., Melander, W., and Molnar, I. (1976). *J. Chromatogr.,* **125,** 129.
11. Karger, B. L., Gant, J. R., Hartknopf, A., and Weiner, P. H. (1976). *J. Chromatogr.,* **128,** 65.
12. Jandera, P., Colin, H., and Guiochon, G. (1982). *Anal. Chem.,* **54,** 435.
13. Karger, B. L., Le Page, J. N., and Tanaka, N. (1980). *High-Perf. Liq. Chromatogr.,* **1,** 113.
14. Horwath, C., Melander, W., Molnar, I., and Molnar, P. (1977). *Anal. Chem.,* **49,** 2295.

15. Snyder, L. R. and Kirkland, J. J. (1979). *Introduction to Modern Liquid Chromatography*. Wiley-Interscience, New York.
16. Lawrence, J. L. (1987). *J. Chromatogr. Sci.,* **17,** 147.
17. Kissinger, P. T. (1983). *Chromatogr. Sci.,* **23,** 125.
18. Stulik, K. and Pacakova, V. (1982). *CRC Crit. Rev. Anal. Chem.,* **14,** 297.
19. Huber, J. F. K., Jonker, K. M., and Poppe, H. (1980). *Anal. Chem.,* **52,** 2.
20. Engelhardt, H. and Neue, U. D. (1982). *Chromatographia,* **15,** 403.
21. Hofmann, K. and Halasz, I. (1980). *J. Chromatogr.,* **125,** 287.
22. Snyder, L. R. (1976). *J. Chromatogr.,* **125,** 287.
23. Deelder, R. W. and Hendricks, P. H. J. (1973). *J. Chromatogr.,* **83,** 343.
24. Scholten, A. H. T. M., Brinkman, N. U. A., and Frei, R. W. (1980). *Anal. Chim. Acta,* **114,** 137.
25. Majors, R. E. (1972). *Anal. Chem.,* **44,** 1722.
26. Kaminski, M., Klawiter, J., and Kowalczyk, J. S. (1982). *J. Chromatogr.,* **243,** 225.
27. Mehdi, S. and Wiseman, S. (1989). *Anal. Biochem.,* **176,** 105.
28. Mori, M. and Tatibana, M. (1978). In: *Methods in Enzymology*, Vol. 51, (ed. P. Hoffee and M. E. Jones), pp. 111–21. Academic Press, San Diego, CA.
29. Soffer, R. L. (1976). *Ann. Rev. Biochem.,* **45,** 73.
30. Cushman, D. W. and Cheung, H. S. (1971). *Biochim. Biophys. Acta,* **250,** 261.
31. Horiuchi, M., Fujimura, K., Tarashima, T., and Iso, T. (1982). *J. Chromatogr.,* **233,** 123.
32. White, P. J. and Kelly, B. (1965). *Biochem. J.,* **96,** 75.
33. Asada, Y., Tanizawa, K., Kawabata, Y., Misono, H., and Soda, K. (1981). *Agric. Biol. Chem.,* **45,** 1513.
34. Weir, A. N. C., Bucke, C., Holt, G., Lilly, M. D., and Bull, A. T. (1989). *Anal. Biochem.,* **180,** 298.
35. Unnithan, S., Moraga, D. A., and Schuster, S. M. (1984). *Anal. Biochem.,* **136,** 195.
36. Hall, P. F. (1986). *Steroids,* **48,** 131.
37. Schatzman, G. L., Laughlin, M. E., and Blohm, T. R. (1988). *Anal. Biochem.,* **175,** 219.
38. Jackson, A. H., Sancovich, H. A., Ferramola, P. M., Evans, N., Games, D. E., Matlin, S. A., Elder, G. H., and Smith, S. G. (1976). *Phil. Trans. R. Soc. Lond. Ser. B.,* **273,** 191.
39. Francis, J. E. and Smith, A. G. (1983). *Anal. Biochem.,* **138,** 404.
40. Kawanishi, S., Seki, Y., and Sano, S. (1983). *J. Biol. Chem.,* **258,** 4285.
41. Busby, W. H., Quackenbush, G. E., Humm, J., Youngblood, W. W., and Kizer, J. S. (1987). *J. Biol. Chem.,* **262,** 8532.
42. Fischer, W. H. and Spiers, J. (1987). *Proc. Natl. Acad. Sci. USA,* **84,** 3628.
43. Bond, M. D., Auld, D. S., and Lobb, R. R. (1986). *Anal. Biochem.,* **155,** 315.
44. Merrifield, R. B. (1963). *J. Amer. Chem. Soc.,* **85,** 2149.
45. Consalvo, A. P., Young, S. D., Jones, B. N., and Tamburini, P. P. (1988). *Anal. Biochem.,* **175,** 131.
46. White, R. L., De Marco, A. C., Shapiro, S., Vining, L. C., and Wolfe, S. (1989). *Anal. Biochem.,* **178,** 399.
47. Stewart, G. R., Mann, A. F., and Fentem, P. A. (1980). In: *The Biochemistry of Plants*, Vol. 5, (ed. P. K. Stumpf, and E. E. Conn), pp. 2271–327. Academic Press, New York.

48. Marques, S., Florencio, F. J., and Candau, P. (1989). *Anal. Biochem.,* **180,** 152.
49. Lindroth, P. and Mopper, K. (1979). *Anal. Chem.,* **51,** 1667.
50. Klein, D. C., Berg, G. R., and Weller, J. L. (1970). *Science,* **168,** 979.
51. Thomas, K. B., Zawilska, J., and Iuvone, P. M. (1990). *Anal. Biochem.,* **184,** 228.
52. O'Donnell, J. J., Sandman, R. P., and Martin, S. R. (1978). *Anal. Biochem.,* **90,** 41.

# 5

# Electrochemical assays: polarography

P. D. J. WEITZMAN and P. J. WATKINS

## 1. Introduction

The technique of 'polarography' was invented nearly seventy years ago by the Czech chemist Jaroslav Heyrovsky and earned him the Nobel Prize in 1959. Polarography is an electroanalytical method in which the current produced at a polarizable micro-electrode immersed in a solution is investigated as a function of the applied potential. It is also sometimes referred to as 'voltammetry' or 'amperometry', but here we shall use 'polarography', the inventor's term. Polarography provides a method of studying substances which are electro-active. Electro-reducible/oxidizable substances respectively accept or surrender electrons at an electrode providing an appropriate potential is applied. The *potential* at which a substance is reduced or oxidized is characteristic of that substance, while the *current* that passes is a reflection of the rate of electrolysis. Under conditions when this rate is determined by the *concentration* of the electro-active substance, changes in current will indicate changes in concentration; these may then be followed by continuous measurement of current at an appropriate fixed potential. This may form the basis for the measurement of the rate of a reaction and thus offers a suitable technique for assaying enzymes.

Polarography has been ignored by most enzymologists, despite a number of attractive features of the technique and some distinct advantages. Curiously, not only has polarography failed to establish a niche for itself within popular enzymology, but most enzymologists are unaware of the technique and its possible usefulness. This is the more surprising in view of the interest which enzymologists have in electron transfer and redox reactions. Some popularity has been gained by a particular type of polarography, the use of the 'oxygen electrode', but the wider field has remained a strictly minority pursuit.

One of us (PDJW) described the application of polarography to enzyme assays a quarter of a century ago (1) and subsequently reported extensions of the technique (2, 3). A literature survey has revealed a number of reports by other investigators of the use of polarography in various enzyme assays. In this book we describe polarographic enzyme assays within the context of a range of other techniques. We hope that the method may gain new adherents and increased popularity.

## 2. Polarographic principles

If an unstirred solution containing an electro-reducible substance, X, is electrolysed between a small cathode and a reference anode, the current–voltage curve, or 'polarogram', will be of the form shown in *Figure 1a*, exhibiting a pronounced 'wave'. As the applied negative potential increases, only a very small current will flow until a potential is reached at which X can be electro-reduced at the cathode, when the current begins to rise. With increased applied potential, the concentration of X at the cathode surface decreases towards zero. Further increase in potential produces no further increase in current, the latter being limited by the rate of diffusion of X from the bulk of the solution to the cathode, where it is instantaneously reduced. This limiting current depends on the concentration of X in the solution and has three components—the residual current, the migration current, and the diffusion current. The residual current is the small current that flows in the complete absence of X. The migration current is caused by the reduction of X attracted to the cathode by electrostatic forces; in practice, this is eliminated by the presence of 'supporting electrolyte' (buffer components or neutral salts). The diffusion current, with which we are concerned here, is due to the reduction of X reaching the cathode by free diffusion under the influence of the concentration gradient between the cathode surface and the bulk of the solution. This limiting diffusion current is therefore directly proportional to the solution concentration of X.

The other important characteristic of a polarographic wave is its 'half-wave potential', $E_{1/2}$, which is the potential at the mid-point of the wave, where the current is half the limiting value. $E_{1/2}$ is independent of concentration and is highly characteristic of the particular electro-reducible species in a particular solution. A polarogram thus provides both qualitative and quantitative information on solution analytes.

The same considerations apply to electro-oxidation of a susceptible species except that, instead of a cathodic wave, an anodic wave is produced (*Figure 1b*). The limiting anodic diffusion current is again proportional to the concentration of the electro-oxidizable species.

## 3. Polarographic techniques

The type of polarography described here is simple direct current (DC) polarography. Other more complex variations have been developed (AC, linear-sweep, oscillographic, square-wave, pulse) but it is only with conventional

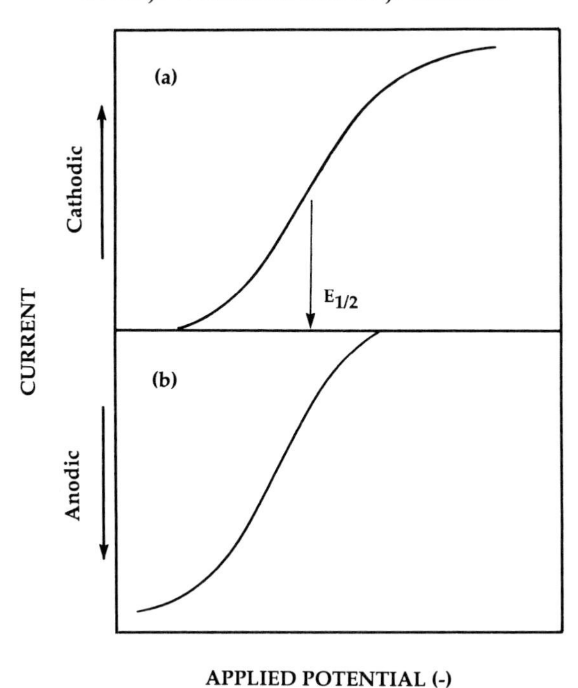


**Figure 1.** Polarograms produced by (a) an electro-reducible or (b) electro-oxidizable species at a micro-electrode.

DC polarography that this article is concerned. The reader is referred to some long-standing established texts for fuller accounts of the technique, its apparatus, and procedures (4–7).

## 3.1 Electrodes

The classic polarographic electrode is the dropping mercury electrode (DME) which is simply a thick-walled glass, fine-capillary tube (internal diameter 0.05–0.1 mm), connected by a flexible tube to a reservoir of mercury and with its open end immersed in the test solution. Mercury passes slowly down the capillary and emerges as a regular succession of identical tiny drops. The height of the reservoir level is adjusted so that a drop of mercury falls every 3–5 s. The presentation of a fresh mercury surface to the solution with each new drop is of particular value when examining solutions containing proteins or other surface-active substances which might otherwise coat, and ultimately block, the electrode surface.

Another beneficial feature of the DME is the high overvoltage of hydrogen on mercury. This enables the DME to be used over a more extended potential range (up to 2 V or more, depending on the pH) than would be the case were $H^+$ ions to be reduced to $H_2$ gas at their thermodynamic reduction potential.

The electro-reduction of $H^+$ ions effectively sets an upper limit on the negative potential for polarographic studies.

A typical polarogram produced with a DME is shown in *Figure 2*. The oscillations reflect the growth of the mercury surface during the lifetime of each drop. Wave heights are best determined by measuring the mean values of the oscillations. It should be noted that wave height depends on drop geometry; it is therefore necessary to calibrate individual electrodes.

The other electrode in the circuit is a non-polarizable reference electrode (i.e. one whose potential remains constant despite changes in current). This is commonly an easily-constructed, saturated calomel electrode (SCE), and electrical connection is achieved by a salt bridge (2% agar, 2 M KCl) which dips into both the reference electrode compartment and the analyte solution under test. *Figure 3* illustrates the experimental set-up. By means of platinum wires making contact with the mercury reservoir of the DME and with the mercury pool of the SCE, the two electrodes are connected to a polarograph—a sensitive instrument capable of applying a controlled variable potential to the DME and measuring the small currents which flow as a response. The current–voltage curve is displayed on a chart recorder.

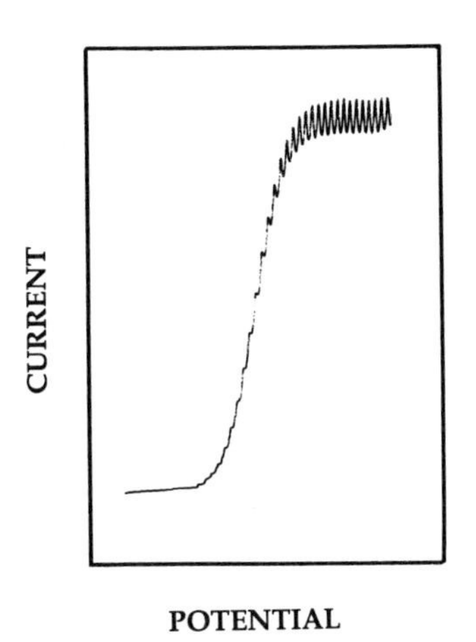


**Figure 2.** Typical polarogram produced by an electro-reducible species at a dropping mercury electrode.

## 3.2 Removal of oxygen

Solutions in contact with air contain dissolved oxygen which is itself electro-reducible at the DME (a two-step process, first to $H_2O_2$ and then to $H_2O$); this is the basis of the 'oxygen electrode', though the cathode generally employed is a platinum electrode. Polarographic waves of oxygen obscure those of other substances, so oxygen must be removed before measurements

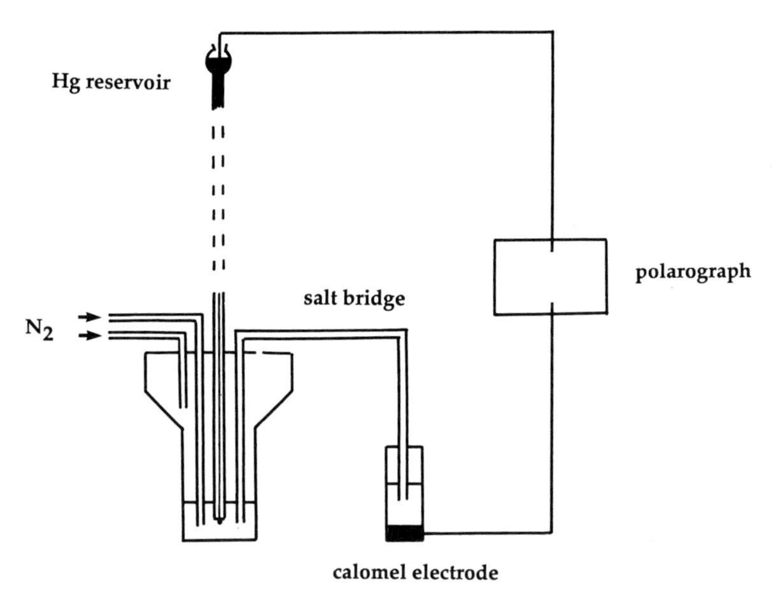


**Figure 3.** Schematic diagram of the experimental arrangement for polarographic measurements with a dropping mercury electrode.

are made. This is done by passing oxygen-free nitrogen through the solution for a few minutes. The nitrogen flow is then diverted over the surface of the solution so as to maintain oxygen-free conditions but allow the solution to become quiescent (*Figure 3*).

## 3.3 Reaction vessel

The choice and design of a cell to contain the test solution allows substantial flexibility. Essentially, all that is required is a small glass container into the top of which a stopper can be placed which carries the DME, salt bridge, and nitrogen inlet and outlet tubes. The stopper should also contain a narrow hole through which a fine drawn-out pipette or syringe needle can be inserted in order to introduce into the assay mixture that component required to start the reaction (generally the enzyme itself). When not in use, this hole may be blocked with a small plastic cap. Thermostatic conditions may readily be achieved by constructing a jacketed cell and pumping through water from a thermostated water bath.

## 3.4 Polarography and spectrophotometry

In view of the widespread use of spectrophotometry in enzyme assays, it is appropriate to compare it with polarography. A polarogram, such as that in *Figure 1*, is analogous to an absorption spectrum. The $E_{1/2}$ value equates to the wavelength of an absorption peak (both are characteristic of the particular species) and the wave height is equivalent to the absorbance (both are dependent on concentration). Just as spectrophotometric assays may be conducted at fixed wavelength by following the change in absorbance with time, so polarographic assays may similarly be performed by monitoring the change in current with time, at a fixed potential. The slope of the trace indicates the rate of the reaction (*Figure 4*).

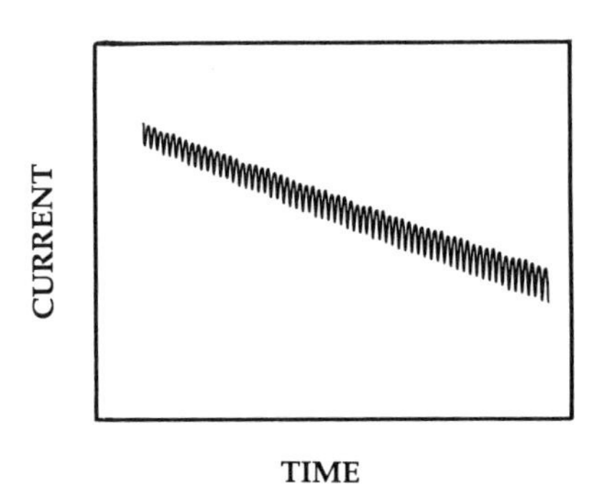


**Figure 4.** Typical polarographic assay 'trace'. The slope of the trace indicates the rate of reaction.

One of the substrates or products of the enzyme must give a polarographic wave in the available potential range; the reaction rate can then be measured by following the increase or decrease of that wave. Like spectrophotometric assays, polarographic assays are ideally conducted in a continuous manner. However, discontinuous assays may have to be performed, particularly if treatment of the reaction mixture or derivatization of one of the components is required before the polarographic wave can be measured. If more than one component of the reaction is electro-active, it may still be possible to assay polarographically if the waves are sufficiently separated on the potential scale.

## 3.5 Summary of requirements for a polarographic assay

In summary, it is necessary to demonstrate:

(a) the display of a polarographic wave by one of the reaction components under the experimental conditions of the assay (buffer, pH, and so on)

(b) the linear dependence of wave height on concentration

(c) the non-interference with the wave by other reaction components

If these conditions are satisfied, the enzyme-catalysed reaction should reveal a change in limiting current at a suitable fixed potential. This current change should enable a straight line to be drawn and a rate to be calculated, using a calibration factor determined in (b) above. Rates so calculated should show a linear dependence on enzyme concentration.

In developing a new polarographic assay, it is advisable to run a current–voltage polarogram at the end of an assay to check that the current change that has been followed really does correspond to an increase or decrease of the expected wave and is not a spurious effect.

## 3.6 Example assay procedure

Into the cell place buffer solution, substrates, and cofactors to a total volume of 0.98 ml. Attach the cell to the stopper so that the electrode, salt bridge, and nitrogen tubes are in place. Pass nitrogen for 2–3 min. Divert the nitrogen to pass *over* the solution, switch on the required potential, and run the current trace for 0.5–1 min to give a base reading. Switch off and divert the nitrogen to bubble through the solution. Using a long-tipped pipette or syringe and long needle, introduce 0.02 ml of enzyme (suitably diluted) into the reaction mixture. After about 20 s, divert the nitrogen to stop the bubbling, switch on the potential and the chart feed, and start recording the changing current.

## 3.7 Summary features and advantages of polarographic assays

Polarography is a straightforward and simple experimental technique and the apparatus required is neither sophisticated nor expensive. High sensitivity may be achieved. As an example, a variety of polarographic assays devised in the authors' laboratory have utilized calibration factors in the range 1.5–6.5 microamps per millimolar concentration of the measured species, and measurements were routinely undertaken at a full-scale deflection of the chart recorder of 0.2 microamps. Commercial polarographs offer a wide range of sensitivities, say from several hundred to one-hundredth microamps full-scale deflection on their chart recorders. Small volumes of reaction mixture are required and the entire measurement may be completed within a few minutes. A wide pH range can be used. Reaction vessels are simple and inexpensive and temperature control is easily achieved. Both continuous and discontinuous methods can be devised. It may be possible to devise a more direct polarographic assay compared with the requirement for a coupled enzyme procedure in the spectrophotometric assay. The method also lends itself readily to coupled enzyme procedures. The technique is unaffected by the colour or turbidity of the reaction mixture, a definite advantage when assaying enzymes in some biological extracts (e.g. cyanobacterial extracts or permeabilized cells).

Finally, newcomers to polarography may fear the toxic effects of mercury on sensitive enzymes, but it is mercury ions, not the metal itself, which can cause inactivation. Properly handled, the dropping mercury electrode is generally perfectly safe to employ in enzyme assays.

# 4. Polarographic enzyme assays

This section describes a variety of polarographic enzyme assays which have been established. They are grouped according to the nature of the polarographically-active substance determined.

## 4.1 Oxygen and hydrogen peroxide

The polarographic activity of molecular oxygen has been referred to earlier. Two waves are produced, corresponding to a two-step reduction—first to $H_2O_2$ and then to $H_2O$. The polarographic determination of oxygen is widely used. As mentioned earlier, oxygen electrodes are nowadays more generally made of platinum. In principle, any reaction which consumes or generates oxygen may be assayed polarographically with an oxygen electrode. Various oxidases have been assayed in this way, for example L-amino acid oxidase, cytochrome oxidase, glucose oxidase, and xanthine oxidase.

An interesting adaptation of this method has been employed for the assay of trehalose (8). Glucose oxidase and catalase were used as coupling enzymes in the system:

$$\text{trehalose} + H_2O \xrightarrow{\textit{trehalase}} \text{glucose}$$

$$\text{glucose} + O_2 + H_2O \xrightarrow{\textit{glucose oxidase}} \text{gluconic acid} + H_2O_2$$

$$H_2O_2 \xrightarrow{\textit{catalase}} \tfrac{1}{2}O_2 + H_2O$$

The net consumption of oxygen was monitored with an oxygen electrode.

In addition to its cathodic reduction wave, hydrogen peroxide also produces an anodic wave, which has been employed in enzymic assay. A platinum electrode is used and various oxidases have been assayed (9).

## 4.2 Sulphur compounds

Thiol compounds give rise to anodic oxidation waves, while various disulphide compounds have been shown to produce cathodic reduction waves. These have both been utilized for polarographic enzyme assays.

### 4.2.1 Coenzyme A

Weitzman (1) proposed the use of polarography for the continuous assay of enzymes which catalyse the cleavage of acyl derivatives of CoA to the free coenzyme. The latter, but not its acylated derivatives, produces a well-defined anodic wave at the DME due to oxidation of its SH group, and the wave height is strictly proportional to concentration. The procedure was first demonstrated with the enzymes citrate synthase and malate synthase which respectively catalyse the reactions:

$$\text{acetyl-S-CoA} + \text{oxaloacetate} + H_2O \rightarrow \text{citrate} + \text{CoA–SH} + H^+$$

and

$$\text{acetyl-S-CoA} + \text{glyoxylate} + H_2O \rightarrow \text{malate} + \text{CoA–SH} + H^+$$

The formation of CoA–SH may be followed at −0.2 V (versus SCE) and the measured rates are directly proportional to enzyme concentration.

The polarographic assay has distinct advantages over other assay methods for these enzymes. One such method monitors the consumption of acetyl-S-CoA at 233 nm. This has the disadvantage that it can only be used with low concentrations of acetyl-S-CoA owing to the high absorbance of this compound, and is not well suited to assays with crude cell extracts as these inevitably contain high concentrations of UV-absorbing materials. Citrate synthase activity is often assayed at 412 nm in the presence of 5,5′-dithiobis-(2-nitrobenzoic acid), which reacts non-enzymically with the reaction product CoA–SH to produce the yellow compound 2-nitro,5-mercaptobenzoic acid. However, incorporation of this reagent into the assay mixture incurs the risk that thiol groups on the enzyme itself may be blocked, leading to inactivation or other modification; for example, desensitization to regulatory effectors. Citrate synthases from some sources are indeed affected in these ways. Malate synthases are also frequently susceptible to inactivation by 5,5′-dithiobis-(2-nitrobenzoic acid). In such cases, the polarographic method may be the assay of choice.

In examining the sensitivity of various citrate synthases in crude extracts to inhibition by the feedback regulator NADH, the anaerobic conditions of polarographic assay prevent the action of NADH oxidase. In normal aerated solutions, NADH oxidase (often present in crude cell extracts) would remove the NADH; this may interfere with a spectrophotometric assay.

Yet another advantage of polarographic assays is their applicability to turbid solutions. An example is the polarographic measurement of citrate synthase and other enzymes in suspensions of permeabilized cells (10, 11). Such assays work perfectly well, even at levels of turbidity which would make spectrophotometric measurements impossible.

The polarographic activity of CoA–SH also forms the basis of an assay for succinate thiokinase (11), which catalyses the reversible reaction:

$$\text{succinyl-S-CoA} + \text{ADP/GDP} + P_i \rightleftharpoons \text{succinate} + \text{ATP/GTP} + \text{CoA–SH}$$

Polarographic assay (−0.2 V versus SCE) may be conducted in either direction, formation or consumption of CoA–SH. Apart from the advantage of being a continuous assay, the polarographic method is much more sensitive than the colorimetric hydroxamate method used for this enzyme by some workers. Another advantage concerns assays with coloured solutions. Thus succinate thiokinase, previously thought to be absent from cyanobacteria, was demonstrated to be present by polarographic measurements, despite the intensely green colour of reaction mixtures (12). In principle, this assay is applicable to any reaction involving the conversion of acyl-S-CoA to CoA–SH. The large number of such reactions invites the wider application of the polarographic assay method.

The feasibility of using the acyl-S-CoA to CoA–SH reaction in a coupled polarographic assay has been demonstrated in the case of aspartate aminotransferase, using a very simple variant of conventional polarographic equipment (13). The reaction scheme is:

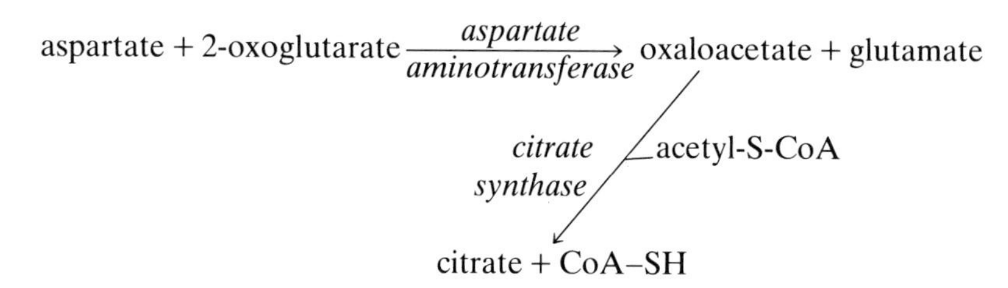


The aminotransferase rate was determined by monitoring the formation of CoA–SH at the DME. Many other similar assays could be devised.

### 4.2.2 Lipoic acid

The polarographic activity of lipoic acid has been described as has also that of lipoamide. Clear complementary cathodic and anodic waves were displayed by the oxidized lipoamide (SS-containing) and the reduced dihydrolipoamide (SH-containing) (14), permitting the design of a polarographic assay for dihydrolipoamide dehydrogenase (15). The enzyme was assayed in the direction of formation of oxidized lipoamide by continuous monitoring of the increase in wave height at −0.75 V (versus SCE). Although dihydrolipoamide dehydrogenase may be assayed at 340 nm, in this particular study it was necessary to examine the inhibition of activity by NADH at concentrations which would have rendered the spectrophotometric assay useless. As $NAD^+$ and NADH do not interfere with the wave of lipoamide at −0.75 V, the polarographic assay is unaffected.

### 4.2.3 Other sulphur compounds

Thiocholine, $(CH_3)_3{\equiv}N^+\text{-}CH_2CH_2SH$, exhibits a well-defined anodic wave which has been used as the basis of an assay for acetylcholinesterase and cholinesterase (16, 17). Acetyl-thiocholine was the substrate and its hydrolysis to thiocholine was followed by monitoring the increasing anodic wave at −0.2 V (versus SCE). The polarographic method agreed well with the spectrophotometric assay method.

Using as substrate the sulphur compound *S*-(4-methoxyphenyl)-thiophosphate, alkaline phosphatase may be assayed by monitoring the increase in the anodic wave of the thiol product (13).

Adenosylhomocysteinase may be assayed polarographically by following the increase in the anodic wave of the homocysteine formed (13).

## 4.3 Carbonyl compounds

Carbonyl compounds contain the reducible >C=O group and should be detectable by polarography. Thus, pyruvate gives a well-defined cathodic wave which may be measured at −1.4 V (versus SCE). This permits a direct assay of the enzyme pyruvate kinase, which catalyses the reaction:

$$\text{phosphoenolpyruvate} + \text{ADP} \rightarrow \text{pyruvate} + \text{ATP}$$

Pyruvate is the only polarographically active species in this reaction and the rate of its formation may be followed continuously. This assay compares favourably with the conventional spectrophotometric assay in which the formation of pyruvate is coupled to the enzyme lactate dehydrogenase:

$$\text{pyruvate} + \text{NADH} \rightarrow \text{lactate} + \text{NAD}^+$$

and the decrease in absorbance at 340 nm is followed. No coupling enzyme is required in the polarographic assay.

Phosphofructokinase (PFK) can also be assayed on the basis of the polarographic activity of pyruvate by coupling to pyruvate kinase (PK) and following the pyruvate wave continuously:

$$\text{fructose-6-phosphate} + \text{ATP} \xrightarrow{PFK} \text{fructose-1,6-bisphosphate} + \text{ADP}$$

$$\text{ADP} + \text{phosphoenolpyruvate} \xrightarrow{PK} \text{pyruvate} + \text{ATP}$$

The reaction catalysed by threonine deaminase also produces a keto compound which may be followed continuously at −1.4 V:

$$\text{threonine} \rightarrow \text{2-ketobutyrate} + NH_3$$

Although citrate synthase may be assayed polarographically by following the appearance of CoA–SH, it may sometimes be desirable to follow the consumption of the substrate, oxaloacetate. This too may be achieved by monitoring the disappearance of the oxaloacetate wave at −1.4 V.

## 4.4 Nicotinamide adenine dinucleotide

$NAD^+$ and $NADP^+$ produce cathodic waves at the DME which may be measured in the range −1.0 to −1.3 V (versus SCE). The reduced nucleotides NADH and NADPH are not electro-active at negative potentials and do not interfere. Assays of NAD(P)-linked dehydrogenases may therefore be performed polarographically. Although the method of choice for such enzymes is clearly the spectrophotometric assay at 340 nm, there may be circumstances when the polarographic assay is useful. An example is the measurement of enzyme activity in a turbid suspension of permeabilized cells (10). A further advantage is that the polarographic assay is performed under anaerobic conditions when any NADH oxidase activity is prevented.

## 4.5 Some other examples

A discontinuous polarographic assay for esterase and lipase activities has been proposed (18) using β-naphthyl esters as substrates (β-naphthyl-acetate and β-naphthylmyristate). The β-naphthol produced by the enzymes is derivatized to α-nitroso-β-naphthol whose polarographic reduction wave may be measured.

Oscillographic polarography has been used to assay the protease-catalysed hydrolysis of *p*-nitrophenyl esters and *p*-nitroanilides (19). Substrates and products (*p*-nitrophenol and *p*-nitroaniline) are polarographically reducible, but at sufficiently distinct potentials to be separately measurable. A polarographic method of measuring *p*-nitrophenyl phosphatase activity has also been described (20) which permits continuous assay of the enzyme in particulate tissue preparations.

Several studies have developed polarographic assays as components of enzyme-linked immunoassay procedures. Thus alkaline phosphatase has been measured by its action on phenyl phosphate to produce phenol which was separated on an octyldecylsilane column and then measured at a carbon paste electrode (21). Another assay for alkaline phosphatase has been based on its action on a novel substrate, (*N*-ferrocenoyl)-4-aminophenyl phosphate (22). The product of enzymic cleavage, (*N*-ferrocenoyl)-4-aminophenol, was monitored polarographically by its oxidation to the ferricinium ion using cyclic voltammetry at a graphite electrode. Finally, alkaline phosphatase used in an immunoassay has been coupled to glucose oxidase (23). The alkaline phosphatase was assayed by allowing it to hydrolyse glucose-6-phosphate to glucose, oxidizing this with glucose oxidase and measuring polarographically either oxygen consumption or hydrogen peroxide formation.

# References

1. Weitzman, P. D. J. (1966). *Biochem. J.,* **99,** 18P.
2. Weitzman, P. D. J. (1969). *Methods Enzymol.,* **13,** 365.
3. Weitzman, P. D. J. (1976). *Biochem. Soc. Trans.,* **4,** 724.
4. Kolthoff, I. M. and Lingane, J. J. (1952). *Polarography,* (2nd edn). Interscience Publishers, New York.
5. Zuman, P. (1964). *Organic Polarographic Analysis.* Pergamon Press, Oxford.

seconds, although modifications have succeeded in achieving response times as low as 1 ms (4).

Dependent on their cathode size and design, all oxygen electrodes have a small residual consumption of oxygen which gives rise to a residual current below which oxygen tensions cannot be measured. For a Clark-type electrode this will be in the order of $1 \times 10^{-10}$ amp which represents an oxygen concentration of 0.0005% or a $PO_2$ of 0.04 mm Hg (3). Electrodes will also drift over a time period and this may be of the order of 0.1 μl $O_2$/hour (2). Both these factors will determine the lower limit of sensitivity at which Clark-type electrodes will operate and consideration must be given to the rate of oxygen utilization by the biological system before embarking on development of a polarographic assay. It is worth noting that oxygen electrodes have a relatively high temperature coefficient and therefore must be operated under conditions of careful temperature control.

Ageing or lack of sensitivity of electrodes is caused by poisoning of the platinum electrode, particularly by biological samples containing phosphates, –SH reagents, and proteins. Much of this is prevented by the thin Teflon membrane covering the electrode, but nevertheless electrodes do age. However, they may be rejuvenated by soaking in 3% (w/v) ammonium hydroxide solution for a few minutes and subsequently rubbing the electrode surfaces with a slight abrasive paper (e.g. fine Emergy paper).

## 5. Calibration

Clearly calibration must be done routinely and under the precise conditions of temperature and media conditions that pertain to the experimental system under investigation. This has been reviewed extensively (4). However, most biochemical assay systems are carried out in dilute aqueous salt solutions in which standard calibration conditions pertain. Either of the following procedures may be used:

(a) Set up the electrode in its chamber and introduce 1 ml of distilled water (equilibrated with air = 100%). Add a few crystals of sodium dithionite ($Na_2S_2O_4$). The oxygen concentration rapidly falls to zero with a consequent response by the recorder. This position should be adjusted on the recorder chart to zero.

(b) Carefully wash out the electrode chamber several times with distilled water to remove any remaining dithionite. Introduce a further ml of distilled water (air equilibrated), allow to equilibrate to the temperature of the electrode chamber (25 °C, electrode response will rise slightly and then plateau). Adjust recorder response to 90% by suitable sensitivity controls. This level now represents 240 μM $O_2$ or 480 ng atoms oxygen/ml at 25 °C. This may be calculated from the dissolved gas constants available in the literature (*Table 1*) (5, 6).

**Table 1.** (a) Volume of oxygen dissolved in aqueous medium (microliters of oxygen per millilitre at 1 atmosphere).

| Temp. °C | Equilibrated with 100% $O_2$ $H_2O$* | Ringer Soln.† | Equilibrated with air (21% $O_2$) $H_2O$* | Ringer Soln.† |
|---|---|---|---|---|
| 15 | 34.2 | 34.0 | 7.18 | 7.14 |
| 20 | 31.0 | 31.0 | 6.51 | 6.51 |
| 25 | 28.5 | 28.2 | 5.98 | 5.92 |
| 28 | 26.9 | 26.5 | 5.65 | 5.56 |
| 30 | 26.1 | 26.0 | 5.48 | 5.46 |
| 35 | 24.5 | 24.5 | 5.14 | 5.14 |
| 37 | 23.9 | 23.9 | 5.02 | 5.02 |
| 40 | 23.1 | 23.0 | 4.85 | 4.83 |

* from Handbook of Chemistry and Physics 40th Ed., Chemical Rubber Pub. Co., Cleveland. 1958–1959.
† recalculated from Umbriet *et al.* (1964). Manometric Methods. 4th Ed. Burgess Pub. Co.

(b) Solubility of $O_2$ in buffered mitochondrial medium equilibrated with air (21% $O_2$).

| Temp. °C | μg atoms $O_2$/ml* | μmoles/ml (mM) |
|---|---|---|
| 15 | 0.575 | 0.288 |
| 20 | 0.510 | 0.255 |
| 25 | 0.474 | 0.237 |
| 30 | 0.445 | 0.223 |
| 35 | 0.410 | 0.205 |
| 37 | 0.398 | 0.199 |
| 40 | 0.380 | 0.190 |

* Solubility of $O_2$ experimentally determined by Chappell (1964). *Biochem J.*, **90**, 225, in a buffered mitochondrial medium containing NADH, inorganic phosphate, and isolated mitochondria.

These tables are taken with permission from the YSI publication on YSI model 5300 Biological $O_2$ monitor manual.

Calibration of the intervening recorder chart may then be carried out assuming linearity or alternatively by carrying out the following procedures:

(a) Pipette 0.95 ml of the experimental buffer solution into the electrode chamber and add 50 μl of a freshly-prepared solution of phenazine methosulphate (2 mg/ml water). Close electrode, allow solution to temperature equilibrate.

(b) Add to electrode chamber 10 μl of 10 mM NADH (i.e. 0.1 μmole NADH). Note that the actual NADH concentration should be determined spectrophotometrically by monitoring the fall in absorbance at 340 nm in the presence of malate dehydrogenase and oxaloacetate.

(c) The PMS will be reduced stoichiometrically by the NADH and the reduced PMS reoxidized by the oxygen in solution. From the amount of NADH injected it should be possible to calculate the theoretical oxygen uptake and hence equate that with the distance fallen on the recorder.

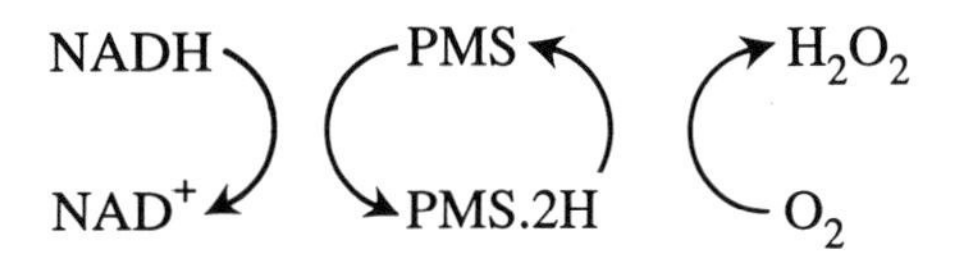


It remains to be stressed, however, that the oxygen electrode measures chemical activity and not the concentration of oxygen present. For this reason the electrode must be calibrated for the reaction medium to be used experimentally and it may not be assumed that one particular calibration holds for a different medium, since the activity coefficient is dependent, amongst other factors, on the ionic strength of the buffer.

## 6. Electrode systems

Several manufacturers supply oxygen electrode systems which consist of the following:

- an oxygen probe (electrode)
- a polarizing, back-off box
- a temperature-compensated incubation chamber with stirrer
- a waterbath
- a potentiometric recorder

The electrode most commonly used is the probe variety available from YSI Inc., Yellow Springs, Ohio, USA (through Clandon Scientific Ltd., Aldershot, Hants, GU12 5QR) although Gilson (through Anachem, Luton, Beds. LU2 0EB) and Rank Bros. Ltd. (Bottisham, Cambs, CB5 9DA) also supply electrode systems. In fact, Rank Bros. manufacture an electrode which is sealed into the base of the incubation chamber. The incubation chambers, made out of perspex or glass, must be water-jacketed and the chamber itself must accommodate a stirrer bar. It must be possible to shut off the atmosphere by means of a sleeve. Suitable injection ports, however, must be available for the introduction of small volumes of samples or solutions.

The polarizing box is a simple circuit carrying out two functions:

- maintaining the electrode at a constant −ve potential
- providing suitable sensitivity and back-off for the electrode output to a recorder

Suitable circuit designs are provided by the manfacturers and, although they may be purchased commercially, any competent electrical workshop will be able to construct one. Most conventional laboratory recorders will be suitable to record the electrode output; ideally they should have a sensitivity better than 20 mV full scale, acceptance of a source impedance of 2K/mV, and a response time of less than 1 s. Most commercially-available electrodes function with a final volume of 3 ml but may be adapted to 1 ml with suitable modifiers. Further reduction to a 0.5 ml final incubation volume is possible with specific modifications, but less than this volume requires special microelectrodes such as are available from Transdyne General Corporation, Ann Arbor, Michigan 48106, USA, or Insted Labs. Inc., Horsham, Penn. 19044, USA (via Clandon Scientific). These systems, however, require sophisticated polarizing and amplification systems and in the author's hands have some stability problems.

# 7. Polarographic assays

## 7.1 Tissue/organelle respiration studies

A major use to which the oxygen electrode has been put is to study oxygen consumption of various biological preparations ranging from whole cells and slices to subcellular fractions such as mitochondria, synaptosomes, and chloroplasts. In view of the wide variety of conditions, incubation media, and so on, which relate to such studies, no attempt will be made to deal with them comprehensively. A typical example of a polarographic study of mitochondrial function will be used to illustrate the use of the electrode in this context. Human skeletal muscle mitochondria were prepared (7) and stored on ice in cold isolation medium (225 mM mannitol, 75 mM sucrose, 10 mM Tris–HCl, 100 μM $K^+$-EDTA, pH 7.2) at a concentration of 10–15 mg mitochondrial protein/ml.

The studies were carried out in a final volume of 1 ml in a thermostatted incubation chamber (25°C) and the assay solution was well-stirred. The respiration medium consisted of 100 mM KCl, 75 mM mannitol, 25 mM sucrose, 10 mM phosphate-Tris, 10 mM Tris–HCl, and 50 μM EDTA, pH 7.4, plus 0.5 mg bovine serum albumin (BSA) and approximately 0.5 mg of mitochondrial protein (see *Protocol 1*).

---

**Protocol 1.** Respiration studies using oxygen electrode

1. Adjust sensitivity of recorder to give full scale deflection equal to the oxygen content of 1 ml water at 25°C; adjust back-off sensitivity to give zero reading in the presence of dithionite (see calibration).
2. Wash electrode chamber carefully to remove all traces of dithionite, add 1 ml of respiration medium (including BSA).
3. Allow electrode to equilibrate, inject mitochondrial sample in a minimal volume (50–100 μl) and allow to re-equilibrate.

4. Add substrates in small aliquots (5 or 10 μl) to give final concentrations; 5 mM pyruvate + 2.5 mM malate or 10 mM glutamate + 2.5 mM malate or 10 mM succinate + 5 μM rotenone, and so on.
5. Allow equilibration and then add 250 nmoles ADP (in 5–10 μl) and measure stimulated rate of oxygen consumption [state 3 (8)].
6. When all the added ADP has been phosphorylated as shown by the slowing down of the respiration rate (state 4) repeat (**5.**) to measure further state 3 respiration.
7. The 'uncoupled' rate of respiration may be measured at the end of the run by adding 1 μM FCCP (5 μl). This will stimulate respiration until all oxygen has been used and will give an additional check on the zero calibration of the recorder.

The respiration rates can be calculated from the calibration, P/O ratios from the oxygen consumed after the addition of a defined amount of ADP, and the respiratory control ratio (measure of mitochondrial integrity) from the ratio of state 3/4 respiration.

Uncouplers (FCCP) and some inhibitors (rotenone, antimycin) are dissolved in ethanol and to remove all traces from the incubation chamber, the chamber and electrode must be rinsed in ethanol. Particular care should be taken that the ethanol does not stay in contact with the electrode membrane from more than a few seconds.

---

*Figure 1* shows a typical oxygen electrode trace involving the study of human skeletal muscle mitochondria. A number of modifications, particularly for small samples, have been reported in the literature and for further details the reader is referred to (9–11). Additionally, in specifically-designed spectrophotometer cells, oxygen electrodes may be used in conjunction with other ion-sensitive electrodes so that simultaneous measurements of parameters such as pH, $K^+$ concentration, and redox state can be made as well as oxygen uptake (4).

## 7.2 Specific enzyme studies

Those enzymes which involve the use or production of molecular oxygen as an obligatory part of their mechanism can be assayed directly by the oxygen electrode. Such enzymes are, however, rather limited; e.g. glucose oxidase, catalase. However, modifications of oxygen electrodes, whereby an oxygen utilizing enzyme (e.g. glucose oxidase) has been impregnated into the membrane associated with the electrode, has allowed an effective glucose-sensing electrode-system to be constructed (4). Such adaptations may well be of considerable use in continuous industrial processes.

### 7.2.1 Glucose oxidase assay

Glucose oxidase (EC 1.1.3.4) catalyses the oxidation of D-glucose to gluconolactone as follows:

$$\text{glucose} + O_2 + H_2O \rightarrow \text{gluconolactone} + H_2O_2$$

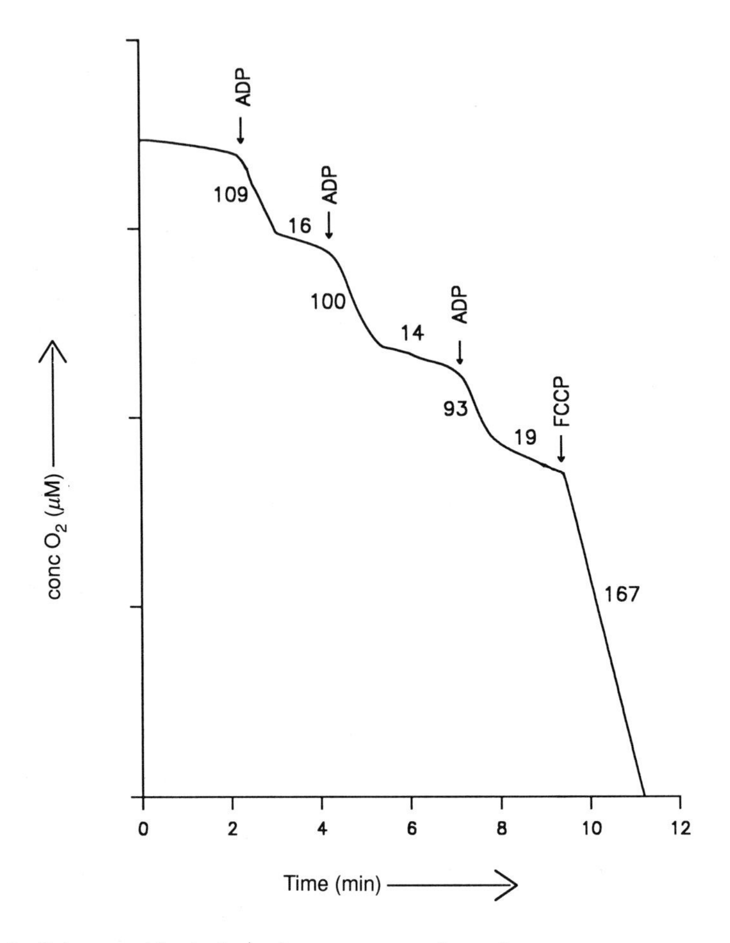


**Figure 1.** *Polarographic studies using an oxygen electrode*
The experiment was carried out at 25°C in a KCl, mannitol, sucrose containing medium (see text) using human skeletal muscle mitochondria (0.75 mg protein). Additions were as follows (final concentrations): malate (2.5 mM), glutamate (10 mM), ADP (250 μM where indicated), FCCP (1 μM). The reaction volume was 1 ml and the time axis units are in minutes. The indicated rates are expressed as ng atoms oxygen consumed per min per mg mitochondrial protein.

The oxygen utilized during the reaction may be used as a measure of the enzyme's activity or alternatively, in the presence of non-saturating glucose concentrations, as a measure of glucose concentration (12). The assay is very simple and may be set up optimally in 50 mM Na acetate buffer, pH 5.6, although other buffers and a wider range of pH may be used without too much loss of activity (pH 3.5–9). Such a system was found to show a linear relationship between either glucose concentration (70 μM–30 mM) or enzyme concentration (0.1–2 units) (12).

### 7.2.2 Catalase assay

Catalase (EC 1.11.1.6) catalyses the breakdown of hydrogen peroxide as follows:

$$2H_2O_2 \rightarrow 2H_2O + O_2$$

It is a widely-distributed enzyme across the animal and plant world and successful analysis of its kinetics was difficult before the establishment of the oxygen electrode system. All assay procedures involve the setting-up of an electrode chamber at constant temperature and with constant stirring of the reaction medium, together with the ability to close-off the reaction system to the atmosphere save for a small injection port for the addition of substrate or enzyme. Such a system has been described (13), the essentials of which are outlined below:

---

**Protocol 2.** Assay of catalase

1. Set up the electrode and add 1 ml of 50 mM $Na_2HPO_4$–$KH_2PO_4$, pH 7, buffer.
2. Allow to equilibrate to 25°C and establish recorder setting near 100%.
3. Bubble buffer system with $N_2$ gas until zero setting is established and close-off reaction chamber to atmosphere.
4. Add 10 μl $H_2O_2$ (33.5 mM final concentration) through injection port and allow to equilibrate.
5. Add 50 μl of suitably-diluted catalase sample and the initial velocity of reaction may be established from the kinetics of the $O_2$ production rate. (Note that any non-enzymatic production of oxygen should be deducted.)

---

It was found (13) that an activity range of 0.01–8.4 μmol $O_2$/min could be measured and further increases in sensitivity could be achieved by improving the electrode/recorder amplification (down to 0.002 μmol $O_2$/min). This permitted the measurement of catalase activities over almost a 1000-fold concentration range and was 20 times more sensitive than other assay systems.

### 7.2.3 Other systems

Considerable use has been made of the oxygen electrode to set up continuous assay procedures for metabolic intermediates along similar lines to those described for the glucose electrode. These are particularly useful in clinical laboratories and include measurement of free and esterified cholesterol by the use of impregnated cholesterol oxidase and hydrolase (14), tyrosine (15), prostaglandin synthesis by the measurement of arachidonic acid (16), and uric acid using urate oxidase and monitoring the oxygen consumption (17). These systems are, however, outside the scope of this article and are only mentioned for completeness.

## References


1. Clark, L. C., Jr. (1956). *Trans. Am. Soc. Artificial Internal Organs,* **2,** 41.
2. Fatt, I. (1976). *'The polarographic oxygen sensor: its theory of operation and its application in biology, medicine and technology' CRC Press, Cleveland, USA.*
3. Davis, P. W. (1962). In: *Biological Research*, Vol. IV, (ed W. L. Nastuk), pp. 137–79. Academic Press, New York.
4. Lessler, M. A. (1982). *Methods of Biochemical Analysis,* **28,** 173.
5. *Handbook of Chemistry and Physics*, (40th edn), 1958–9. Chemical Rubber Co., Cleveland.
6. Chappell, J. B. (1964). *Biochem. J.,* **90,** 225.
7. Holt, I. J., Harding, A. E., Cooper, J. M., Schapira, A. H. V., Toscano, A., Clark, J. B., and Morgan-Hughes, J. A. (1989). *Annals. Neurol.,* **26,** 699.
8. Chance, B. and Williams, G. R. (1956). *Adv. Enz.,* **17,** 65.
9. Nakamura, M., Nakamura, M. A., and Kobayashi, Y. (1978). *Clin. Chim. Acta,* **86,** 291.
10. Lessler, M. A. and Scoles, P. V. (1980). *Ohio J. Sci.,* **80,** 262.
11. Pappas, T. N., Lessler, M. A., Ellison, E. C., and Carey, L. C. (1982). *Proc. Soc. Exp. Biol. Med.,* **169,** 438.
12. Hertz, R. and Barenholz, Y. (1973). *Biochim. Biophys. Acta,* **330,** 1.
13. Del Rio, L. A., Gomez Ortega, M., Lopez, A. L., and Lopez Gorge, J. (1977). *Anal. Biochem.,* **80,** 409.
14. Dietschy, J. M., Delente, J. J., and Weeks, L. E. (1976). *Clin. Chim. Acta,* **73,** 407.
15. Kumar, A. and Christian, G. D. (1975). *Clin. Chem.,* **21,** 325.
16. Lord, J. T., Ziboh, V. A., Blick, G., Poitier, J., Kursonoglu, I., and Penneys, N. S. (1978). *Brit. J. Dermat.,* **98,** 31.
17. Nanjo, M. and Guilbault, G. G. (1974). *Anal. Chem.,* **46,** 1769.

# 7

# Electrochemical assays: the pH-stat

KEITH BROCKLEHURST

## 1. Introduction

A pH-stat (stationary pH) device in its most literal sense (which is the subject of the present chapter) is used to monitor the progress of chemical reactions in which protons (or, strictly, hydrons) are liberated or taken up. This is achieved by measuring the quantity of base or acid that needs to be added at various times to keep the pH essentially constant. The technique was developed in 1923 by Knaffle-Lenz (1, 2) in connection with his work on esterases and has been applied widely in many biochemical systems (3). In a more general sense, 'stat' techniques are used to monitor an even greater variety of reactions in which a chemical species that is either generated or consumed can be detected by using electrodes. Reactions involving metal ions can be monitored by using ion-selective electrodes and reactions involving redox changes by using platinum electrodes (4).

The pH-stat method of determining reaction rates has wide applicability and complements spectrophotometric methods. It requires a change in proton binding sites during reaction rather than a change in chromophoric character. Particular advantages are that it can be used to study kinetics in non-buffered solution and in stirred suspension. A consequence of the latter is that the technique is readily used to study turbid cellular extracts and immobilized enzymes or cells. In such systems in particular it may be necessary to establish conditions in which reaction rate is independent of the rate of stirring of the heterogenous reaction mixture, or at least for comparative studies to keep the stirring rate constant (5, 6). Variants on the standard pH-stat methodology include electrolytic titrant generation (7, 8) and spectrophotometric monitoring of pH (8).

When using a simple pH-stat assembly with a 0.5 ml burette, 0.02–0.05 M titrant, and a reaction time restricted to a few minutes to attempt to obviate problems of enzyme instability in dilute solution, it is possible to record enzyme activities as low as *c.* 0.1–0.5 μmol of substrate transformed per min. When this limiting sensitivity is not required, however, it is common practice to use assay mixtures such that reaction can be monitored by using 0.1 M titrant. When the titrant is sodium hydroxide, use of this concentration ensures that absorption of residual carbon dioxide does not seriously interfere with the assay and permits the inclusion of very dilute buffer ($\leqslant$1 mM) to assist in establishment of the required assay pH without perceptible influence on the measured value of the steady state rate.

# 2. Principles and theoretical basis of pH-stat methodology

## 2.1 General principle

A pH stat is a type of autotitrator that can be used to maintain a constant pH in a non-buffered solution during a reaction that involves the production or uptake of protons by addition of a solution of acid or base of known concentration (9). The amount of titrant added to maintain the pH is recorded as a function of time to provide a progress curve for the reaction which may be subjected to kinetic analysis. In many kinetic studies reaction volumes and concentrations of reactants and titrant are arranged such that there is negligible volume during a kinetic run and thus standard kinetic equations are applied. The more general situation in which concentrations of substrate and titrant are comparable, and thus reaction is accompanied by significant volume change, has been treated analytically (10).

## 2.2 The nature of pH-stat components and their functions

A pH-stat consists of:

- a thermostated magnetically-stirred reaction vessel
- a glass electrode
- a KCl bridge
- a reference (calomel) electrode
- an electrode shield
- a controller
- a motor-driven burette
- a recorder

The principles underlying the function of each of these components have been discussed in detail by Jacobsen *et al.* (3). The electrochemical cell, comprising the glass electrode immersed in the reaction mixture, the KCl bridge and calomel electrode, is connected to the pH meter which is itself connected to the controller. The electrode assembly needs to be accurate and stable and well shielded from the surroundings. Even with such precautions the wearing of clothing made from artificial fibres by the experimenter should be avoided to minimize electrostatic perturbations. The controller starts the motor-driven burette when the output potential of the pH meter changes from its fixed initial value and stops it when the initial potential is re-established. The rate of addition of acid or base (titrant) necessary to effectively maintain the pH and to provide a good estimate of the reaction rate depends upon a variety of factors (3), notably the buffer capacity and size of the sample as well as the titrant concentration. The control functions that determine the rate of addition of titrant, therefore, are of considerable importance. The traditional pH-stat uses an incremental technique in which the progress curve is produced in a stepwise manner through a series of alternate reagent additions and pauses. Until recently regulation of the size of each increment of titrant and each pause has usually been achieved by a proportional control known as a proportional band. The limitations of this approach are:

- overshooting or titration lag
- a steady-state pH that is offset from the required value
- inability to 'remember' how the system reacted previously to a given offset and subsequent addition of titrant

In the current state-of-the-art pH stat, the Radiometer TitraLab (Radiometer Ltd., Crowley, West Sussex, RH10 2PY, UK), a considerable improvement in proportional control has been achieved which eliminates the offset from the desired pH value and provides a fast response to changes in the reaction mixture (11). The reaction kinetics mode of the equipment incorporates proportional integral derivative control (PID). The PID-algorithm has two adjustable parameters, controller gain and time constant. The controller gain regulates the titrant addition rate in accordance with the reaction rate and the volume size and buffer capacity of the sample and the time constant adapts this regulation to the response time of the electrode and to the stirring efficiency.

## 2.3 Some limitations and sources of error

To obtain satisfactory results in pH-stat experiments it is necessary to arrange for reliable constant stirring, stable electrode systems, adequate shielding, and an effective system for the prevention of absorption of atmospheric carbon dioxide. Other problems are existence of an unknown liquid junction potential between saturated KCl and the reaction mixture, and the tendency of the liquid from the burette tip (situated below the surface of the reaction solution) to leak. It is assumed (3) that the diffusion potential is small in dilute solutions between pH 3 and pH 11, being of the order of 0.1 mV (0.002 pH unit) in a solution containing 1 mM HCl and 0.1 M KCl and that, in such solutions, its variation may be neglected. With more concentrated solutions, determination of whatever non-enzymic rate may be observed prior to addition of enzyme is considered to deal effectively with a variety of extraneous effects, including whatever change in diffusion potential may be occurring. Leakage from the submerged tip of the burette is minimized by making the density of the titrant lower than that of the reaction mixture. A potential systematic error that may arise in pH-stat assays of enzymes in haemolysates is described in Section 5.

# 3. Commercial and custom-made pH-stat assemblies: automation, computer control, and special applications

## 3.1 The range of equipment

The simplest pH stat, sometimes used in undergraduate laboratories, consists of a pH meter and electrode system, a reaction vessel mounted on a magnetic stirrer, a manually-operated burette, and a stop-clock. Automatic burettes were developed in the 1930s (12, 13) and autotitration devices (14) led to the development of the automated pH stats built at the Carlsberg Laboratory in the 1950s (15). In the 1950s and 1960s many pH-stat experiments were conducted by using the Radiometer TTT1 autotitrator, an SBU1 burette, a reaction vessel such as the TTA31, and an SBR2 recorder, and reports of experiments conducted with this type of equipment and its successors (e.g. TTT2, TTA80, TIM90, and VIT90 autotitrators) continue to appear in the literature. The state-of-the-art Radiometer TitraLab is described in Section 2.2. Some custom-made pH-stat systems that have been reported in the literature, constructed to meet special requirements in some cases, are described below.

## 3.2 Some pH-stat systems described in the literature

### 3.2.1 Inexpensive systems

Warner *et al* (16) describe an inexpensive pH-stat autotitrator which can be used for kinetic experiments with small reaction volumes (e.g. 2–3 ml). A digitally-controlled burette adds titrant to a reaction mixture at a rate proportional to the differences between the solution pH and the specified stat pH. Hamilton glass syringes (250 μl–2.5 ml) deliver titrant at rates from 100 nl/min to *c.* 1.5 ml/min. The volume of titrant and the pH are displayed digitally and can be sent as an analog signal to an external chart recorder. The difference between the potential provided by the pH-measuring electrode and the stat potential generator is converted to a proportional frequency. A counter accumulates the incremental additions of titrant and provides a digital display. The digital signal is converted back to an analog signal and any number of incremental additions can be displayed full scale on the chart recorder. The digital burette employs a stepping motor connected to the micrometer drive by a keyed shaft which slides into a slightly larger slotted tube. The micrometer drive (0–2.5 cm travel) is used to push the plunger of the Hamilton syringe containing the titrant. Advantages of this inexpensive pH-stat system include the following:

(a) proportional control of the digital titrant dispensing system provides a large range of rates of titrant addition
(b) the nature of the comparator and the proportionality of titrant addition

permit relatively rapid reactions to be followed accurately without overshooting

(c) the digital nature of the system provides for relatively low noise and permits easy interfacing with a computer

Detailed descriptions of the circuits and the theory of operation are given in (16). In addition, an equation relating reaction parameters (reaction volume, titrant concentration, buffer capacity, and kinetic constants) to tracking accuracy is derived and discussed.

An inexpensive, simple, comparator/control unit for a pH-stat containing solid-state circuitry is described by Job and Freeland (17). This can be used with a pH meter and a solenoid burette (18) to construct a general purpose pH-stat-autotitrator system.

### 3.2.2 Automated systems dealing with multiple samples

Keijer (19) describes equipment that can be used in an automatic procedure for large numbers of pH-stat assays of a given enzyme in sequence and illustrates its performance by a cholinesterase assay. In conventional pH-stat procedures for the determination of enzyme activities, the following operations need to be performed: the controls of the titrator and titrigraph are set; the titration assembly is brought to an initial position with the syringe burette filled with titrant, and the recorder pen set to a zero position; the pH of the substrate solution is brought to the required value and the reaction is started by addition of enzyme solution, often as the recorder pen crosses a marked line on the chart. After recording the progress curve, the reaction vessel and electrodes are rinsed, the burette is refilled, and the recorder pen repositioned in readiness for the next reaction. The automation of this procedure described by Keijer (19) involves the following. The samples containing enzyme are stored in U-shaped tubes fitted with siphons and located in a turntable. Each sample in succession is mixed with a particular volume of substrate solution and flushed into the reaction vessel via a delivery funnel making use of valves and motors controlled by a programming unit. The reaction is followed for a previously selected time and the volume of titrant delivered as a function of time is recorded. During this time, the substrate solution required for the next reaction is transferred to its location above the appropriate valve by the dosing pipette. At the end of the reaction, the pen motor is switched to rapid reverse mode to bring the pen to the starting position and the syringe burette is refilled simultaneously. Whilst these operations are being carried out, the sample holder, delivery funnel, and reaction vessel are automatically rinsed three times with distilled water. When the syringe filling, pen setting, and rinsing have been completed (at least 3 min) the syringe is connected to the titrant delivery tube, the next sample is moved over the delivery funnel, and the cycle is repeated.

A more flexible automated pH-stat system involving an alternating stop and flow system, which permits the changeover from one type of enzyme assay (i.e. one type of substrate) to another, is described by Vandermeers *et al* (20). The operation of the system is illustrated by assays of hydrolases in pancreatic homogenates and intestinal juice (lipase, trypsin, and chymotrypsin).

### 3.2.3 Multiple pH-stat systems: linked pH-stat systems for maintenance of substrate concentration and data acquisition and processing for several pH stats simultaneously

A method preventing substrate depletion during a kinetic run at low substrate concentrations, by coupling an automatic titration burette to a second burette containing an equimolar solution of the substrate, is described by Konecny (21). Rousseau and Atkinson (22) describe the application of digital data logging equipment in the recording of the operations of multiple pH-stats and the development of software for the conversion of digital data to titrimetric data and then into kinetic parameters. The data acquisition and processing system is based on a digital data logger linked to individual pH-stat assemblies and generates a computer-compatible record of titrimetric data. An external module causes an operating autotitrator to activate the logger to perform a continuous scan until all autotitrators return to the passive mode. The operating time, the real time, and the channel identification are recorded when any one autotitrator executes a pH correction. The data generated by the logger are manipulated by computational algorithms. These identify the operative titrators and produce data for computer manipulation on cumulative titrant volumes added as a function of time. The system allows experimental data to be sampled, collected, and analysed subsequently by off-line computer methods.

### 3.2.4 Computer-controlled systems

Innovations in reagent delivery systems and computer automation now provide for rapid, accurate, automated kinetic determinations by pH-stat methods. Of particular value are the development of a digital computer-controlled titrant delivery system capable of producing μl aliquots (23) and the description of a pH-stat under real-time computer control (24). Real-time computer-controlled processing eliminates time consuming calculations and permits real-time modification of parameters according to experimental demands. Feedback from experiments eliminates pH overshoot or lag and allows dynamic alteration of sampling frequency.

The system comprises a microlitre titrant addition system, a sensitive pH meter, a reaction vessel, and a mini-computer with graphics display and disk storage. The pH meter continuously monitors the pH of the reaction mixture. At a selected frequency the computer reads the output and compares it with a stored reference potential to determine the pH change. It then calculates the rate of addition of titrant necessary to return the pH to its command value.

The design of the microlitre titrant addition system is based on the generation and selection of charged droplets by an electric field rather than on a conventional syringe mechanism. The addition system has no moving parts, no burettes that require refilling, and a rapid response time to a demand for titrant addition. The design and operation of this system are described extensively in (23) and (25) and updated in (24).

The pH-stat adds titrant at rates selected by the computer, measured in terms of the number of droplets charged per second. Charging is achieved by the computer interface comprising a computer-controlled frequency divider, pulse-delay and pulse-width adjustment circuitry, and a high voltage transistor switch. The rate at which droplets are charged and thus deflected into the reaction vessel is loaded by the computer into its common instrument bus which communicates with negative logic with the frequency divider.

The computer program controlling the pH stat is written in FORTRAN IV and operates under the PDP-11 RSX-11M V3.1 real-time operating system. The program is subdivided into individual concurrent tasks to achieve modular program design and efficient use of the CPU. Communication between tasks is by use of a shared data region. Synchronization, where required, is achieved by the use of event flags and suspend–resume constructs. An important feature of the pH-stat described in (24) is the algorithm used to calculate a new titrant addition rate. This determines the necessary incremental change from the current titrant addition rate. It makes use of two factors in the calculation of the change in the rate of addition of titrant; the difference between the current pH and the control pH, and the difference between the pH when the reaction was last sampled and the current pH. The first factor provides for the reaction to be held at the control pH, and the second determines the direction in which the pH is currently changing and the potential.

### 3.2.5 A pH-stat with spectrophotometric pH monitoring and electrolytic titrant generation

Conventional pH-stat systems utilize volumetric addition of titrant and potentiometric detection of deviations from the control point. Electrolytic titrant generation provides high sensitivity and accuracy and eliminates the need for standard solutions. It is particularly useful for kinetic studies because rates are followed continuously without dilution errors. Spectrophotometric monitoring of the operating pH can provide greater sensitivity to pH change in alkaline media and faster response times.

Karcher and Pardue (26) describe equipment for this type of pH-stat with the following characteristics: the operating point is monitored spectrophotometrically, a pulsed electrolysis current source is used rather than a continuous source, and the data are processed automatically to yield reaction rates or chemical concentration data directly. The use of pulsed electrolysis current makes the instrument more directly compatible with digital equipment and eliminates the problem of overshoot inherent in continuous current generation. The instrument was developed for the assay of acetylcholinesterase. Acetic acid produced in the reaction is monitored at 404 nm by using 4-nitrophenolate as indicator. Deviation of the transmittance at 404 nm by a preselected amount from an adjustable control point triggers a pulse of current which generates base and titrates the acetic acid produced. Electrolysis current and pulse duration are constant during each pulse of titrant and so a given number of pulses corresponds to a constant reaction interval for each rate measurement.

### 3.2.6 An electrochemical pH-stat and controlled current coulometric acid-base analyser

Potential problems with using spectrophotometric pH detection as described in Section 3.2.5 are that some pH indicator dyes undergo redox changes or bind to proteins, both of which can cause spectral changes. A totally electrochemical device, therefore, is attractive. The major problem encountered when using such devices, both for the measurement of pH and for its restoration following a perturbation, is the interaction between the two processes; for example, the coulometric acid-base titrator described by Johansson (27) has a slow response and is susceptible to oscillation about the end-point of a titration. This device measures the electrode signal directly, amplifies it, and applies the signal from which a preset voltage is subtracted to a pair of generating electrodes. Adams *et al.* (28), prompted by the development by Brand and Rechnutz (29) of a high impedance differential potentiometric circuit, reported the construction and evaluation of equipment in which an isolation amplifier is used to eliminate difficulties that derive from a common ground for pH-measuring and current-generating systems. Their totally electrochemical pH-stat has a minimum time constant of 2–3 s at which no overshoot is observed, and has a current efficiency of almost 100% over a wide range of concentrations.

### 3.2.7 A temperature-scanning pH stat

Thermal denaturation of protein in non-buffered solution is accompanied by change in pH (30) and this phenomenon can be investigated at constant pH in a device consisting of a pH-stat, a programmable heating unit, and a temperature measuring and recording system (31).

### 3.2.8 A flow-through pH stat for studies on enzymes (such as lipases) that are optimally active at low pH

Taylor (32) describes a continuous pH-stat method in which conditions for the lipase-catalysed reaction and the titration of the fatty acid products are separately controlled. This allows the use of optimal conditions for titration of fatty acids, despite their inhibitory effects on most lipases, and optimal conditions for the lipase-catalysed reaction. In this method, enzyme and substrate are pumped into a stirred emulsion reactor where the catalysed reaction occurs. The reaction mixture is then allowed to flow to a second stirred vessel for product titration.

# 4. General pH-stat procedure and chemical principles and experimental protocols for some individual enzymes

## 4.1 A general pH-stat procedure

In essence the procedure for carrying out pH-stat assays is common to most reactions. A general protocol is described in *Protocol 1* in terms of reactions such as ester hydrolysis in which protons are liberated. This is readily adapted for use with other reactions in which protons are taken up, or in which special requirements such as a supply of oxygen is required and where more specialized pH-stat systems, such as those involving automation or computer control, are used. A number of other protocols are then described (*Protocols 2–8*) to illustrate the range of reaction types to which the pH-stat assay technique has been applied. In these specific protocols the reaction conditions (notably temperature and concentrations of substrate and titrant) used in the references cited are given. These usually relate to pH values in the range associated with optimal enzyme activity, and values of substrate concentration that are sufficiently larger than relevant $K_m$ values to provide rates that approximate to $V_{max}$ values. The reaction may be started by addition of either substrate or enzyme depending upon the stabilities of the reactants at the pH employed. Starting a reaction by addition of enzyme has the advantage that conformationally-labile enzymes are not kept in dilute solution prior to reaction, and that any (slow) non-enzymic background rate of reaction of substrate can be determined in the same kinetic run prior to addition of enzyme. It may be necessary to stabilize the pH of solutions prior to starting the reaction by using very dilute buffer (Section 1). When using immobilized enzymes it may be necessary to start the reaction by addition of substrate to permit some types of solid support (such as carboxymethylcellulose) to equilibrate at the desired pH before starting the reaction. In the study of immobilized enzymes, a convenient way of delivering aliquots of stock suspension to the reaction vessel is by means of a calibrated graduated 1-ml pipette from which the narrow tip has been removed. The protocols described here are readily adapted to suit other reaction conditions, e.g. for use in experiments over a range of substrate concentrations.

Many pH-stat assays are based on the production of weak acid–conjugate-base equilibrium mixtures as, for example, in ester hydrolysis:

$$\mathrm{R{-}C({=}O){-}OR' + H_2O \rightarrow R'OH + R{-}C({=}O){-}OH \rightleftharpoons R{-}C({=}O){-}O^- + H^+}$$

When the kinetic run is carried out at pH $\gg$ p$Ka_{(RCO_2H)}$, (i.e. such that the pH is at least 2 units greater than the p*K*a) the reaction results in production of protons essentially stoichiometric with the substrate consumed, which are titrated by base in the pH-stat. At pH values nearer to the p*K*a, experimentally-determined reaction rates will need to be corrected for the incomplete ionization of the acid product. The correction factor will depend on the difference between the pH and the p*K*a of the proton-producing product such that

$$\text{rate}_{\text{true}} = \text{rate}_{\text{observed}} \times [10^{\text{pH}-\text{p}Ka}/(10^{\text{pH}-\text{p}Ka} + 1)]$$

---

**Protocol 1.** General pH-stat assay for reactions producing protons as products

*Reagents*

- sodium hydroxide solution (1 M)
- sodium hydroxide solution (0.1 M)
- potassium chloride solution (0.5 M)
- EDTA (di-Na salt) solution (0.1 M)
- Substrate solution
- Enzyme solution or suspension

*Procedure*

**1.** Set up the pH-stat comprising autotitrator, burette (e.g. 0.5 ml capacity), and recorder.

**2.** Load the delivery syringe with NaOH solution of the appropriate concentration (say 0.1 M).

**3.** Maintain the reaction vessel at the desired temperature (e.g. 25°C) by means of a water jacket and water bath.

**4.** Pass a stream of oxygen-free nitrogen through a wash bottle containing water at the reaction temperature, and then over the surface of the fluid in the reaction vessel to exclude carbon dioxide and oxygen.

**5.** Prepare a reaction mixture (minus substrate or enzyme) at the required pH (e.g. by addition of small volumes of 1 M NaOH from an Agla syringe to the stirred mixture).

**6.** Switch on the pH-stat and record any non-enzymic reaction of substrate (for enzyme started reactions) before addition of the other reactant (say enzyme) in solution at the same pH as the reaction pH.

**7.** Record the progress curve.

**8.** From the settings of titrant delivery and chart speed calculate the initial rate in mol $l^{-1}s^{-1}$ (for zero-order reactions from linear traces or tangents at t = zero), or the rate constant for first-order reactions by regression analysis of conventional logarithmic plots.

---

## 4.2 Oxidoreductases (EC 1)

### 4.2.1 Glucose oxidase (EC 1.1.3.4)

*Reaction*

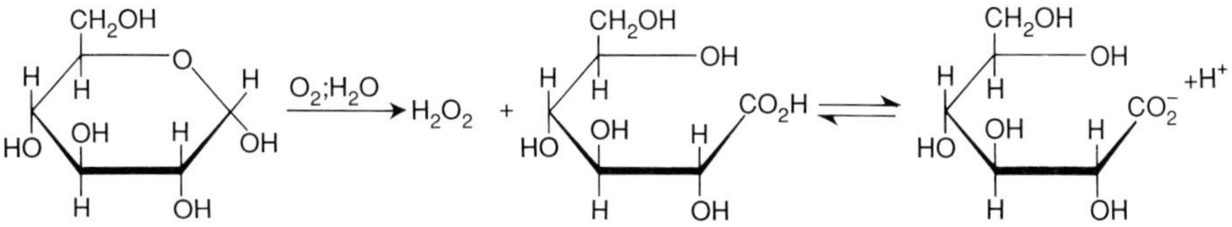

Oxidation of glucose catalysed by glucose oxidase produces gluconolactone which undergoes hydrolysis to gluconic acid (p*K*a 3.6). References to reports of other techniques for assay of glucose oxidase activity by determination of gluconic acid using polarimetry, GC, electronic spectroscopy, and iodometric titration are given in (33).

---

**Protocol 2.** pH-stat assay for glucose oxidase (33) [see also (24) for a glucose oxidase assay by a computer-controlled pH-stat]

**1.** Set up the pH-stat with:

(a) the stirred reaction mixture (10 ml) containing glucose [e.g. 0.16 M to provide approximately saturating ($V_{max}$) conditions], and enzyme in deionized water, pH 5.6 (or other pH as required)

(b) Temperature = 30°C

(c) a supply of pure oxygen to the bottom of the reaction vessel and an oxygen probe to check for appropriate agitation and oxygen flow rate (approx 0.5 volumes of oxygen per volume of reaction mixture per min)

(d) 5 mM sodium hydroxide solution in the burette

**2.** Record titration of released protons for 5 min as a zero-order (initial rate) linear progress curve.

---

### 4.2.2 Dihydrofolate reductase (DHFR) (EC 1.5.1.3)

*Reaction*

$$\underset{\text{(DHF)}}{\text{Dihydrofolate}} + \text{NADPH} + H^+ \rightarrow \underset{\text{(THF)}}{\text{tetrahydrofolate}} + \text{NADP}^+$$

Addition of strong acid to maintain the pH provides a means of monitoring the reaction. Advantages of the pH-stat assay (34) over the spectrophotometric assay involving decrease in $A_{340}$ consequent upon conversion of NADPH to $NADP^+$ are:

(a) a 10 times greater sensitivity
(b) crude preparations or biological fluids with high absorbance can be assayed directly (buffer upper limit equivalent to 2 mM phosphate, pH 7.4)
(c) speed of assay (5 min)

---

**Protocol 3.** pH-stat assay for dihydrofolate reductase (34)

**1.** Set up the pH-stat as follows:
  (a) de-gas stock solutions of DHF and NADPH (each approx. $10^{-4}$ M) and maintain under nitrogen
  (b) temperature = 28°C
  (c) use 0.1 M $Na_2SO_4$ as electrolyte support to avoid adsorption phenomena
  (d) place 2 ml of each of these stock solutions in a thermostatted reaction vessel
  (e) initiate reaction by injection of 1.50 µl of DHFR solution previously adjusted to assay pH (e.g. 7.4)

**2.** Maintain the pH as reaction proceeds by automatic addition of $H_2SO_4$ ($\geq 5 \times 10^{-5}$ M) and record the addition as a function of time (for approx. 30 s in thoroughly stirred solution).

**3.** If the solutions are not de-gassed or if a nitrogen is not used, take account of the baseline drift (approx. $10^{-10}$ mol $H^+$ $min^{-1}$) to correct the initial value.

---

### 4.2.3 Hydrogenase (EC 1.18.99.1) (previously designated EC 1.12.1.1)

*Reactions*

$$2H^+ + \text{reduced mediator} \rightleftharpoons H_2 + \text{oxidized mediator}$$

Ferridoxin is the physiological mediator of hydrogenases from *Clostridium pasteurianum* and *Megasphera elsdenii* whereas cytochrome $c_3$ appears to be the mediator for the *Desulfovibrio* sp. enzyme (35). For *in vitro* assays viologen dyes such as methyl viologen (MV) and benzyl viologen (BV) are efficient mediators, where the following reactions may be monitored by titration with base in a pH-stat.

$$S_2O_4^{2-} + 2H_2O + 2MV_{ox} \rightleftharpoons 2HSO_3^{2-} + 2MV_{red} + 2H^+$$

$$2HSO_3^- \rightleftharpoons 2SO_3^{2-} + 2H^+$$

$$2H^+ + 2MV_{red} \xrightleftharpoons{hydrogenase} 2MV_{ox} + H_2$$

i.e.

$$S_2O_4^{2-} + 2H_2O \rightleftharpoons 2SO_3^{2-} + H_2 + 2H^+$$

or

$$H_2 + 2BV_{ox} \xrightleftharpoons{hydrogenase} 2H^+ + 2BV_{red}$$

**Protocol 4.** pH-stat assay for hydrogenase from *Desulfovibrio gigas* (35)

*Equipment*

(a) Use a pH-stat (e.g. Radiometer) with separate electrodes.

(b) The cover of the reaction vessel holds the two tightly-fitted electrodes, a bubbler, a gas outlet, and the burette tip, and a motor-driven stirrer is fitted to the reaction vessel with springs to make an air-tight seal.

*Preparation for reaction under anaerobic conditions*

1. Introduce solutions by inserting syringes through a septum-capped hole in the cover of the reaction vessel.
2. Prepare solutions of MV containing sodium dithionite by transferring weighed salts to a tube, subsequently sealed with a rubber stopper.
3. Flush argon gas through the tube using syringe needles as gas inlet and outlet and introduce an appropriate volume of 0.1 M NaCl solution, previously deoxygenated by bubbling argon, using a syringe.
4. Renew this solution every 4 h.
5. Prepare BV solutions, if required, in the same manner.
6. Remove air from the titration vessel by flushing with argon or hydrogen, depending on the direction of the hydrogenase-catalysed reaction that is to be studied, after first passing the gases through a deoxygenating cartridge and subsequently through a bubbler flash partly filled with 15 mM sodium dithionite solution containing 1 mM MV (to saturate the gases with water and to provide a test for the absence of oxygen).
7. Use air-tight tubing to conduct the gases.

**Protocol 4.** *Continued*

8. Use 0.2 M NaOH solution as titrant and deoxygenate by continuous bubbling of argon saturated with water.
9. For studies of catalysis in the direction of hydrogen oxidation, maintain hydrogen bubbling; for studies in the direction of proton reduction, maintain argon bubbling.

*Rate assay*

1. Introduce 2 ml of a previously-deoxygenated solution of 0.1 M NaCl into the reaction vessel and allow it to equilibrate with the gas phase during 5 min.
2. Add 1 ml of substrate solution (MV containing dithionite or BV).
3. Adjust the mixture to the required pH in the pH-stat (during 2–3 min when using BV-dithionite, and during approx. 15 min when using MV-dithionite because of the slow oxidation of dithionite).
4. Prepare enzyme solution as a 10-fold dilution in 0.1 M NaCl of a 50 mM Tris–HCl or 50 mM phosphate buffered solution, pH 7.6, containing 1 mg of purified hydrogenase per ml.
5. Adjust the diluted enzyme solution to the required pH by addition of 0.1 M NaOH or 0.1 M HCl and subject it to $H_2$ bubbling for 90 min at room temperature, followed by argon bubbling for 10 min at 4°C under an argon atmosphere.
6. Start the reaction by addition of a few μl of the enzyme solution prepared as described.
7. Record the linear, zero-order, progress curve.
8. Calculate the number of moles of $H_2$ evolved ($A$) from the number of equivalents of NaOH consumed ($n\mathrm{OH}^{-2}$) by using the equation [see (3)]

$$A = \frac{n\mathrm{OH}^-}{2} \times \frac{1 + 10^{\mathrm{pH}-\mathrm{p}Ka}}{10^{\mathrm{pH}-\mathrm{p}Ka}}$$

where p$Ka$ (7.2) is that of the $HSO_3^-$ produced in the reaction. For the reverse reaction, calculate the number of moles of $H_2$ oxidized as 0.5 of the number of equivalents of NaOH added without further correction because the buffer capacity of $BV_{ox}$ may be neglected and does not change consequent upon reduction (36).

To obtain an instantaneous response to the addition of enzyme in BV reduction it is essential to flush the enzyme preparation with oxygen-free $H_2$ for at least 40 min. With untreated enzyme, activity appears only after a lag phase of 30–60 min duration and is less than that observed with treated enzyme.

Other methods for the assay of hydrogenases (by mannometry, GC, amperometric measurements of $H_2$ evolution or consumption, and spectroscopy) are cited in (35). Advantages of the pH-stat method are:

(a) it avoids the use of buffers which can have a considerable effect on hydrogenase activity (37)

(b) the $H_2$-evolving activity is free of limitations derived from gas–liquid equilibration inherent in the manometric method

(c) by changing the gas passing through the titrant and the reaction vessel ($H_2$ or Ar) it can be used to follow product formation in both directions of the hydrogenase-catalysed reaction which is not possible in the manometric or spectrophotometric methods.

## 4.3 Hydrolases (EC 3)

### 4.3.1 Triacylglycerol lipase (3.1.1.3)

Reaction:

$$\begin{array}{l} CH_2{-}O{-}\overset{\displaystyle O}{\overset{\|}{C}}{-}R_1 \\ CH{-}O{-}\overset{\displaystyle O}{\overset{\|}{C}}{-}R_2 \\ CH_2{-}O{-}\overset{\displaystyle O}{\overset{\|}{C}}{-}R_3 \end{array} + H_2O \rightarrow \begin{array}{l} CH_2OH \\ CH{-}O{-}\overset{\displaystyle O}{\overset{\|}{C}}{-}R_2 \\ CH_2{-}O{-}\overset{\displaystyle O}{\overset{\|}{C}}{-}R_3 \end{array} + R_1{-}CO_2H \rightleftharpoons R_1CO_2^- + H^+$$

---

**Protocol 5.** pH-stat assay for triacylglycerol lipase (39) [see also (32) for a procedure using a flow-through pH-stat (Section 3.2.8) and (40) for a procedure for assay of lipoprotein lipase (EC 3.1.1.34)]

*Equipment and reaction conditions*

- use a conventional pH-stat with a 25 ml reaction vessel and a 250 μl burette
- temperature 30°C

*Substrate emulsion*

1. Add 171.4 ml of hydroxypropylmethyl cellulose solution, previously cooled to 4–8°C, to 28.6 ml of triolein or purified olive oil in a high-speed blender.

**Protocol 5.** *Continued*

2. Blend at high speed for 5 min, cool at 4°C for 1 h, and repeat the emulsification process for 5 min.
3. Store the substrate emulsion (which should be stable for 5 days) in a tightly-stoppered bottle at 4°C.
4. Check adequacy of the surface area of the substrate emulsion by ensuring that the activity of a lipase preparation (≥4000 U/l) has the same value with emulsified substrate concentration 150 ml/l as with concentration 100 ml/l.

*Sodium glycocholate/ $CaCl_2$ solution*

1. Dissolve 6.095 g of sodium glycocholate and 0.4463 g of calcium chloride dihydrate in approx 80 ml of $CO_2$-free water with constant stirring.
2. Transfer the solution to a 100 ml volumetric flask and make up to mark with $CO_2$-free water.
3. Store at 4–8°C in a stoppered bottle protected from the light with Al foil (for up to 6 weeks if required).

*Colipase*

1. Prepare a stock solution of colipase at a concentration of 1200 mg/l in sterile isotonic saline (assuming that lyophilized preparations of colipase are approx 60% colipase).
2. Store the solution in appropriate aliquots at −70°C (for up to 1 year if necessary) or at 4°C (for up to 2 weeks).
3. Prepare a secondary stock solution of colipase for immediate use by mixing 1 volume of stock solution with 1 volume of sterile saline.

*Final assay concentrations*

- substrate 100 ml/l
- sodium glycocholate 35 mmol/l
- $CaCl_2$ 8.5 mmol/l
- colipase 6 mg/l

*Assay procedure*

1. Introduce 7.0 ml of the substrate solution into the reaction vessel.
2. Add 2.8 ml of the sodium glycocholate–$CaCl_2$ solution.
3. Add 0.10 ml of the colipase solution (600 mg/l).
4. Conduct a gentle, constant stream of nitrogen gas over the surface of the reaction mixture through a tube in the port of the cover of reaction vessel to remove $CO_2$ gas and prevent $CO_2$ absorption.

5. Adjust the pH to 9.0 with NaOH solution (3 mM) and equilibrate for 5 min.
6. Add 0.10 ml of NaCl solution (155 mM) in the case of blank run, and 0.10 ml of enzyme sample to the assay mixture.
7. Readjust the pH with either HCl (50 mM) or NaOH (30 mM) until the pH meter reads between 8.99 and 9.00, and record the addition of the NaOH titrant as a function of time to provide a linear progress curve over, say, 5 min.
8. Determine blank rates in duplicate at the beginning of the day and at intervals during the day and carry out electrode maintenance if variation is noted.

*Calculation*

Calculate lipase activity as:

$$150 \times (T/t_T - B/t_B) \text{ U/litre}$$

when U = micromoles of fatty acid produced per min; $T$ and $B$ are the volumes (μl) of a 15 mM solution of NaOH needed to titrate the test and blank mixtures over times $t_T$ and $t_B$ (min) for the test and blank mixtures, respectively; the factor of 150 derives from 0.015 (the molar concentration) of titrant and $1 \times 10^4$ which converts the 100 μl sample volume to 1 litre.

---

### 4.3.2 Acetylcholinesterase (EC 3.1.1.7)

Reaction

$$CH_3{-}C(=O){-}O{-}CH_2{-}CH_2{-}N^+(CH_3)_3 + H_2O \rightarrow HO{-}CH_2{-}CH_2{-}N^+(CH_3)_3 + CH_3CO_2^- + H^+ \rightleftharpoons CH_3CO_2H$$

---

**Protocol 6.** pH-stat assay for acetylcholinesterase (38) [see also (26) for a procedure for this enzyme using a pH-stat with spectrophotometric pH monitoring and electrolytic titrant generation described in Section 3.2.5]

*Reaction conditions*

- acetylcholine chloride as substrate at conc 0.11 M.
- pH 8.0
- temperature = 37°C
- 1 mM sodium hydroxide solution as titrant

*Assay*

- use a conventional pH-stat and standard procedure (see *Protocol 1*)

---

### 4.3.3 Peptide hydrolases (peptidases and proteinases) (EC 3.4)

*i. General considerations*

*Reactions*

Many members of this large group of enzymes possess esterase activity in addition to the ability to catalyse the hydrolysis of peptide bonds. This facilitates the application of pH-stat assays because in approximately neutral media where the pH is much greater than the p*K*a of the carboxylic acid product, hydrolysis results in essentially stoichiometric release of protons.

$$\mathrm{R{-}\underset{\underset{O}{\|}}{C}{-}O{-}R' + H_2O \rightarrow R'OH + R{-}CO_2H \rightleftharpoons RCO_2^- + H^+}$$

By contrast, hydrolysis of simple amides at pH values near to neutrality is not accompanied by proton release because the p*K*a values of the carboxylic acid and amine products will usually be approximately 4 and 9 respectively.

$$\mathrm{R{-}\underset{\underset{O}{\|}}{C}{-}NH{-}R' + H_2O \rightarrow R{-}CO_2^- + R'{-}NH_3^+}$$

Even in weakly alkaline media, pH-stat assays of such substrates will be very insensitive. With peptide substrates, the lower value of the p*K*a of an α-amino group (approx. 7.6) does permit the reaction to be monitored in a pH-stat if the stability of the enzyme and its active-centre characteristics allow the reaction to be studied in alkaline media (41, 42). If several peptide bonds within the substrate molecule are cleaved at different rates, however, meaningful interpretation of the kinetic data will be difficult or impossible.

*ii. Serine proteinases (3.4.21) and cysteine proteinases (3.4.22)*

*Reactions:*

As indicated above, the use of ester substrates in which there is only one susceptible bond facilitates the assay of these enzymes by the pH-stat technique. The particular substrate to be used is chosen by making use of known molecular recognition (specificity) characteristics of a particular enzyme. Commonly-encountered examples of the serine proteinase family (43) are chymotrypsin (EC 3.4.21.1) and trypsin (EC 3.4.22.4). Chymotrypsin is readily assayed by using *N*-acetyl-L-tyrosine ethyl ester (ATEE), which makes use of the need for a hydrophobic occupant for the $S_1$ subsite of chymotrypsin.

Trypsin is similarly assayed by using α-*N*-benzoyl-L-arginine ethyl ester (BAEE), which makes use of the need for a cationic side chain of the

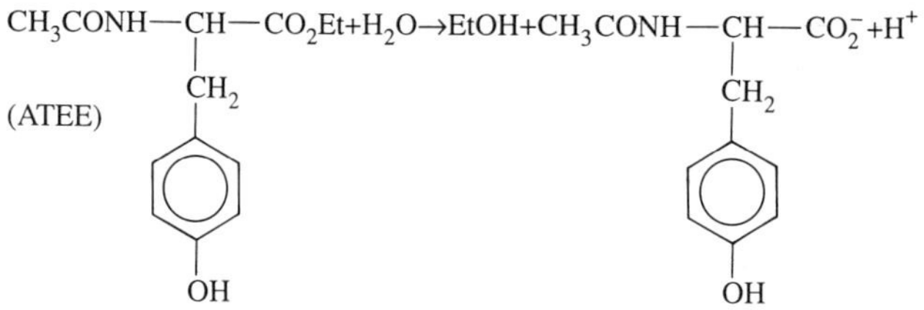

$$CH_3CONH-CH(CH_2C_6H_4OH)-CO_2Et + H_2O \rightarrow EtOH + CH_3CONH-CH(CH_2C_6H_4OH)-CO_2^- + H^+$$

(ATEE)

substrate to bind in the $S_1$ subsite of trypsin. Many members of the cysteine proteinase family (44), such as papain (EC 3.4.22.2), require a hydrophobic substituent in the substrate to bind in the $S_2$ subsite of the enzyme and a good, sensitive, substrate for these enzymes which could be used in pH-stat assays is *N*-acetyl-L-phenylalanyl-glycine ethyl ester:

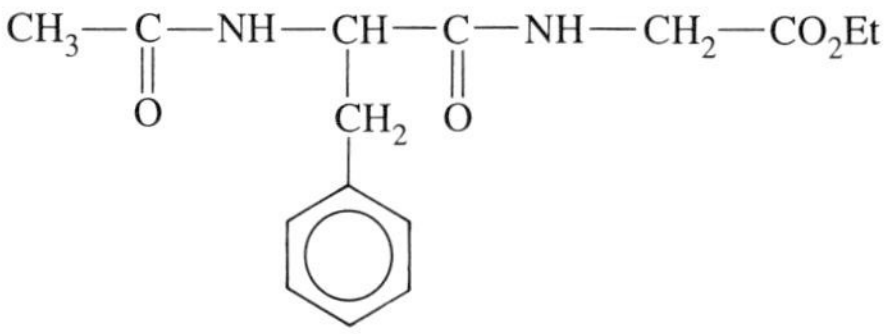

$$CH_3-C(=O)-NH-CH(CH_2C_6H_5)-C(=O)-NH-CH_2-CO_2Et$$

BAEE, which is available commercially, is commonly used as a substrate for papain and many other cysteine proteinases because the α-*N*-benzoyl group fortuitously binds near to the hydrophobic $S_2$ subsite of papain.

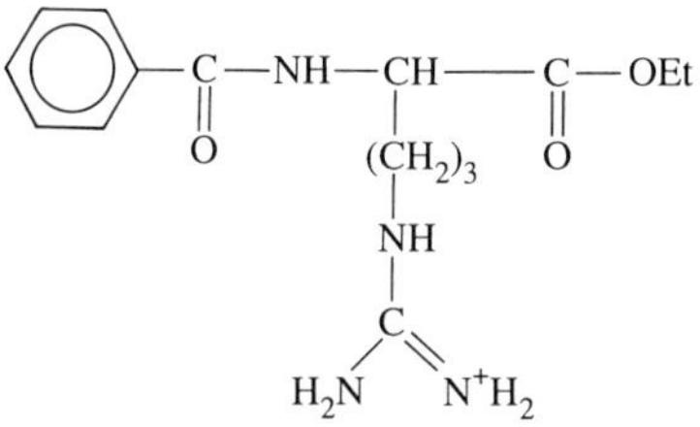

$$C_6H_5-C(=O)-NH-CH[(CH_2)_3-NH-C(=N^+H_2)NH_2]-C(=O)-OEt$$

(BAEE)

---

**Protocol 7.** pH-stat assay for cysteine proteinases and trypsin using BAEE as substrate, and for chymotrypsin using ATEE as substrate, with the enzymes either in solution or covalently immobilized (CI), e.g. on CM-cellulose (CI-enzymes) (45).

1. Use a conventional pH-stat with 0.5 ml burette and micro-reaction vessel.
2. Maintain the reaction vessel at 25 °C and pass a stream of $O_2$-free nitrogen

**Protocol 7.** *Continued*

first through a wash bottle containing water at 25°C, and thence over the surface of the reaction vessel to exclude $CO_2$ and $O_2$.

3. Use a reaction volume of 5 ml and 0.1 M NaOH as titrant.
4. To obtain approximate value of $V_{max}$, use the following concentrations in the stock solutions of substrate: 0.25 M BAEE for papain, ficin (EC 3.4.22.3), bromelain (EC 3.4.22.4), and trypsin, and 0.5 M ATEE in 50% (v/v) aq. methanol for chymotrypsin.
5. For assays with BAEE, place the following in the reaction vessel: 1 ml of 0.5 M KCl, 0.5 ml of 0.1 M EDTA, $x$ ml of standard suspension of a CI-enzyme (e.g. 1 ml containing *c.* 50 mg of CI-bromelain or 5–10 mg of CI-trypsin) or enzyme solution, and $1.95 - x$ ml of deionized water, and equilibrate to pH 7 using 1 M NaOH added from an Agla micrometer syringe.
6. When equilibration is complete (*c.* 5 min) add 2.0 ml of the substrate solution and rapidly readjust the pH to 7.0 using the Agla syringe.
7. Switch on the pH-stat and record for sufficient time to obtain a linear trace (with CI-enzymes, a degree of curvature may occur in the early stage of reaction due to final re-equilibration, particularly with CM-cellulose derivatives).
8. For assays with ATEE, follow the same procedure as for BAEE except use 1.0 ml of substrate solution and $2.95 \times x$ ml of water.
9. Use chart speeds of 1–4 cm/min.

### 4.3.4 Urease (EC 3.5.1.5)

*Reaction*

$$(H_2N)_2C{=}O + 2H_2O \rightarrow 2NH_3 + H_2CO_3$$

Acid needs to be added to a reaction mixture at pH 7.0 to maintain the pH. Under the assay conditions described the initial rate is linear with respect to urease concentration (46) and the decomposition of the ammonium carbamate formed is not rate-determining. It is present in steady-state concentration and does not affect the rate measurement (46).

It is necessary to correct the experimentally determined uptake of acid at pH 7.0 for the ionization of the cationic acid and ammonia products.

$$H_2CO_3 \rightleftharpoons H^+ + HCO_3^- \underset{}{\overset{pK_2\ pK_1}{\rightleftharpoons}} H^+ + CO_3^{2-}$$

$$NH_4^+ \overset{pK_3}{\rightleftharpoons} H^+ + NH_3$$

$$-d[\text{urea}]/dt = (d[H^+]/dt)/(2f_n - f_c)$$

where
$$f_c = (1 + 2K_2/[H^+])/(1 + [H^+]/K_1 + K_2/[H^+])$$
$$f_n = [H^+]/(K_3 + [H^+])$$

---

**Protocol 8.** pH-stat assay for urease (46)

1. Use a conventional pH-stat with a 0.5 ml burette and 0.01 M HCl as titrant and thoroughly-cleaned glassware.
2. Standardize the electrodes frequently at the assay temperature with 0.05 M phosphate buffer (pH 7.0).
3. For routine assays, use 10 ml aliquots of 0.05 M recrystallized urea in boiled-out distilled water, thermostatted at 38 ± 0.1 °C for at least 15 min in 20 ml glass vessels.
4. Place a combination electrode (used only for urease assays), a glass stirrer, and the burette tip into the reaction vessel.
5. Rinse the above three items in buffer A (0.02 M phosphate buffer, pH 7.1, 1 mM EDTA, and 1 mM 2-mercaptoethanol) between assays.
6. Add 20 μl of 1 mM dithiothreitol containing 0.5 mM EDTA just prior to addition of urease solution to the reaction mixture.
7. Start the reaction by addition of urease in buffer A and record the uptake of acid at pH 7.0.
8. For a convenient rate, use of 1 IU of enzyme activity (1 IU of urease activity is that amount of enzyme that causes the decomposition of 1 μmol of urea/min under the standard assay conditions of 38 °C, pH 7.0, 0.05 M urea, 2 μM dithiothreitol).

---

### 4.3.5 β-Lactamases (EC 3.5.2.6)

*Reactions:*

This group of enzymes catalyses the hydrolysis of β-lactams, some acting more effectively on penicillins and some more effectively on cephalosporins. The former were formerly designated as EC 3.5.2.8. Release of protons consequent upon hydrolysis of a β-lactam ring in neutral media is illustrated overleaf. Methodology involving a standard pH-stat system with a 2.5 ml burette and 20 ml reaction volume is described in (47).

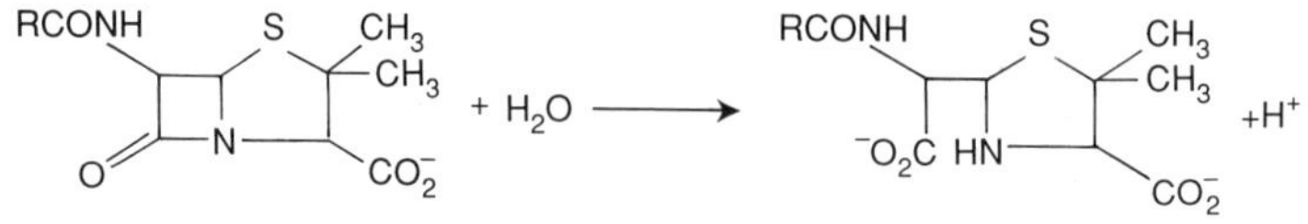

## 4.4 Guanine deaminase (EC 3.5.4.3) and adenosine deaminase (EC 3.5.4.4)

*Reactions*

For guanine deaminase:

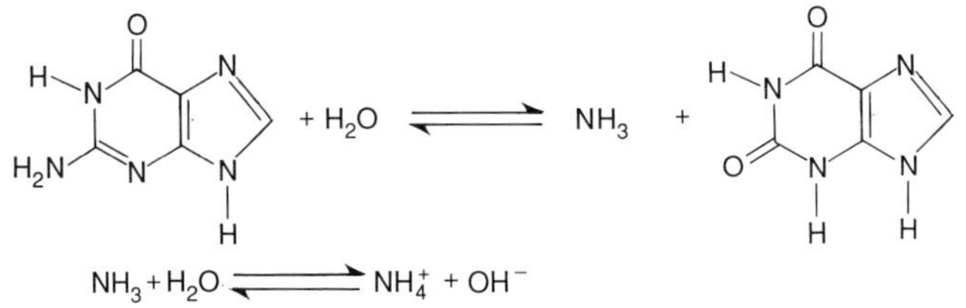

$$NH_3 + H_2O \rightleftharpoons NH_4^+ + OH^-$$

Methodology involving a standard pH-stat system with a 0.5 ml burette, a 5 ml reaction vessel, 0.1 mM HCl, and a reaction pH of 6.6 is described in (48).

Methodology for assay of adenosine deaminase is described in (49).

## 4.5 Carboxy-lyases (EC 4.1.1): glutamate decarboxylase (EC 4.1.1.15) and lysine decarboxylase (EC 4.1.1.18)

*Reactions*

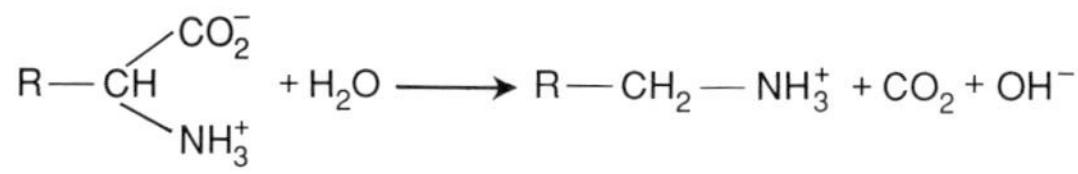

$$R{-}CH(CO_2^-)(NH_3^+) + H_2O \longrightarrow R{-}CH_2{-}NH_3^+ + CO_2 + OH^-$$

The stoichiometry of release of $OH^-$ depends upon the pH, the p*K*a of the α-carboxy group of the amino acid substrate (e.g. 2.18 for lysine), and the first p*K*a of carbonic acid (6.37).

Methodologies involving standard pH-stat systems are described for glutamate decarboxylase in (50) and for lysine decarboxylase in (51).

### 4.6 Acid-ammonia ligases (EC 6.3.1): aspartate-ammonia ligase (asparagine synthetase) (EC 6.3.1.1) and glutamate-ammonia ligase (glutamine synthetase) (EC 6.3.1.2)

*Reactions*

$$\text{Asp} + \text{ATP} + NH_3 \rightarrow \text{Asn} + \text{AMP} + \text{PPi}$$

$$\text{Glu} + \text{ATP} + NH_3 \rightarrow \text{Gln} + \text{ADP} + \text{Pi}$$

The various p*K*a changes in these reactions (52) provide for net proton release which can be titrated with base to provide pH-stat assays for these enzyme-catalysed reactions. Methodology is described in (53).

### 4.7 Some other ATP-utilizing enzymes

p*K*a changes in reactions involving ATP catalysed by the following enzymes result in proton 'signals' in the pH range 5.5–8.5 that permit the use of pH-stat assays described in the references cited:

- hexokinase (EC 2.7.1.1) (54)
- phosphofructokinase 1 (EC 2.7.1.11) (55, 56)
- creatine kinase (EC 2.7.3.2) and adenylate kinase (EC 2.7.4.3) (57)
- glucose-6-phosphate isomerase (phosphoglucoisomerase) (EC 5.3.1.9) (58).

### 4.8 Miscellaneous biochemical applications of pH-stat techniques

Procedures for the application of pH-stat techniques in the following are described in the references cited:

- microbial protease activity (59)
- rhodanese activity (60)
- protein denaturation (ovalbumin in acid solution) (61)
- chemical modification of proteins, e.g. inactivation of trypsinogen by methanesulfonyl fluoride (62)
- determination of thermodynamic quantities characteristic of ATP–Mg complexes (63)

## 5. A systematic error in pH-stat assays of enzymes in haemolysates

In many pH-stat procedures, a stream of $O_2$-free nitrogen is continuously passed over or through the reaction mixture to prevent the absorption of atmospheric carbon dioxide. Newman and Nimmo (64) pointed out that a problem arises when this procedure is applied to reaction mixtures containing oxyhaemoglobin, such as when erythrocyte acetylcholinesterase is being determined. In such circumstances, the oxyhaemoglobin slowly loses its oxygen ligand and becomes more basic. This decreases the amount of base needed to neutralize the acid liberated in the reaction being assayed. One answer for such systems is to use a stream of oxygen instead of nitrogen to prevent absorption of carbon dioxide.

## 6. Concluding comment

The pH-stat assay is a convenient and sensitive technique that is applicable to a wide range of enzyme-catalysed reactions in which there is a net change in hydrogen ion concentration. Equipment is available, both commercially and described in the literature, that permits automation of the technique at a number of levels, and its computer control. The technique is applicable not only to reactions in solution but also to reactions in stirred suspension such as those involving cellular extracts and immobilized enzymes and cells.

## Acknowledgements

It is a pleasure to thank Professor Eric Crook and his group of enthusiasts (Garth Kay, Bob Bywater, P. V. Sundaram, and Chris Wharton) for introducing me to this technique in the 1960s when it was used extensively, in the Biochemistry Department of St. Bartholomew's Hospital Medical College, University of London, for the assay of a variety of enzymes attached to solid supports. Thanks are due also to Liz Beech for searching and assembling relevant examples from the literature, Joy Smith for the rapid production of the typescript at short notice, and the Science and Engineering Research Council for project grants and Earmarked Research Studentships involving kinetic studies on enzyme reactions.

## References

1. Knaffle-Lenz, E. (1923). *Arch. Pathol. Pharmakol.*, **97,** 242.
2. Knaffle-Lenz, E. (1923). *Medd. f. K. Ventenskapsakademiens Nobelinst.*, **6,** No. 3.
3. Jacobsen, C. F., Leonis, K., Linderstrom-Lang, K., and Ottesen, M. (1957). In: *Methods of Biochemical Analysis*, (ed. D. Glick), Vol. IV, pp. 171–210. Interscience Publishers Inc., New York.
4. Jenkins, M. R. (1986). In: *A Biologist's Guide to Principles and Techniques of Practical Biochemistry* (ed. K. Wilson and K. H. Goulding), (3rd edn), pp. 345–380. Edward Arnold Ltd., London.
5. Hornby, W. E., Lilly, M. D., and Crook, E. M. (1966). *Biochem. J.*, **98,** 420.
6. Lilly, M. D., Regan, D. L., and Dunnill, P. (1974). *Enzyme Eng.*, **2,** 245.

7. Matsen, J. M., and Linford, H. B. (1962). *Anal. Chem.*, 34, 142.
8. Karcher, R. E. and Pardue, H. L. (1971). *Clin. Chem.*, **17,** 214.
9. Bucher, T., Hofner, H., and Romayrere, J-F. (1974). In: *Methods of Enzymatic Analysis* (ed. H. U. Bergmeyer and K. Gawehn), (2nd English edn), pp. 254–61. Academic Press, Inc., New York.
10. Tsibanov, V. V., Loginova, T. A., and Neklyudov, A. D. (1982). *Russian Journal of Physical Chemistry*, **56,** 718.
11. Saunders, I. (1990). *Laboratory Products Technology, Feb.*, 10.
12. Whitnah, C. H. (1933). *Ind. Eng. Chem. Anal. Ed.*, **5,** 352.
13. Longsworth, L. G. and MacInnes, D. A. (1935). *J. Bacteriol.*, **29,** 595.
14. Lingane, G. (1949). *Anal. Chem.*, **21,** 497.
15. Jacobsen, C. F. and Leones, J. (1951). *Compt. rend. Lab. Carlsberg, Ser. chim.*, **27,** 333.
16. Warner, B. D., Boehme, G., Urdea, M. S., Pool, K. H., and Legg, J. I. (1980). *Anal. Biochem.*, **106,** 175.
17. Job, R. and Freeland, S. (1977). *Anal. Biochem.*, **79,** 575.
18. Ke, B. (1975). *Bioelectochem. Biochem. Bioenerg.*, **2,** 93.
19. Keijer, J. H. (1970). *Anal. Biochem.*, **37,** 439.
20. Vandermeers, A., Lelotte, H., and Christophe, J. (1971). *Anal. Biochem.*, **42,** 437.
21. Konecny, J. (1977). *Biochem. Biophys. Acta*, **481,** 759.
22. Rousseau, I. and Atkinson, B. (1980). *Analyst*, **105,** 432.
23. Hieftje, G. M. and Mandarano, B. M. (1972). *Anal. Chem.*, **44,** 1616.
24. Lemke, R. E. and Hieftje, G. M. (1982). *Anal. Chim. Acta*, **141,** 173.
25. Forensen, J. K. and Stockwell, P. B. (1975). *Automatic Chemical Analysis*, pp. 45–54, Wiley, New York.
26. Karcher, R. E. and Pardue, H. L. (1971). *Clinical Chemistry*, **17,** 214.
27. Johansson, G. (1965). *Talanta*, **12,** 111.
28. Adams, R. E., Betso, S. R., and Carr, P. W. (1976). *Anal. Chem.*, **48,** 1989.
29. Brand, M. J. D. and Rechnitz, G. A. (1969). *Anal. Chem.*, **41,** 1185.
30. Bull, H. B. and Breese, K. (1973). *Arch. Biochem. Biophys.*, **158,** 681.
31. Eigtved, P. (1981). *J. Bioche. Biophys. Methods*, **5,** 37.
32. Taylor, F. (1985). *Anal. Biochem.*, **148,** 149.
33. Iturbe, F., Ortega, E., and Lopez-Munguia, A. (1989). *Biotechnology Techniques*, **3,** 19.
34. Gilli, R., Sari, J. C., Sica, L., Bourdeaux, M., and Briand, C. (1986). *Anal. Biochem.*, **152,** 1.
35. Aquirre, R., Hatchikian, E. C., and Monson, P. (1983). *Anal. Biochem.*, **131,** 525.
36. Prince, R. C., Linkletter, S. J. G., and Dutton, P. L. (1981). *Biochim. Biophys. Acta*, **635,** 132.
37. Glick, B. R., Martin, W. G., and Martin, S. M. (1980). *Canad. J. Microbiol.*, **26,** 1214.
38. Garcia-Lopez, J. A. and Monteoliva, M. (1988). *Clin. Chem.*, **34,** 2133.
39. Tietz, N. W., Astles, J. R., and Shuey, D. F. (1989). *Clin. Chem.*, **35,** 1688.
40. Chung, J. and Scanu, A. M. (1974). *Anal. Biochem.*, **62,** 134.
41. Milhalyi, E. (1978). In: *Applications of Proteolytic Enzymes to Protein Structure Studies*, (ed. R. C. East), Vol. 1, p. 129. CRC Press, Cleveland, USA.

42. Rothenbuhler, E. and Kinsella, J. E. (1985). *J. Agric. Chem.*, **33**, 433.
43. Polgar, L. (1987). In: *Structure and function of serine proteases in hydrolytic enzymes* (ed. A. Neuberger and K. Brocklehurst), pp. 159–200. Elsevier, Amsterdam.
44. Brocklehurst, K., Willenbrock, F., and Salih, E. (1987). In: *Cysteine proteinases in hydrolytic enzymes* (ed. A. Neuberger and K. Brocklehurst), pp. 39–158, Elsevier, Amsterdam.
45. Crook, E. M., Brocklehurst, K., and Wharton, C. W. (1970). *Methods Enzymol.*, **19**, 963.
46. Blakeley, R. L., Webb, E. C., and Zerner, B. (1969). *Biochemistry*, **8**, 1984.
47. Hou, J. P. and Poole, J. W. (1972). *J. Pharmaceutical Sci.*, **61**, 1594.
48. Bieber, A. L. (1971). *Anal. Biochem.*, **43**, 247.
49. Garth, Von H. and Zoch, E. (1984). *J. Clin. Chem. Clin. Biochem.*, **22**, 769.
50. Salvadori, C. and Fasella, P. (1970). *Ital. J. Biochem.*, **19**, 193.
51. Vienozinskiene, J., Januseviciute, R., Paulinkonis, A., and Kazlauskas, D. (1985). *Anal. Biochem.*, **146**, 180.
52. Albert, R. A. (1969). *J. Biol. Chem.*, **244**, 3290.
53. Wedler, F. C. and McClune, G. (1974). *Anal. Biochem.*, **59**, 347.
54. Hammes, G. G. and Kochavi, D. (1962). *J. Amer. Chem. Soc.*, **84**, 2076.
55. Dyson, J. E. and Noltmann, E. A. (1965). *Anal. Biochem.*, **11**, 362.
56. Lorenz, I. (1972). *Math-Naturwiss.*, **21**, 551.
57. Mahowald, T. A., Noltmann, E. A., and Kuby, S. A. (1962). *J. Biol. Chem.*, **237**, 1535.
58. Dyson, J. E. and Noltmann, E. A. (1968). *J. Biol. Chem.*, **243**, 1401.
59. Alkanhal, H. A., Frank, J. F., and Christen, G. L. (1985). *J. Food Protection*, **48**, 351.
60. Cannella, C., Pensa, B., and Pecci, L. (1975). *Anal. Biochem.*, **68**, 458.
61. Ottensen, M. and Wallevik, K. (1968). *Biochim. Biophys. Acta.*, **160**, 262.
62. Morgan, P. H. and Hass, G. M. (1976). *Anal. Biochem.*, **72**, 447.
63. Sari, J.-C., Ragot, M., and Belaich, J.-P., (1973). *Biochim. Biophys. Acta*, **305**, 1.
64. Newman, P. F. J. and Nimmo, I. A. (1980). *J. Clin. Pathol.*, **33**, 1009.

# 8

# Detection of enzymatic activity after polyacrylamide gel electrophoresis and agarose gel isoelectric focusing

OTHMAR GABRIEL and DOUGLAS M. GERSTEN

## 1. Introduction

Many options are currently available to perform electrophoretic protein separations (1–4). Unique problems arise in the context of enzyme electrophoresis because there is the additional requirement to demonstrate functionally active enzyme after the electrophoretic separation is completed. Separations are usually carried out under 'native' conditions which attempt to avoid denaturation and inactivation of the enzyme protein. However, separations under denaturing conditions are possible, followed by renaturation and subsequent demonstration of enzymatic activity. Readers who are interested in comprehensive aspects of polyacrylamide gel electrophoresis (3) and isoelectric focusing (4) are referred to the references listed.

### 1.1 Electrophoretic separation techniques

There is an abundance of products and equipment for the methods described, that are available from a variety of manufacturers. The quotation of any product, equipment or manufacturer in this article is not intended as an endorsement or a negative statement about an unmentioned competing product. It merely reflects our experience and familiarity with a particular product.

Electrophoretic separations of proteins can be performed in free solution or in a variety of anti-convection media such as paper, membranes, or gels (5). We will confine our comments to analytical applications using gels as supporting medium. Polyacrylamide and agarose are available in high purity. The characteristics and advantages of these matrices are given for polacrylamide in (4) and for agarose in (6).

For zymography, slab gels are the preferred geometric format; electrophoresis of multiple samples is performed more conveniently in slabs than in cylindrical gels, samples can be run side by side for ready comparison, and extrusion of fragile gels from a glass tube is unnecessary. Coated glass or plastic backing sheets for slab gels are now widely available, providing mechanical stability to otherwise fragile gels. To demonstrate enzymatic activity, overlays containing the necessary substrates and cofactors required can be easily brought into direct contact with gel slabs. 'Blotting' of enzymes from slab gels onto membranes in a manner which faithfully preserves the electrophoretic pattern is possible. Protein or 'Western' blotting is the subject of a recent textbook (7).

Charged molecular species migrate in an electrical field, and electrophoretic mobility can be used to resolve proteins in two fundamentally different ways, either by using their intrinsic charge differences or by imparting an extrinsic charge.

Separations based on intrinsic charge generally use low ionic strength electrolytes and attempt to maintain the proteins in their 'native' conformation in order to preserve enzymatic activity (8, 9). A different approach based on a protein's intrinsic charge is isoelectric focusing. In this method, a pH gradient is established across the electric field and proteins will migrate to the point of the pH gradient where their net charge is zero; individual proteins will be separated according to their isoelectric point along the pH gradient (4). Electrofocusing is attractive for zymograms since isozymes with small differences in their isoelectric points can be clearly resolved (4) and since conditions of separation frequently involve maintenance of proteins in their native state.

The most common form of electrophoresis employs the application of a large extrinsic charge by interaction of proteins with sodium dodecylsulphate (SDS)(10). In this method, the hydrophobic tail of SDS interacts with the protein's hydrophobic domains, imparting a negative charge. The separation is based on the size of the proteins, which determines their relative migration rates through the gel matrix. Interaction of the protein with SDS causes denaturation and inactivation of most enzymes, but some will still retain activity even in the presence of SDS. Usually after electrophoretic separation, steps are required to restore native conformation to the protein before one can demonstrate enzymatic activity.

## 2. Preparation of sample and material

The following guidelines selected for the preparation of materials and sample for electrophoretic separation are based on representative and commonly-used methods. This should provide a start to examine the feasibility of the separation. It will alert the investigator to possible problems and difficulties one has to anticipate. For a more comprehensive treatment of electrophoretic separations the reader is referred to (3). A 'quantitative' approach that explores 'optimization' of separation conditions has been published by Chrambach (11).

## 2.1 Sample preparation

Prior to electrophoresis, determination of the protein concentration is essential. About 1–10 μg of a purified protein is optimal, while mixtures of proteins can contain between 50–100 μg. Too little protein may limit detection of the enzyme. Overloading of gels will result in tailing and distortion of protein bands, and could interfere with the banding pattern of adjacent sample lanes. Isoelectric focusing is even more sensitive to overloading. Here, too much protein will distort the pH gradient. The protein should be applied in a small volume in continuous buffer systems. For vertical gels, 10–20 μl in 0.75-mm-thick and 7-mm-wide sample well should be used. It is less critical in discontinuous buffer systems as long as the height of the sample volume is kept to less than one half of the stacking gel height. The buffer should only contain the trailing ion and its counter-ion. If the sample is to be layered under a buffer, 10–30% (w/v) sucrose or glycerol are added. The mixture with a tracking dye such as bromphenol blue helps to monitor sample application. For horizontal gels we recommend a 10 μl sample volume if direct loading is used. Wicks can accommodate 25 μl and a sample applicator mask can accommodate up to 125 μl (12).

In consideration of maintaining enzymatic activity, it should be noted that factors like presence of divalent cations, coenzymes, detergents and ionic environment, pH, and absence of oxidizing agents cannot be neglected. In addition, tracking dyes may react irreversibly with enzyme protein resulting in inactivation of the enzyme. Thus caution is required during sample preparation.

Before any attempt for separation and localization of an enzyme sample is carried out, it is imperative to obtain the best available information about the enzyme under investigation. Even when one has essential information about enzyme stability, requirements of cofactors, and other significant data, it is sometimes useful to make small pilot experiments. One such experiment is to cast acrylamide gels on slides and to apply samples of the enzyme in question to a small well. The sample is permitted to diffuse into the gel matrix and the detection method and limits are tested without the electrophoretic run. This pilot assay will in most cases assess whether or not recrystallization of acrylamide monomer, pre-electrophoresis, or other strategies will be required.

## 2.2 Recrystallization of the acrylamide monomer prior to polymerization

Not more than what can be used in one week should be recrystallized and the material should be stored desiccated and protected from light at −20 °C. This procedure eliminates a number of contaminants that accumulate during storage of the monomer (13).

Dissolve completely 45 g of acrylamide monomer in 90 ml of ethanol-free chloroform brought to 60°C in a hot water bath. (The use of ethanol-free chloroform is essential for the detection of oxidoreductases, as discussed in Section 11.1.2. Commercial chloroform contains 1–2% (v/v) ethanol as stabilizing agent and it can be removed by washing the chloroform with water in a separatory funnel just prior to use. The washed chloroform can be directly used for recrystallization.) Filter through a funnel kept at 60° using Whatman No. 1 filter paper. The filtrate is cooled to 22°C and the crystals are filtered immediately *in vacuo* on Whatman #1 filter paper, collected, air dried, and stored as described above. This procedure must be carried out in a fume hood, wearing plastic safety gloves. (See also Section 3.1.)

### 2.3 Removal of oxidants

Gels that are polymerized by the addition of peroxides will contain residual peroxides which are detrimental to many enzyme proteins. There are several ways to eliminate peroxides. Pre-electrophoresis of the gel alone (14) or inclusion of sulfhydryl group components such as cysteic acid into the buffer system can be used (15). Direct addition of sulfhydryl reagents to the polymerization mixture is not recommended as chain termination of polymerization will occur, resulting in prevention or delayed polymerization of gels. Thin gels (1 mm and less) can be soaked with buffers containing diothiothreitol to remove oxidizing agents.

## 3. Preparation of slab gels for zymography

### 3.1 General considerations

Inhibition and/or loss of enzymatic activity due to the presence of contaminants in the gel is unpredictable. Consequently, we suggest starting with the highest purity acrylamide monomer then recrystallizing the acrylamide again prior to use. Both acrylamide and bisacrylamide monomers are neurotoxins and repeated, extended exposure must be avoided. Handling of solutions should be carried out wearing plastic safety gloves and no pipetting by mouth is permitted. As mentioned in Section 2.2, recrystallization of acrylamide monomer must be carried out in a fume hood.

### 3.2 Preparation of gels

Where possible, we recommend that the flat-bed gels be covalently cast onto a plastic sheet or onto a siliconized glass plate. This permits easier handling of gels during staining, especially when using large pore gels with poor mechanical stability. Large pore gels provide advantages due to their reduced molecular sieving and electrophoretic separation time. These gels permit ready diffusion of coenzymes and substrates into the gel following electrophoresis. To silanize glass plates, mix 2 ml 3-methacryloxypropyltrimethoxysilane (Polifix 100, Serva Fine Biochemicals) with 1 l of 50% (v/v) ethanol, and allow to stand for 30 min. Immerse the scrupulously clean glass plates for 3–5 min, then allow to drip-dry vertically. Plates coated in this way can be stored indefinitely.

An alternative to siliconized glass is the use of polyester plastic backing sheets. These can be obtained in large sizes or precut to fit commercial electrophoresis units as GelBond-PAG (FMC Bioproducts) or Gel-Fix for PAGE (Serva). The backing sheets must be in contact with the glass electrophoresis plate without air bubbles, or the gel will have uneven thickness when polymerized. Cut the plastic sheet to the precise size of the glass. Hold the edge of the sheet at 45° to the edge of the glass plate. If the plastic has both sides coated (Gel-Fix), place a drop of water between the glass and the plastic backing. If the plastic has only one side coated (GelBond), the side facing the glass will be hydrophobic. In this case use 0.1% Triton X-100 as the wetting agent. Lower the plastic sheet onto the glass and expel excess liquid from between glass and plastic by rolling firmly with a rubber roller or a test tube wrapped with a paper towel. Make sure there are no air bubbles trapped between the surface of the glass and the plastic backing.

Wherever possible, we recommend the use of rehydratable gels for isoelectric focusing and applications that do not use SDS (16, 17). These gels can be prepared well in advance, kept stored until required, and are much less likely to inactivate enzymes due to unpolymerized acrylamide or unreacted polymerization initiators such as ammonium persulphate. The strategy for the preparation of rehydratable gels is to cast the gels with a minimum of buffer salts and ampholytes on a glass or plastic backing. Once polymerized, remove the covering plate, and soak the polymerized gel and its backing in water. Soaking time depends on gel thickness. A 0.75-mm gel will require about 20 min soaking. We suggest the use of AcrylAide (FMC) rather than bisacrylamide as the cross-linker for rehydratable gels because, in our experience, gels dried on plastic sheets tend to curl less with this cross-linker. The dried paper-thin gel can be stored for months. Prior to use, the gel is soaked in the appropriate electrolyte or ampholyte solution. The extent of rehydration varies with time and %T.* A convenient way to follow the extent of rehydration is to weigh the gel before drying and to check rehydration by weight. We have had occasional difficulties with dried gel separating from the backing sheet. If this occurs, reduce the thickness to 0.25 mm.

## 3.3 Casting the gel

To cast the polyacrylamide slab gel, either a vertical or horizontal set-up is possible. If plastic backing is used, be certain that the plastic does not buckle during assembly, as this will cause the gel to have uneven thickness. For

* The pore size of polyacrylamide gels is defined by two parameters: %$T$ refers to the total monomer concentration (w/v); %$C$ refers to the amount of cross-linking agent (bisacrylamide or AcrylAide) as a percentage of the monomer concentration.

vertical casting, assemble the plates with a U-form spacer, such that the spacer is near the edge of the plates. Glass plates much larger than the spacer will flex once clamped together, causing buckling of the plastic. If silanized glass plates are used, be certain that only one of the two plates is siliconized! Clamp the plates and spacer together using binder or bulldog clips. Clamps should completely surround the perimeter to prevent leakage. If leakage occurs, disassemble and grease the spacer slightly with Celloseal (Fisher) on both sides before clamping the plates together. Calculate the appropriate volume for the gel, leaving 0.5 cm at the top for the water layer. Pour or pipette the acrylamide mixture into the U-form, then carefully place (use a syringe with plastic-tubing that has a small diameter or a peristaltic pump) a 3-mm-layer of water or water-saturated butanol on top of the acrylamide solution.

To cast the gel in the horizontal format by the sliding technique (not recommended for thick gels), place the siliconized or plastic-backed plate on a level surface. Grease two spacers, which are the size of the long dimension, on one side each, and place them at the edges of the plate with the greased side facing the glass. Take the non-siliconized plate and hold it parallel to the level surface, so that about 1 cm is resting on the spacers and the rest of the plate extends over the edge of the lower plate. Pour the acrylamide solution into the space where the plates overlap, while constantly sliding the upper plate along the spacers along the lower plate until the two plates overlap and are completely lined-up. The ends are left open. Gels cast in this manner will polymerize to within a few mm of the ends. An alternative method to the sliding technique is the flap technique (17).

Agarose is the medium of choice for many applications, e.g. isoelectric focusing. Only recently has the use of rehydratable agarose gels been reported (18), therefore experience is limited. Agarose can be obtained in different variations, which melt and solidify at different temperatures. Usually, the melting temperature is considerably higher than the gelling temperature. To cast an agarose gel weigh out the appropriate amount of agarose, add water, and melt the agarose in a hot water bath or microwave oven at the temperature suggested for the particular brand of agarose. If heat-sensitive substances are to be included in the gel, allow the gel to cool to within 10 degrees of the gelling temperature before adding them to the molten agarose. As with polyacrylamide, tight adherence of the gel to a glass plate is not always assured. (Polishing and cleaning of glass plates and removal of fingerprints help.) Plastic backing sheets for agarose are available (GelBond and Gel-Fix). In the event that fluorescence detection of enzymatic activity is to be used, the non-fluorescent GelBond-NF (FMC) should be used. If plastic backing sheets are used, roll them onto the glass plates as noted above. If glass plates are to be used, dispense approximately 0.5 ml of molten agarose per 250 $cm^2$ of area onto the plate. Wipe the agarose over the plate with a no-lint tissue, and allow the surface to air-dry. Agarose gels are usually cast horizontally and rarely cast between parallel plates. For gels of 1-mm thickness or less, surface tension is sufficient to hold the molten agarose in place until it solidifies. Place the plate on a levelling table, and pour or pipette the calculated volume of agarose solution onto the centre of the plate after it has cooled to near its gelling temperature. If necessary, spread the agarose to the edges with the end of the pipette. Agarose gels thicker than 1 mm require a wall around the plate. Prepare such a wall by application of waterproof masking tape around the perimeter of the plate. Agarose plates may be prepared in advance of use, but they must be maintained in a humidified chamber until use.

## 4. Gel formulations

There are many variations of gel recipes described in the literature—far too many to be considered here. We have selected four main types of gels for illustrative purposes: electrophoresis under 'native' conditions in Tris–glycine polyacrylamide gels, SDS polyacrylamide gel electrophoresis, and isoelectric focussion in polyacrylamide gels and in agarose gels. For a more complete discussion of various forms of protein separations, the reader is referred to *Gel Electrophoresis of Proteins: A Practical Approach* in this series (3).

One of the attractive features of polyacrylamide gels is the ability to achieve superior resolution by matching the pore size of the gels with the size of the proteins to be resolved. This is readily accomplished by controlling the acrylamide monomer content as a percentage of acrylamide plus cross-linking agent (%$T$), and the degree of cross linking (%$C$) as a percentage of the total monomer. It is usual practice to prepare a stock solution of acrylamide monomer and cross-linker, and adjust the pore size by diluting the stock solution appropriately.

### 4.1 'Native' Tris–glycine gels (8, 9)

Native Tris–glycine gels are prepared as described in *Protocol 1*.

---

**Protocol 1.** Casting of a 12 ml vertical slab of 10%$T$, 2.66%$C$.

*Stock solutions*

(a) Stock solution of acrylamide/AcrylAide* cross-linker or acrylamide/bisacrylamide cross-linker: 30 g acrylamide, 0.8 g cross-linker in 69 ml water. (These solutions can be stored at room temperature with the exception that AcrylAide must be refrigerated.)

(b) Stacking gel buffer: 5.98 g Tris base, 0.46 ml TEMED (N, N, N′, N′-tetramethylethylene diamine) in 100 ml of 0.48 M HCl. Adjust to pH 6.7 if necessary.

**Protocol 1.** *Continued*

(c) Separating gel buffer: 36.6 g Tris base, 0.23 ml TEMED in 100 ml of 0.48 M HCl. Adjust to pH 8.9 if necessary.

(d) *i.* 5% ammonium persulphate (w/v) in water or
*ii.* riboflavin (0.1%, w/v): 4 mg riboflavin dissolved in 4 ml of water.

Riboflavin-initiated polymerization is required only if residual persulphate interferes with enzymatic activity.

(e) Running (tank) buffer: 6.0 g Tris base, 28.8 g glycine to 1 l, adjust to pH 8.3.

*Procedure*

1. Assemble the plates and backing sheets.
2. To make 12 ml of gel solution, mix 4.0 ml of acrylamide/AcrylAide, 1.5 ml separating gel buffer, and 6.4 ml water.
3. Mix gently and de-gas the solution in a suction flask connected to a vacuum source (12–20 Torr).
4. Add 100 μl of persulphate solution, pour between the plates, and add a water layer carefully without mixing of the solutions. Completion of polymerization is noted by the appearance of a sharp boundary between the polymerized gel and the water layer.
5. Pour off the water layer and replace with separating gel buffer.
6. Leave the gel overnight before polymerization of the stacking gel. If there is evidence that indicates that the residual peroxide in the gel is detrimental to the enzyme, a remedy such as pre-electrophoresis must be carried out prior to polymerization of the stacking gel.
7. After completion of polymerization of the separating gel, pour off the buffer layer.
8. Prepare a 4% *T*, 2.66% *C* stacking gel by mixing 1.33 ml of acrylamide/bisacrylamide solution, 1.25 ml stacking gel buffer, 6.9 ml water, 25 μl riboflavin or 50 μl persulphate. If rapid polymerization is desired, also add an additional 50 μl of TEMED. It should be noted that in stacking gels bisacrylamide, not AcrylAide, is used as the cross-linker.

* AcrylAide, persulphate and riboflavin solutions should be stored at 4°C. Riboflavin requires storage in the dark. All other solutions may be stored at room temperature.

---

To achieve a contiguous interface between the separating and stacking gel, it will be necessary to wash the separating gel surface with a small amount of stacking gel mixture. Discard the wash and fill the space between the plates to the top with stacking gel solution; then insert the well-forming comb. If riboflavin is used as the initiator, polymerization of the gel is carried out by illumination with a fluorescent light.

## 4.2 SDS polyacrylamide gel electrophoresis: Laemmli technique (10)

SDS polyacrylamide gels are prepared as described in *Protocol 2*.

**Protocol 2.** Casting a vertical 10%*T*, 2.66%*C*, 12 ml SDS slab gel

*Solutions*

- stock solution of acrylamide/AcrylAide* or acrylamide/bisacrylamide: 30 g acrylamide, 0.8 g cross-linker dissolved in 69 ml water
- buffer for separating gel: 226.9 g Tris base, 1.25 ml TEMED, 5.0 g SDS, adjust to pH 8.8 and bring total volume to 1 l
- buffer for stacking gel: 9.85 g Tris–HCl, 0.125 ml TEMED, 0.5 g SDS, adjust to pH 6.8 and bring volume to 100 ml
- running (tank) buffer: 3.027 g Tris base, 14.41 g glycine, 1.0 g SDS, adjust to pH 8.3 and bring volume to 1 l.
- 5% (w/v) ammonium persulphate

*Procedure*

1. Assemble the plates as described in *Protocol 1*.
2. To make 12 ml of gel solution, mix gently to avoid frothing, 4.0 ml acrylamide solution, 2.4 ml separating gel buffer, and 5.5 ml water. If the gel is thin (less than 0.5 mm), de-gas the mixture. Add 100 μl ammonium persulphate.
3. Pour or pipette between the plates, add water layer, and wait for polymerization to occur.
4. When the separating gel is polymerized, pour off the water layer and prepare a 3% *T*, 2.66%*C* stacking gel by mixing 1.0 ml of acrylamide solution, 2.0 ml stacking gel buffer, 6.9 ml water and 50 μl persulphate. If rapid polymerization is desirable, also add an additional 50 μl TEMED.
5. Fill the space between the plates to the top, then insert the well-forming comb.

* AcrylAide and persulphate solutions should not be stored at room temperature.

## 4.3 Isoelectric focusing in polyacrylamide (16)

Isoelectric focusing, as other electrophoretic procedures, has many variations. Here we describe a procedure using carrier ampholytes, the most common mode of electrofocusing. In this procedure, the pH gradient is formed by the field-directed movement of soluble amphoteric compounds (the most common are polyaminopolycarboxylic acids) that can buffer at their isoelectric points (4). There are two main variations: (a) buffer electrofocusing substitutes a mixture of 47 buffer salts for ampholytes (19); (b) the pH gradient is cast into the gel when acrylamide polymerizes (20).

Zymography, following isoelectric focusing in rehydrated polyacrylamide gels, has been reported to give better results than the same procedures following focusing in conventional gels (16).

---

**Protocol 3.** Casting a 10 ml, 5%*T*, 2.66%*C* rehydratable gel

*Solutions*

- stock solution of acrylamide/AcrylAide: 30 g acrylamide, 0.8 g AcrylAide dissolved in 69 ml water
- 5% (w/v) ammonium persulfate
- 0.5 M Tris–HCl, pH 8.0
- ampholyte solution*—this varies as the desired range of the gradient to be used varies. There is also variation in ampholyte solutions provided by different manufacturers and between lots from the same manufacturer. In general, when a narrow range (e.g. pH 5–7) is needed, prepare an aqueous solution which is 4% (v/v) 5–7 LKB Ampholine, 10% (v/v) glycerol. If a wide range is needed, (e.g. pH 3–10), substitute a blend of 0.8% (v/v) 3–10 LKB Ampholine and 3.2% 5–7 LKB Ampholine containing 10% (v/v) glycerol.

*Procedure*

1. Prepare a U-form gasket by cutting a piece of polyvinylchloride plastic or silicone rubber of suitable thickness (usually 1 mm or less) with a 'guillotine-style' paper cutter. The width of the gasket should be at least 5 mm.
2. Assemble the plates and backing sheets; calculate the required volume.
3. To make 10 ml of gel, mix 0.5 ml of acrylamide/AcrylAide solution, 1.0 ml Tris buffer, 0.1 ml persulphate solution, and 8.35 ml water. Vacuum degas, then add 0.05 ml TEMED and mix gently. Pour the solution between the plates and overlay with a water layer.
4. Following polymerization, remove the gel bonded to the plastic and soak it, face down, in a dish of water for 20–30 min, depending on the thickness of the gel.
5. Allow the gel to air-dry. It may be necessary to weigh down the edges of the plastic backing sheet to prevent curling. Dried gels may be stored for several months.

6. Rehydrate the gel. Some authors advocate rehydrating the gels in a cassette especially designed for this purpose in order to be economical with the ampholyte solution (16). Others recommend overlaying the precise amount of solution on top of the dried gel, and allowing it to soak in (21). A simpler, if slightly more expensive approach, is the following. If the gel is cast on a plastic sheet, float the gel, face down, in a dish containing ampholyte solution. If the gel is cast on a siliconized glass plate, put the plate face up in a dish containing ampholyte solution. Monitor the progress of rehydration by weight.
7. To prevent drying after completion of rehydration, cover the gel with gel bond or with plastic wrap until use. Before use, be certain to remove excess ampholine from the gel surface. This is best accomplished by carefully drawing a piece of polyester backing sheet across the surface in the manner of a windshield wiper.

* Ampholyte solutions should be prepared fresh. Persulphate should be stored at 4°C. Other solutions can be stored at room temperature.

## 4.4 Isoelectric focusing in agarose (22)

Several different agarose preparations with different electroendosmotic properties are commercially available. Only those specifically designated for isoelectric focusing should be used (e.g. Agarose-IEF, Pharmacia or Isogel, FMC).

**Protocol 4.** Preparation of a 15 ml agarose-isoelectric focusing gel

1. Assemble the plates and plastic backing sheets.
2. Weigh out 0.075 g Isogel agarose, add 15 ml of water and heat in a boiling waterbath or microwave oven until the agarose has completely dissolved. No beads of residual undissolved agarose should be visible.
3. Allow to cool to 70°C. Add 225 µl LKB 3.5–10 Ampholine, 75 µl LKB 2.5–4 Ampholine, and 75 µl LKB 5–8 Ampholine and gently mix.
4. Pour onto plate. If necessary, spread the liquid to the edges of the plate with a pipette tip. When the solution has gelled, put the gel into the refrigerator for 1 h.

# 5. Electrophoresis

The overall objective is to perform the electrophoretic separation in the shortest possible time. The rate of the electrophoretic migration depends on the field strength. The major limitation to applied voltage is Joule heating

(23) which, in extreme cases, can cause boiling or melting of the gel. Therefore, some manner of heat exchanger is required for high-voltage electrophoresis. For vertical gels this is most frequently accomplished by immersion of the plates in a tank containing an electrolyte, with cold water circulating through a cooling coil. For flat-bed gels the gel and backing sheet are placed on a platen having a circulating coolant, or one which is a Peletier cooling device. A water layer (similar to that described for gel casting on backing sheets) is interposed between the gel backing-sheet and platen to ensure good contact. Failure to maintain good contact will lead to hot spots in the gel. Constant temperature in the gel is an absolute requirement for isoelectric focusing experiments because the isoelectric point and viscosity are temperature dependent.

The input voltage depends on the efficiency of cooling and the thickness of the gel. The electrical resistance of the gel, hence the heating, decreases as the cross-sectional area increases. This is balanced by the decrease in heat dissipation as thickness increases.

For discontinuous electrophoresis systems [Ornstein and Davis (8, 9) or Laemmli (10)], the resistance of the gel changes as the run progresses because the leading ion occupies more and more of the separating gel. Two parallel slabs, 0.75 mm thick, immersed in electrolyte, with cooling, running in parallel in the vertical position, should be run at 200 V using a constant voltage power setting.

For flat-bed isoelectric focusing systems, 0.5 mm thick, using efficient cooling, a starting applied voltage of 100–200 volts/cm (interelectrode distance) is appropriate (16). Electrode wicks for a 3.5–10 pH gradient should be soaked with 1.0 M NaOH and 1.0 M $H_3PO_4$ for the cathode and anode, respectively. For other gradients see (24), and for immobilized gradients see (25). The electrical resistance of an isoelectric focusing gel changes rapidly once the current has been applied, since fewer and fewer species remain charged as the separation proceeds. Therefore the best separations are achieved using power supplies which can be operated at constant power. The power limit is 1–10 Watts, and depends on the cooling efficiency and width (not thickness!) of the gel.

## 6. Staining

If gels must be stained following enzyme assay, or if it is necessary to monitor the efficiency of blotting and elution, the gel should first be immersed in a fixing solution containing, per litre, 300 ml absolute methanol, 34.5 g sulfosalicylic acid, and 115 g trichloroacetic acid. A 0.75-mm gel requires at least 1 h. The gels should be then stained in a staining solution containing, per litre, 250 ml absolute ethanol, 80 ml glacial acetic acid, and 1.0 g Coomassie Brilliant Blue R-250 (also called Serva Blue R-250). Filter the stain immediately before use. Destain in the same solution without Coomassie dye.

**Protocol 5.** Silver staining

We recommend the procedure of Guevara *et al.* (26).

1. Fix the gel by incubating in 200 ml of a solution containing 20% (v/v) absolute ethanol, 5.0% (v/v) glacial acetic acid, and 2.5% (w/v) sulfo-salicylic for at least 12 h with one solution change.
2. Wash the gel in 20% ethanol three times for 20 min each.
3. Incubate the gel in 200 ml of 20% ethanol containing 5.9 mmoles $AgNO_3$, 31 mmoles $NH_4OH$, and 2.8 mmoles NaOH for 1 h. Ammoniacal silver solution is prepared by mixing 2.1 ml 14.8 M $NH_4OH$, 0.113 g NaOH, 40 ml absolute ethanol, and 148 ml water. Dissolve 1.0 g $AgNO_3$ in 10 ml water and then add dropwise to the ammoniacal ethanol solution.
4. Rinse the gel with water then repeat step 2.
5. Develop the image by incubating the gel in 200 ml of 20% ethanol containing 0.01% (w/v) citric acid and 0.037% formaldehyde for 15 min or until the pattern is clear.
6. Stop the development by incubating the gels in 20% ethanol, 5% acetic acid for 30 s.
7. Wash the gel in running tap water for 45 min.

## 7. Troubleshooting the electrophoresis

The majority of difficulties encountered in gel electrophoresis are due to polymerization problems and dirty plates. Sticky or lumpy gels with poor mechanical stability are usually incompletely polymerized. The most common reason is that the acrylamide, and most frequently the persulphate solutions, are old or need de-gassing.

Uneven migratory patterns or 'smiling' bands are usually caused by uneven heat dissipation in which the temperature increase makes the pores of the gel larger in the middle. In this case reduce the applied voltage or cool the gel to a lower temperature. Bands that streak are usually caused by dirty plates. Wash the plates first in acetone to remove oils, grease from the gaskets, and fingerprints, then in cleaning acid before rinsing and drying.

In vertical systems, problems are sometimes encountered due to short circuits where there is electrolyte leakage from the upper to the lower electrolyte reservoir. In these cases the tracking dye runs slower than usual because the current is shunted away from the gel to the short circuit. If the short circuit is intermittent, as with dripping electrolyte, the voltmeter will be intermittently fluctuating. Test for electrolyte leaks by adding bromphenol blue to the upper electrolyte reservoir. To remedy this type of short circuit, grease the spacers and gaskets with 'Celloseal' as described in Section 3.3.

The analogous problem in horizontal systems, uneven voltage, is usually due to poor contact between the electrode and the electrolyte-soaked filter wick, or between the wick and the gel. Occasionally, the wick will dry out. Should this be the case, start with a thicker wick.

# 8. Preparation of electrophoretically-separated enzymes for detection

The gel containing the separated components must be worked up immediately following the electrophoretic run to localize the enzyme. The proteins are located in narrow bands and fixation of proteins that usually precedes staining would result in enzyme denaturation and can not be carried out. To avoid diffusion of the separated bands, speed is essential to retain the electrophoretic pattern and minimize re-mixing of separated components. For enzymes that can retain activity during freezing and thawing the gels can be kept frozen until detection of enzymatic activity can begin.

# 9. Blotting/elution/renaturation

It is intuitively obvious that the most desirable scenario for demonstration of enzymatic activity following electrophoresis is one in which the separation can be performed under 'native' conditions, loss of activity is negligible, and activity is so high that the amount of protein at the surface of the slab gel is sufficient to effect very rapid substrate conversion. Such a situation would allow incubation of the gel in a substrate solution, or abutting the gel against a substrate-impregnated matrix, in a manner such that the reaction is confined to a tight band, before the resolution attained by the separation is lost to diffusion. Separation could also be performed in SDS gels if the enzyme maintains its activity with SDS present. More enzymes are in this category than originally envisioned, but the vast majority are not and frequently require elution from the gel, renaturation to their native conformation, tethering to a solid-phase matrix, or combinations of the above.

## 9.1 Blotting

Protein blotting (27) has become a popular technique because the pattern of the electrophoretic separation is maintained after the proteins have been eluted from the gel. The original application of protein or 'Western' blotting was for locating the position of proteins using antibodies, not enzymatic activity, and was a variation of the 'Southern' technique (28), first used for DNA transfers. Capillary action is used to draw the proteins out of the electrophoresis gel onto a solid-phase membrane matrix. The matrices originally used were DBM (diazobenzyloxymethyl) paper or nitrocellulose membranes (29), but fibreglass, nylon, and polyvinylidene difluoride sheets are also used (30).

The transfer of the proteins to the matrix by capillary blotting is an overnight procedure (29) but there are modifications that accelerate the transfer. It can be hastened by using a vacuum table (an ordinary vacuum gel dryer) (31), or the transfer can be driven by electrophoresis (32).

The decision to blot or not to blot is generally made on the basis of time and activity considerations. The ideal condition described above would have enough activity at the surface of the gel to ensure rapid substrate conversion. In the absence of this, blotting serves to concentrate the protein from the three dimensions of the gel onto an essentially two-dimensional matrix. If this manoeuvre, by itself, does not give enough active enzyme for rapid substrate conversion, the immobilized enzyme can be incubated with substrate for very long periods of time without detachment from the membrane or diffusion. Another advantage of blotting (33) is that the blotting matrix can be incubated in more than one substrate. Hence the same separation can be tested for different activities.

Each of the matrices and transfer modes has its advantages and disadvantages, which should be evaluated in consideration of the enzymes and substrates in question; for example, not all proteins bind to all matrices. Preliminary experiments are required in which the crude mixture is spotted, fixed to the matrix, and tested for enzymatic activity. Alternatively, some matrices will bind everything, thereby causing a high background. Such supports can be blocked following transfer by incubation in a solution of irrelevant proteins. These include BSA, fish gelatin, powdered non-fat milk, and Irish cream liqueur. DBM paper must be blocked following transfer by incubation in ethanolamine and gelatin (29). Such treatments may serve to inactivate the enzymes. Some electroblotting protocols require transfer in the presence of denaturants, such as acetic acid (32) or methanol (30). Electroblotting and capillary blotting require that both sides of the slab gel be open. Gels cast on backing sheets can only be transferred by vacuum blotting. Overnight transfer may be too long to maintain the activity of some enzymes. Nitrocellulose membranes are very brittle and sometimes difficult to handle, but they are now available with fabric reinforcements. Some membranes can be prewetted with the substrate solution. As enzymes change shape in order to function, the effect of their fixation to a membrane is unknown. Careful kinetic studies of membrane-bound enzymes have generally not been performed, but the effect is probably not far different from that for enzymes tethered to beaded matrices. Not all proteins transfer with the same efficiency; transfer is frequently less complete for larger proteins. Generally, more protein is eluted from the gel by electroblotting than by capillary blotting.

## 9.2 Capillary blotting to nitrocellulose

This is one of the easiest and the least expensive methods and does not require denaturing electrolytes.

**Protocol 6.** Capillary transfer from a polyacrylamide slab gel to nitrocellulose

1. Pre-wet an Immobilon-NC membrane by floating it in a dish containing 0.02 M phosphate buffer, pH 8.0. Do not immerse the membrane since this may cause air to become trapped inside the pores.
2. Pre-wet 6 sheets of Whatman No. 1 filter paper in the same buffer.
3. Fill a square or rectangular glass dish with the same buffer, and cover the dish with a glass plate which extends over two of the four sides. Drape the 6 thicknesses of filter paper over the plate such that both ends are immersed as wicks in the buffer.
4. Position the slab gel on the filter paper, being careful to avoid air bubbles between the gel and the paper.
5. Use plastic wrap or paraffin film to 'frame' the gel. This is to ensure that the only fluid flow is through the gel and not around it. The entire surface of the filter paper, except that occupied by the gel, should be covered by the plastic.
6. Place the saturated membrane on the gel, again avoiding air bubbles.
7. Cover the membrane with two sheets of dry filter paper (S and S#597).
8. Prepare a mat of dry paper towels, 25 mm thick, and place these on top of the dry filter papers.
9. Cover the stack with a glass plate, and put a 3 kg weight on the glass. Incubate overnight.
10. Following incubation, one may wish to stain the gel with Coomassie blue to determine whether any of the protein remains. Alternatively, stain the membrane by incubation for 3 h in 0.1% (w/v) Naphthol blue–black in water. Destain by incubation in water.

## 9.3 Elution

There are several reasons why enzymatic activity is determined in free solution, rather than in the gel or on a blotting matrix. In our opinion, the two major reasons are that suitable coloured, insoluble, cleavage products are unavailable, and that the enzyme must be re-natured in order to restore activity. The first problem has been partially solved by the use of overlay gels impregnated with substrate. The second, re-naturation, is discussed in Section 9.4.

Over the years, there have been several different schemes devised for elution of the protein from the gel. Two brief summaries of these can be found (34, 35). Generally, they fall into three categories; mechanical, diffusional, and electrical. In the mechanical category, gels have been crushed, homogenized, or dissolved; agarose gels with low gelling temperatures have been melted. Diffusion by itself, in which the band (or spot from two-dimensional gels) is merely excised and incubated in an elution buffer, is decidedly the least efficient approach. As such, it is almost always combined with mechanical disruption of the gel. Naturally, the more finely the gel is ground, the more rapid the diffusional recovery. In consideration of this, homogenization of the gel in a glass/teflon tissue grinder has been reported (34, 36).

Many arrangements for electroelution have also been described. Electroelution of the proteins from the excised band into a dialysis bag was first described by Lewis and Clark (37), and many variations using a dialysis bag or membrane have been reported since (38–41). Electroelution into sucrose gradients (35), hydroxylapatite (42), conductivity gradients (43), capillary matrices (44), and by steady-state stacking (39) have also been used.

In many of the published applications electroelution in the presence of 0.1% (w/v) SDS is advocated to keep the proteins in solution and to ensure that they migrate to the anode. A few examples of applications which do not use SDS are *N*-ethylmorpholine:acetic acid (40), steady-state stacking (39), and capillary matrices (44).

## 9.4 Renaturation

Enzymes to be re-natured must potentially recover from the untoward effects of two harsh treatments. These are the use of detergents/reducing agents/polar solvents (sometimes required during solubilization, electrophoretic separation, and electroelution) and acid/alcohol/high salt conditions (sometimes required during fixation and staining to determine the location of the protein in the gel). Furthermore, it is not clear whether enzymes entrapped at high local concentration in gels behave more as soluble or immobilized proteins. A general review of re-naturation of immobilized enzymes was published (45).

We have cautioned the reader above that great care must be taken throughout a zymography experiment to preserve enzymatic activity. Assuredly, in the absence of prior knowledge regarding enzyme stability, this must be so. At first approximation it is difficult to envision enzymology under anything but 'native' conditions. None the less, a scan of the available literature indicates that many more enzymes than originally envisioned can maintain activity under what would generally be considered very harsh conditions; for example (Section 11.1), tyrosinase activity can be demonstrated in an SDS gel. Other well-known examples of enzymes which function in the presence of SDS are proteinase K which is used commonly in nucleic acid extractions (46), and V8 protease, chymotrypsin, and papain used in the Cleveland procedure (47). Enzymes which modify nucleic acids appear to be particularly hardy in terms of their retention of activity in SDS or their recovery from SDS treatment. They are discussed briefly in Section 11.7. In a certain sense, enzymatic activity in the presence of SDS is not surprising since SDS is biodegradable by bacteria, particularly Pseudomonas (48).

Enzymatic activity in the presence of non-ionic detergents is generally considered to be milder treatment than SDS. Heegaard (49), however, points out that some enzymes which are inactivated by Triton X-100 are active in Tween 20. Even though activity can sometimes be demonstrated in their presence, such detergents must bind very tightly since they are routinely used to displace SDS pre-bound to the proteins. A good discussion of the effect of non-ionic detergents on immunoblotting (not enzymoblotting) can be found in Bjerrum *et al.* (50). An example of an equally harsh technique is the demonstration of phosphodiesterase activity at the surface of a gel containing 8 M urea (51). Observations such as these lead us to the inescapable conclusion that, in more instances than one would expect, full renaturation is not required for satisfactory localization of enzymatic activity after electrophoresis.

In light of this, many applications merely rely on soaking the SDS gel for a short time in a buffer without detergents as the only renaturation step. At best, this is a variable procedure, because proteins can differ in their rates of spontaneous re-folding, from fractions of seconds to days (52). On the one hand, some authors have had good success with simple soaking, being able to demonstrate activity for amylases and proteases (53) for example. However, those results could not be readily duplicated (54). On the other hand, attempts to renature multimeric dehydrogenases, composed of identical subunits without interchain disulphides, by simple soaking, yielded less than 1% of their original activity (53).

Given that renaturation in the gel by rapid (and perhaps incomplete) buffer exchange may not be entirely satisfactory, alternative methods have been reported. A general method was proposed by Weber and Kuter (55) which elutes the protein and uses 6 M urea to displace the SDS. SDS in solution is then adsorbed to Dowex 1, the sample is diluted slowly to reduce the urea to 0.6 M, and residual urea is removed by equilibrium dialysis. Other authors advocate acetone precipitation of the proteins to ensure complete removal of SDS (34). The precipitates are re-dissolved in 6 M guanidine–HCl, then diluted.

The addition of sulphydryl agents, such as 1 mM mercaptoethanol, during the renaturation of enzymes in the gel has been reported (56). This stems from the notion that renaturation is a multi-step process that requires the sequential formation and breakage of intermediate disulphides before the least-energy conformation is reached. Renaturation, of course, can be catalysed by disulphide interchange enzyme (57) but, to our knowledge, this has not been applied to enzymes in gels. Thioredoxin, however, also catalyses the re-folding of proteins (58), and this has been used for the renaturation of enzymes in gels (59).

Recovery of activity by any of the above approaches appears to vary from protein to protein, suggesting that no universal method of renaturation is appropriate. This means that attempts to compare the activity in two bands of the same gel, or in different gels, are difficult; they necessarily reflect both the amount of protein present and the extent of renaturation. Moreover, for some proteins, no manner of renaturation seems to be satisfactory. In those cases the antibody capture approach described in Section 12 might be considered.

Finally, some workers have found it necessary to elute proteins from blotting membranes (59). In these procedures the elution required 2% SDS and/or 1% Triton X-100, so the same considerations as above apply regarding renaturation.

## 10. Detection of enzymes in gels

The detrimental effects of reagents used to prepare gels (e.g. oxidizing agents, incomplete reaction of polymerizing agents) and conditions dictated by the requirements for protein separations (e.g. pH, buffer ions) must be recognized and appropriate steps must be taken to minimize enzyme inactivation. Ampholytes used for isoelectric focusing interfere with some enzyme detection methods and removal by soaking in appropriate buffers may be required. In most instances a compromise must be established between optimal conditions for the enzyme and requirements for protein separation.

### 10.1 *In situ* localization of enzymes: principles and quantitation

The separation of enzymes in thin-layer gels by electrophoresis and isoelectric focusing provides a restrictive matrix for the enzyme and easy access of low-molecular-weight substrate to the enzyme. This permits *in situ* localization with little diffusion of the enzyme protein. The thinner the gel, the more rapid equilibration to optimal enzymatic reaction conditions will take place. This will result in faster colour development and less diffusion time for the protein band. There are several basic types of reactions and reagents suitable for *in situ* localization of enzymes. All these methods are modifications of procedures originally described by histochemists.

#### 10.1.1 Simultaneous capture of substrate product

This technique has many applications with enzymes that remain active in the presence of the staining reagents. The product of the enzymatic reaction couples with a reagent present in the incubation mixture to result in formation of an insoluble coloured product. The major advantage of this method is the immediate formation of a product that precipitates into the gel matrix. Consequently, there is no diffusion of substrate product.

### 10.1.2 Post-incubation coupling of substrate product

Incubation of enzyme and substrate yields the product of the reaction. The product of the enzymatic reaction in turn is converted with a suitable reagent to yield an insoluble coloured product. In contrast to the above mentioned method, diffusion of the low-molecular-weight substrate product during the first incubation period results in broadening of bands. This method should only be used when the reagents required to form the insoluble coloured stain are not compatible with the enzyme and lead to its inactivation.

### 10.1.3 Autochromic methods

The progress of an enzymatic reaction can be directly observed when there are changes of optical properties of either the substrate or the product. There are a number of substrates and/or coenzymes that change their optical properties during enzymatic reactions in the visible or ultraviolet light. Changes of fluorescence properties are included as very sensitive indicators of autochromic methods. During the progress of the reaction, a photograph is taken to obtain a permanent record.

### 10.1.4 Indicator–matrix assays

The use of components that cannot be accommodated in the separating gel, such as auxiliary enzymes or high-molecular-weight substrates, will require the overlay of the gel with an 'indicator–matrix' after electrophoresis. Sandwich-type incubation of separating and indicator gel permit localization of enzymatic activity.

### 10.1.5 Copolymerization of substrate into separating gel

High-molecular-weight substrates such as carbohydrate polymers, proteins, or nucleic acids are sometimes included in separating gels and may present special problems. The separation will be influenced by the presence of these macromolecules. During electrophoresis, action of the enzyme on its substrate can be prevented by inclusion of inhibitors, chelating agents, or other methods. Following restoration of optimal conditions for enzymatic activity, the enzyme can then be detected.

## 10.2 General comments about *in situ* detection methods

The localization of enzymatic activity by any of the above described methods will be followed by a method to detect the presence of proteins. In this way a correlation between the protein stain and the appearance of one or more bands of enzymatic activity will give the investigator important experimental evidence about the purity, the presence of isozymes, or other significant information about the enzyme of interest. One of the most important facts about staining for enzymatic activity (that is many times neglected) is the vigorous experimental verification that the stain indeed represents the activity of the enzyme of interest. An extensive battery of control experiments is essential. There are some obvious questions. Is the stain substrate dependent? Do specific inhibitors of the enzyme prevent appearance of stain? Is denatured enzyme inactive in the assay procedure employed? Does the omission of essential cofactors or metals abolish enzymatic activity? These controls should be performed especially when staining techniques new to the investigator are employed.

In general, *in situ* staining procedures will necessitate a compromise between the conditions required for electrophoretic separation and the conditions that provide optimal activity for the enzymatic process that results in localization of the enzyme in a gel matrix. In many instances, quantitative retention of enzymatic activity during the incubation period should not be assumed without experimental verification. This fact should be remembered when attempts are made to derive quantitative information from enzyme staining patterns. The kinetic events that occur in the gel matrix during staining are complex and will require substantial experimental evidence to establish that the quantity of the enzyme protein is the only rate-limiting factor and that enzymatic activity does not diminish due to partial inactivation. For most systems it is not advisable to derive quantitative conclusions from the intensity of an enzymatic stain.

## 11. Practical examples for enzyme detection

Below are some practical applications for the staining of various enzymes. In each group examples are presented that relate to the various separation methods discussed and with typical staining protocols. The examples were selected to present, wherever possible, methods that may find applications to a whole group of enzymes. There will be instances where a satisfactory technique for a given application cannot be found. A list of applications with literature references was published earlier by us (1, 2). Another valuable source for uncharted territory that we have used on many occasions is a text of histochemistry (60, 61). Generally, histochemical methods are applicable with few modifications to the staining of polyacrylamide or agarose gels.

A division into major groups of enzymes as defined by the International Union of Biochemistry was considered to be useful because enzymes that catalyse related reactions lend themselves to the application of similar detection methods.

### 11.1 Oxidoreductases

The substrate being oxidized is considered a hydrogen or electron donor. The classification is based on 'donor:acceptor oxidoreductase', and wherever possible, the recommended name is 'dehydrogenase'. Many dehydrogenases use $NAD^+$ or $NADP^+$ as coenzymes.

### 11.1.1 Detection of NAD$^+$ or NADP$^+$ requiring enzymes by direct visualization in ultraviolet light

Reduced pyridine-nucleotides exhibit yellow fluorescence under ultraviolet light, while the oxidized form of the coenzymes appear as black bands. Advantages of this method are that the assay works in either direction of the enzymatic reaction (oxidation or reduction of substrate). The recognition of 'nothing dehydrogenases' that lead to oxidation or reduction of coenzyme in the absence of substrate is another useful feature of the method (see example below). A disadvantage of the method is that as an autochromic method, diffusion of coenzyme leads to broadening of separated components, and for this reason it is best to obtain a photographic record at different time intervals during the progress of the enzymatic reaction.

Electrophoresis of most oxidoreductases requires protection of critical sulphydryl residues and conditions that retain the enzyme protein in its native conformation. It is recommended to use the conditions of native Tris–glycine discontinuous gels (8, 9). The stacking gel polymerization should be carried out with riboflavin as a catalyst. Pre-electrophoresis of the separating gel is recommended prior to casting the stacking gel. Analytical gels are run in 0.5–0.75 mm thick vertical gel slabs as described in Section 4.1. Before application of the sample to the gel, the protein samples are diluted 1:1 with a sample buffer containing 10 mM Tris pH 7.5, 30% (v/v) glycerol and 0.01% (w/v) bromophenolblue as the tracking dye. The gels are run at 4°C at constant current towards the anode.

Following electrophoresis (62) the gels are soaked for a couple of minutes in a 1 mM coenzyme solution. If the reaction catalyses the transfer of hydride ion from NADH or NADPH to substrate, the coenzyme (1 mM) is dissolved in 0.2 M bis-Tris buffer, pH 7.0. By contrast, if NADH or NADPH are generated by the reaction, resulting in oxidation of substrate, the buffer containing the coenzyme is 0.2 M Tris buffer, pH 8.0 or 9.0. In each instance, the coenzyme solution is prepared just prior to use. After incubation of the gel with the appropriate coenzyme–buffer solution, the gel is briefly rinsed with water to remove excess coenzyme and is viewed in an ultraviolet light box (Chromato-Vue Trans Illuminator, Model TL-33, Ultra-Violet Products, Inc., San Gabriel, CA). After a short time, oxidoreductases that mediate coenzyme–hydrogen transfer independent of substrate are visible. The gel is overlaid with Whatman No. 1 filter paper soaked with a solution containing substrate and other enzyme requirements (e.g. metal ions). Depending on the amount of active enzyme present, substrate-dependent bands of enzymatic activity become visible with time. Photographs are taken to keep a permanent record (e.g. Polaroid MP-4 Land Camera with type 667 Polaroid film).

### 11.1.2 Detection of $NAD^+$ or $NADP^+$ requiring enzymes by coupling to formazan formation

The second type of technique to detect pyridine nucleotide requiring oxidoreductases depends on the formation of reduced coenzymes (NADH or NADPH). The reducing equivalents are transferred to various tetrazolium dyes, a process which is mediated by phenazine methosulphate and which results in formation of a deeply-coloured insoluble formazan. The main advantage of the method is the fact that the reaction can be carried out in a single incubation mixture.

Unfortunately, this method is sensitive to light and oxygen exposure and can lead to high background and non-specific staining. It is essential to demonstrate substrate dependence of the staining process. As mentioned in the previous example, 'nothing dehydrogenases' were reported by several investigators. It should be noted that several explanations for the appearance of these 'nothing dehydrogenases' have been offered; presence of substrate (most commonly ethanol) as an impurity of the system, presence of sulphydryl groups of proteins with sufficiently low redox potential to reduce pyridine nucleotide, and substrate so tightly bound to the enzyme that it was not separated from the enzyme protein. More recently, the enzyme alcohol dehydrogenase was identified as 'nothing dehydrogenase'. The source of alcohol was found in commercial samples of acrylamide (Section 2.2). Technical grade chloroform contains ethanol as a stabilizing agent, and therefore chloroform free of ethanol must be used for recrystallization of acrylamide (63).

*i. Example: α-glycerophosphate dehydrogenase (64)*

The electrophoretic separation is carried out in a native discontinuous gel system. Make sure that the enzymatic reaction proceeds in the direction of substrate oxidation. The incubation is carried out in the dark and reagents are prepared just prior to use. 50 ml of the incubation mixture contain 25 mg of $NAD^+$, 15 mg of nitro blue tetrazolium (NBT), 1 mg of phenazine methosulphate, 5 ml of 1 M sodium α-glycerophosphate, pH 7.0, 10 ml of 0.2 M Tris–HCl, pH 8.0, and 35 ml of water.

Incubate the gel in the dark at 37°C until dark-blue bands appear. Rinse the gel with water and fix in ethanol–acetic-acid–glycerol–water (5:2:1:4). Be aware of artificial staining caused by light and oxygen and make sure that the stain is substrate-dependent.

This method can be modified by appropriate changes to detect a variety of $NAD^+$ and $NADP^+$ requiring dehydrogenases. There are also a number of various other tetrazolium salts available that provide advantages for the detection of oxidoreductases (3).

### 11.1.3 Pyridine-nucleotide independent oxidoreductases

A third example of an oxidoreductase, tyrosinase, an enzyme which does not use a pyridine-nucleotide as a coenzyme, is presented here. Mammalian tyrosinase is an enzyme that retains activity even in the presence of SDS. Therefore, although the procedure below describes the removal of SDS after the electrophoretic separation, for very active samples this step is not necessary. The staining technique employs the simultaneous capture method. The enzyme catalyses the conversion of L-tyrosine to L-dopa (3,4-dihydroxy phenylalanine) to dopaquinone. Dopaquinone converts spontaneously and rapidly to dopachrome, resulting in formation of a brown–black-coloured product (65). The first step of this reaction sequence is very slow, so the method described here uses dopa as the substrate, and is based on work originally published by Holstein *et al.* (66).

---

**Protocol 7.** Demonstration of tyrosinase activity in an SDS gel

1. Cast a 10% *T*, 2.66% *C* SDS separating and stacking gel as described in *Protocol 2*.
2. Mix the sample with the following SDS-containing buffer: 9.87 g Tris–HCl, 20.0 g SDS, 75 ml glycerol, adjust to pH 6.8 and dilute up to one litre. It should be noted that boiling and addition of reducing agents such as mercaptoethanol will inactivate the enzyme. Add a tracking dye [up to 5% of the sample volume, such as 0.1% (w/v) bromphenol blue].
3. Load the sample and apply the current. For example, a 140 × 140 × 0.75 mm slab gel can be subjected to electrophoresis at 100 V, constant voltage. Stop the run when the tracking dye reaches the bottom of the gel slab. Pry the plates apart, remove the gel and the remaining backing sheet.
4. Equilibrate the gel and backing sheet for 15 min at room temperature in 0.1 M phosphate buffer, pH 6.8. Agitate occasionally.
5. Incubate the gel and backing sheet at room temperature in a solution containing 0.1 mg L-dopa/ml phosphate buffer. An active sample will produce a brown precipitate after about 5 min. Samples with weaker enzymatic activity may require incubation times up to 2 h. To detect the enzyme protein as well as other proteins present in the sample, the proteins are fixed and stained by the procedure described in Section 6.
6. To preserve stained gels cross-linked with AcrylAide incubate the destained gel in 5% (v/v) glycerol and air dry. Alternatively, the gel can be rapidly dryed in a microwave oven (67).

---

An example of the method is shown in *Figure 1*.

## 11.2 Transferases

Transferases transfer a group such as a methyl-, glycosyl-, or phosphate-group from a donor to an acceptor. Classification of this type of enzymes is

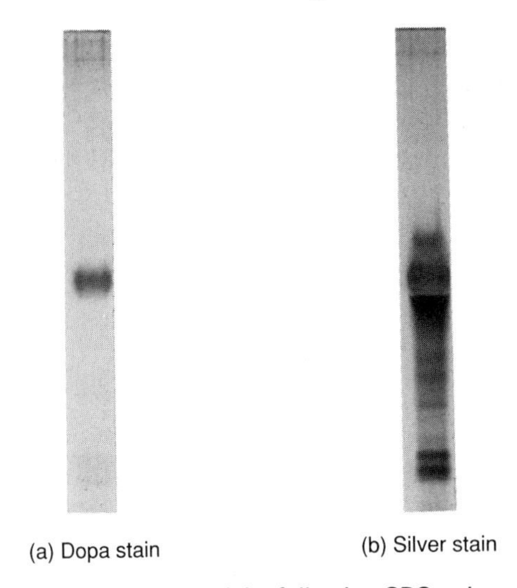


**Figure 1.** Demonstration of tyrosinase activity following SDS-polyacrylamide gel electrophoresis. Specimen is an extract of B16 murine melanoma cells. Lane a—Dopa stain indicating a single band of enzymatic activity. Lane b—same preparation using silver stain to demonstrate total protein. Figure courtesy of Dr V. J. Hearing, reference 65.

based on the scheme 'donor:acceptor grouptransferase'. In many instances, the donor is a coenzyme that contains the group to be transferred. Usually, the detection of these enzymes involves coupling of the reaction product to an auxiliary enzyme which in turn can be visualized by a direct capture reaction. As an example, in the detection of hexokinase (ATP:glucose phosphotransferase), the enzyme is first incubated with ATP, $Mg^{2+}$, and glucose to yield the product glucose-6-phosphate. Glucose-6-phosphate is in turn detected with the auxiliary enzyme glucose-6-phosphate dehydrogenase, an NADP-requiring enzyme. The formation of NADPH is then used to follow the technique described for oxidoreductases which leads to formation of insoluble deeply-coloured formazan. The application of auxiliary enzymes is usually carried out with overlay techniques using a paper or an agar overlay in a sandwich-type incubation.

### 11.2.1 Detection of hexokinase

The incubation mixture is prepared just prior to use. The gel is either directly incubated in the solution or with a filter paper soaked in the solution and placed in a sandwich-type incubation. In either case, the incubation is carried out shielded from light. The incubation mixture contains 0.2 mM $NADP^+$, 10 mM $MgCl_2$, 8 mM ATP, 0.16 mg/ml PMS, 0.1 mg/ml NBT, 0.4 units/ml glucose-6-phosphate dehydrogenase, 0.5 mM glucose, 0.05 M Tris buffer, pH 8.0. Incubation is carried out at 37°C. Termination of incubation and washing of the gel is carried out in 10% acetic acid. Store in the dark.

### 11.2.2 Detection of glutamate-pyruvate transaminase

It should be noted that the strategy employed in this reaction is useful for all NAD-requiring dehydrogenases with formazan formation as the mode of detection. The glutamate–pyruvate transaminase uses L-alanine and alpha-keto glutaric acid as the substrate with pyridoxal phosphate as the coenzyme of the reaction. L-glutamic acid, one of the products of the reaction, is coupled with glutamic dehydrogenase. The substitution of $NAD^+$ in the reaction by its 3-acetylpyridine analogue, 3-acetylpyridine adenine dinucleotide (APAD), changes the equilibrium of NAD-dependent reactions. The more positive redox potential of $APAD^+$/ADAPH as compared to $NAD^+$/NADPH displaces the equilibrium of dehydrogenase reactions in favour of substrate oxidation. This provides the proper conditions for the oxidation of L-glutamate resulting in formation of APADH. Phenazine methosulphate mediates hydrogen transfer from APADH to a tetrazolium salt [3-(4,5-dimethylthiazol-2-yl)-2,5-diphenyltetrazolium bromide (MTT)] which leads to the formation of the corresponding formazan.

A summary of the coupled reactions involved is shown below:

$$\text{L-alanine} + \alpha\text{-ketoglutarate} \rightarrow \text{pyruvate} + \text{L-glutamate}$$

$$\text{L-glutamate} + \text{APAD}^+ \rightarrow \alpha\text{-ketoglutarate} + \text{APADH} + \text{NH}_4^+$$

$$\text{APADH} + \text{MTT} \xrightarrow{\text{PMS}} \text{APAD}^+ + \text{Formazan}$$

The gel is incubated at 37°C in a container shielded from light, covered with filter paper soaked with a solution containing the components described below. The rate of colour formation depends on gel thickness and enzyme concentration.

Staining solution prepared just prior to use: 4 ml of 0.4 M Tris–HCl buffer, pH 8.2, containing 60 mg of L-alanine, 7 mg of α-ketoglutaric acid, 1.5 mg of $APAD^+$, 0.4 mg of PMS, 0.6 mg of MTT, 0.2 mg of pyridoxalphosphate, and 2 units of L-glutamic dehydrogenase. After staining, the gel is washed with 10% (v/v) acetic acid and stored in the dark.

### 11.2.3 Detection of aspartate transcarbamylase (69)

Gels are incubated in 0.1 M glycine/NaOH buffer, pH 10, containing 1 mM aspartate, 0.2 mM carbamyl phosphate, and 2 mM $CaCl_2$. This is an example of transferase localization that is applicable to various other enzyme classes such as decarboxylases, synthases, phosphatases, and pyrophosphatases, and is based on the release of phosphate, pyrophosphate, or $CO_2$ as reaction product. Calcium ions are used to form insoluble salts. The visual detection of the calcium precipitates can be further enhanced by Alizarin red (68).

## 11.3 Hydrolases

These comprise a large group of enzymes which catalyse the addition of the elements of water to various linkages, resulting in the cleavage of the bond. Many chromogenic substrates are commercially available and the progress of cleavage can be followed directly in the visible or ultraviolet light. Substrates or products that change their fluorescence are also widely used and render this a very sensitive technique. Some of these enzymes release inorganic phosphate, inorganic pyrophosphate, or $CO_2$ as their products and can be observed by precipitation with $Ca^{2+}$. (See also examples Section 11.2.3.) Generally, hydrolases are more stable than any of the other groups of enzymes described so far.

### 11.3.1 Neuraminidases

---

**Protocol 8.** Detection of neuraminidases (sialidases) (70)

Gels are stained with 2′-(methylumbelliferyl)-α-D-N-acetylneuraminic acid (Koch-Light Ltd.). Staining for enzymatic activity can be carried out in several ways.

*A*

1. After the electrophoretic separation, incubate the gel in buffer (0.2 M acetate buffer, pH 5.0, containing 5 mM $CaCl_2$) for 15–20 min.
2. Spray the methyl umbelliferyl derivative of neuraminic acid (1 mM dissolved in the buffer just described) onto the gel.
3. Observe the appearance of a fluorescent band by UV radiation at 366 nm.
4. To intensify low fluorescence, incubate the gels with 0.133 M glycine buffer, pH 10, containing 0.06 M NaCl and 0.04 M $Na_2CO_3$ for 2 min.
5. Photograph the fluorescent band, then stain the gel with Coomassie brilliant blue.

*B*

1. Preincubate the gel with the acetate buffer as just described and then cover with a nylon sheet (Sartorius SM 20006, Goettingen, FRG) presoaked in 5% (v/v) glycerol.
2. Dry and immerse in a solution of the above acetate buffer containing 0.05 mM methyl umbelliferyl neuraminic acid derivative and dry again.
3. Roll this sheet on the gel and incubate for 1–3 min at 70°C.

---

It should be noted that the fluorogenic method permits detection of neuraminidase in the microunit range, far below the detection limit for proteins by conventional staining.

The same technique can be used for various glycosidases, sulfatases, esterases, and phosphatases. Many umbelliferyl derivatives are commercially available for this purpose (Sigma. Chem. Co., St. Louis, MO, USA). The staining technique can be carried out in an analogous manner to the one just described or as follows: Soak Whatman Number 1 filter paper in 0.1 M sodium citrate buffer, pH 4.0, containing the umbelliferyl derivative (0.5 mg/ml). Cut the paper to the same size as the gel and cover the gel with the filter paper and wrap into a saran wrap. The gel is viewed in the UV light during incubation at room temperature. Fluorescence can be enhanced by spraying with ammonia.

### 11.3.2 Detection of alkaline phosphatase enzymes (71)

*i. Indoxyl phosphate-formazan staining*

Stain the gels in a solution of 50 ml 0.1 M Tris, containing 5 mM $MgSO_4$ and 500 μg/ml of indoxyl phosphate (5-bromo-4-chloro-3-indolyl phosphate) and 100 μg/ml MTT [3-(4,5-dimethylthiazolyl-2)-2,5-diphenyltetrazolium-bromide]. Incubate at 37°C for up to 16 h. Acceleration of the staining process can be obtained by the addition of phenazine methosulphate (50 μg/ml). Under these conditions the reaction is carried out in the dark. Enzymatic activity appears as blue bands.

*ii. Alpha naphthyl phosphate–Fast Red (diazo) staining*

The gels are placed in a 0.05 M bicarbonate/carbonate buffer, pH 9.6, containing 0.001 mM $ZnCl_2$, 0.1 mM $MgCl_2$, 20 mM sodium α-naphthyl phosphate and Fast Red (5-chloro-2-toluenediazonium chloride) 1 g/l (contains only about 15% active dye). The gels are kept at 4°C overnight. Bands of enzymatic activity are red–brown.

*iii. Silver-staining*

---

**Protocol 9.** Beta-glycerophosphatase-silver staining

In this procedure, steps 3–8 must be carried out in a dark room with a yellow safelight.

1. Incubate gels at 37°C in a closed box in 0.1 M Tris containing 0.185 M sodium β-glycerophosphate, 20 mM $MgSO_4$, and 60 mM $CaSO_4$.
2. Wash gels successively for 15, 30, and 45 min, using each time a 200-fold volume of water in a rotary shaker.
3. Immerse the washed gels in 0.2 M silver nitrate for one 1 h.
4. Repeat step 2.
5. Expose the gels to long-wave UV light or bright sunlight for 5 min. Areas of enzymatic activity stain as brown–black bands.
6. Intensification and increased sensitivity can be accomplished by immersion

of gels for 20 min into a freshly-prepared solution of 10 ml of water containing 500 mg of powdered gum acacia and 10 ml of water containing 5 mg of silver lactate, 100 μl of 15 mM dodecylamide, and 100 μl of 0.05% (v/v) Triton X-100.

7. Remove the gels and place into 2.5 ml of 2 M citrate buffer, pH 4.0. Add 1 ml of 0.6 M quinol and keep the gels in this developer for 2 min.
8. Wash the gels in running water for 10 min, after which they can be exposed to light.

---

The diazo technique is the least sensitive but the simplest technique. The formazan technique is more sensitive and is almost as simple to carry out. The silver technique is the most sensitive but the many steps required makes it less attractive. In addition, samples with very low enzymatic activity require long incubation time with β-glycerophosphate (up to 100 h).

Examples for the use of sandwich type incubation with agar overlays are described below

### 11.3.3 Lipases

---

**Protocol 10.** Detection of lipase (72)

1. Dissolve 0.4 g of agar by heating in a water bath in 20 ml of 0.1 M sodium succinate buffer, pH 6.0.
2. Cool the mixture to 60–65 °C, and add 4 mg of rhodamine B and 0.5 g of triolein and emulsify for 1 min in a homogenizer.
3. De-gas the mixture and pour it into a pre-heated chamber consisting of two polytetrafluoroethane-coated glass plates (200 $mm^2$), separated at the edges by a flexible silicone rubber, 7 mm wide and 0.7 mm thick. The chamber is held together with spring clips.
4. When the gel is set, carefully remove the glass plates and replace with transparent plastic sheets.
5. Layer carefully the substrate slab on top of the electrophoretogram avoiding formation of air bubbles.
6. Saran wrap (cling film) on top to prevent drying.
7. Incubate the sandwich at 37°C. Light pink bands (fluorescent at 366 nm) appear on a deep red gel.

---

### 11.3.4 Proteases

---

**Protocol 11.** Detection of protease (73)

1. Suspend agar, 1.5% (w/v) and gelatin 0.5% (w/v) in 20 mM Tris–HCl buffer, pH 8.8, containing 1 mM $NiCl_2$ and 0.02% (w/v) $NaN_3$.

**Protocol 11.** *Continued*

2. Heat the mixture in a boiling bath until melted and subsequently degassed.
3. Pour the solution in a pre-heated chamber (100°C), as described in Protocol 10, step 3.
4. Incubate for 30 min at 37°C the sandwich formed between the electrophoretic gel and the indicator gel.
5. Immerse the indicator gel into saturated $(NH_4)_2SO_4$ for 5–10 min.
6. Place between flexible clear polypropylene sheets and photograph against a dark background.

## 11.4 Lyases

Lyases catalyse the cleavage of bonds other than by the addition of water. In some instances, the altered chemical property of the cleavage product lends itself to specific coupling with a chromogenic reagent.

### 11.4.1 Argininosuccinate lyase (74)

This method illustrates the coupling of several enzymatic steps to the indicator enzyme glutamate dehydrogenase as follows:

$$\text{Argininosuccinate} \xrightarrow{\textit{Argininosuccinase}} \text{fumarate} + \text{L-arginine}$$

$$\text{L-arginine} \xrightarrow{\textit{Arginase}} \text{L-ornithine} + \text{urea}$$

$$\text{Urea} + H_2O \xrightarrow{\textit{Urease}} CO_2 + 2NH_3$$

$$\alpha\text{-ketoglutarate} + NH_4^+ + \text{NADH} \xrightarrow{\textit{Glutamate dehydrogenase}} \text{L-glutamate} + H_2O + NAD^+$$

**Protocol 12.** Detection of argininosuccinate lyase

1. The staining solution: 50 g of barium argininosuccinate, 1 mg (40 units) of arginase, 2 mg (20 units) of urease, 25 mg of α-ketoglutarate, 10 mg of NADH, and 50 μl of glutamate dehydrogenase (500 units/ml) in 5 ml of 0.1 M Tris–HCl, pH 7.6.
2. Soak the solution onto a filter paper (Whatman 3MM) and place on the surface of the slab gel.
3. Incubate the sandwich at 37°C and examine for non-fluorescent quenched bands when illuminated with UV light.
4. Keep a photographic record.

5. Glutamate dehydrogenase and lactic acid dehydrogenase bands may appear in crude extracts. These bands are visible on the gel before they can be seen on the paper, in contrast to argininosuccinase bands.

### 11.4.2 dTDP glucose 4,6-hydro-lyase

The method described below is a general method, applicable to various enzymes that convert a nonreducing substrate into reducing products. The example chosen converts dTDP-glucose to dTDP-4-keto-6-deoxy glucose (75).

**Protocol 13.** Detection of dTDP glucose 4, 6-hydro-lyase

1. Cover the slab gel with a filter paper soaked with a solution containing 6 mM dTDP-glucose and 20 mM Tris buffer, pH 8.0.
2. Incubate the sandwich for 10 min at 37°C.
3. Rinse the gel with water and immerse into a freshly-prepared solution of 0.1% (w/v) TCC (2,3,5-triphenyltetrazolium chloride) in 1 M NaOH.
4. Keep the gel at room temperature for 5–10 min in this solution and keep exposure to light during the staining to a minimum.
5. Wash the gel extensively in 7.5% acetic acid. After removal of TCC, the same gel can be stained for protein with Coomassie blue.

## 11.5 Isomerases

As indicated by the name, these enzymes result in formation of isomers that involve a molecular rearrangement (e.g. glucose-6-phosphate isomerase, conversion between a keto- and aldehydo-hexose phosphate). These enzymes have recently attracted additional interest since many isozyme forms exist that have genetic and clinical importance.

### 11.5.1 Glucose-6-phosphate isomerase

The method is suitable for the rapid staining of enzymes in gels. It requires higher concentrations of reactants to enhance the reaction and diminish diffusion, thereby increasing the resolution. The method can be readily modified for the visualization of several enzymes (76). The reaction is coupled to glucose-6-phosphate dehydrogenase as in *Protocol 14*.

$$\text{Fructose-6-phosphate} \rightarrow \text{glucose-6-phosphate}$$
$$\text{glucose-6-phosphate} + NADP^+ \rightarrow \text{6-phosphogluconolactone} + H^+ + NADPH$$

$$NADPH + MTT \xrightarrow{PMS} NADP^+ + \text{formazan}$$

**Protocol 14.** Detection of glucose-6-phosphate isomerase

1. Dissolve 18 g of Tris and 1 g of $NaN_3$ in 1 l of water.
2. Take 1.6 ml of this solution and dilute with 6.0 ml of water.
3. Take one half of the solution (3.8 ml) and dissolve 60 mg of agarose by heating in a water-bath.
4. In the other half of the solution (3.8 ml) dissolve 350 mg of fructose-6-phosphate, 50 mg of $NADP^+$, 10 mg of PMS, 25 mg of MTT, and add 0.1 ml of glucose-6-phosphate dehydrogenase (1 mg/ml).
5. Cool the heated agarose gel solution to 60°C and mix with the substrate–cofactor solution just prepared.
6. Pour the mixture on a GelBond sheet and allow to set.
7. Overlay the indicator gel on the separating gel and incubate for 2–4 min at 60°C.
8. Remove the indicator gel and dry. Bands of enzymatic activity appear blue-violet.

## 11.6 Ligases

Ligases are enzymes that establish a new bond between two molecules at the expenditure of a nucleoside triphosphate.

### 11.6.1 Pyruvate carboxylase (77)

The reaction is carried out in the presence of a positive allosteric effector, acetyl-CoA, and leads to the formation of oxaloacetic acid. Coupling of the reaction product oxaloacetate with Fast Violent B (6-benzamido-4-methoxy-*m*-toluidine diazonium chloride) results in a red colour.

**Protocol 15.** Detection of pyruvate carboxylase.

1. Incubate the electrophoresis gel at room temperature in 100 mM Tris–HCl, pH 7.8, 5 mM sodium pyruvate, 2 mM ATP, 5 mM $MgCl_2$, 50 mM $KHCO_3$, 0.3 mM acetylCoA.
2. Rinse the gel in water and transfer into a solution of Fast Violet B (1 mg/ml) for 15 min in the dark.
3. Minimize exposure to light, wash gel with water.

## 11.7 Enzymes that modify nucleic acids

There has been a recent surge of interest in electrophoresis of enzymes that modify nucleic acids. We shall use these as an example of an enzyme assay employing radioactivity. In order to demonstrate enzymatic activity in the gel following electrophoresis, an overlay gel containing nucleic acid substrate is sometimes used (78), or the electrophoresis gel itself is pre-cast to contain high-molecular-weight DNA. Those pre-impregnated gels in which enzyme assays are to be performed have been termed 'activity gels'. They usually use SDS-based separations, thus renaturation is frequently necessary.

Bertazzone *et al.* (79) have reviewed the different types of nucleic-acid-modifying enzymes which have been studied using activity gels. They include DNA polymerase, primase, ligase, methyltransferase, terminal transferase, DNase, RNase, RNA polymerase, and alkaline phosphatase (DNA substrate). The assays generally use either radioactive nucleic acid in the gel and measure release into an acid-soluble form, or use radioactive precursors and measure incorporation into an acid-insoluble form.

---

**Protocol 16.** Detection of DNA polymerase in an activity gel (80, 81)

1. Cast 12 ml of a 7.5% *T*, 2.66% *C* SDS gel. Use the recipe given in *Protocol 2*, with the following modifications:
   (a) Mix 3.0 ml of acrylamide solution, 2.4 ml of separating gel buffer, 5.0 ml of water, 0.5 ml of 'gapped' DNA solution. Do not de-gas. Add 100 μl of persulphate solution and pour between the plates.
   (b) Prepare the solution of 'gapped' DNA by treating calf thymus DNA (3 mg/ml) with bovine pancreatic DNase I, according to (81). After DNase cleavage, shear the DNA further by filtering through a 0.45 micron millipore filter. Dilute the DNA to a final concentration of 0.6–1.8 mg/ml in water. Use 0.5 ml of this in step 1 (a).
2. Perform the electrophoresis
3. Renature the DNA polymerase by soaking the gel in 50 mM Tris–HCl, pH 7.5, 10 mM 2-mercaptoethanol at 37°C for 60 min to remove SDS. Then incubate in 50 mM Tris–HCl, pH 7.5, 5 mM 2-mercaptoethanol, 1 mM EDTA at 4°C for 30 h.
4. Incorporate radioactive precursors into high-molecular-weight DNA by incubating the gel in 70 mM Tris–HCl, pH 7.5, 7 mM $MgCl_2$, 10 mM 2-mercaptoethanol, 45 μCi $\alpha^{32}$P-dTTP (3000 Ci/mmol), 14 μM each of dATP, dCTP, dGTP at 37°C for 18 h.
5. Wash out unincorporated radioactivity by incubating in 5% (w/v) TCA, 1% (w/v) sodium pyrophosphate. This should require at least three changes until the count rate in the wash reaches background.
6. Dry the gel.
7. Expose the radiofluorograph by placing a sheet of Kodak XAR-5 (or its equivalent) film between the dried gel and an intensifying screen (Dupont Extra Lite, or Cornex, or equivalent). Store the cassette at −70°C until development.

---

One recent study uses a non-radioactive overlay gel to assay DNase isozymes (78). In that study, an agarose gel impregnated with high-molecular-weight DNA is abutted against the separating gel. DNA which has been degraded will not intercalate ethidium bromide, hence there is an absence of fluorescence above the bands containing active enzyme.

## 12. Other enzymology-after-electrophoresis techniques

The primary concern of this chapter has been the identification of proteins separated using electrophoresis by demonstration and analysis of their enzymatic activity. For completeness, some other techniques which come under the umbrella of enzymology after electrophoresis should be included.

The most commonly used method is the Cleveland technique (47). In this technique, peptide mapping is performed on a single component of a mixture of proteins resolved by SDS polyacrylamide gel electrophoresis. Following SDS electrophoresis, the gel is stained, and the band containing the protein of interest is excised. This band is re-equilibrated in SDS-containing buffers and inserted into the well of a stacking gel on top of a second SDS separating gel. The excised block is overlaid with a solution containing proteolytic enzymes (V8 protease, chymotrypsin, or papain). The target protein and the enzyme are driven into the stacking gel by the current. The power is turned off, and digestion proceeds in the gel. Following digestion the current is re-applied, and the cleavage products are resolved in the separating gel.

A second variation of enzymology after electrophoresis uses glycosyltransferases to identify those glycopeptides in an SDS gel which have the appropriate carbohydrate moieties to serve as acceptors for β (1–4) galactosyltransferase (82). In this application glycoproteins separated in SDS gels are transferred to nitrocellulose by electroblotting. The nitrocellulose is incubated with galactosyltransferase in the presence of tritiated UDP-galactose, and radioactivity is then located by autoradiography. Thus those bands in the pattern which can serve as acceptors are identified.

A third variation of enzymology after electrophoresis has been used in those situations where enzymatic activity following separation in SDS gels cannot be restored to the proteins by renaturation. This procedure uses specific antibodies in the manner of an immunoblot. Nitrocellulose membranes containing the blotted enzyme are incubated in a large excess of antibody. The antibody excess ensures that antibody is attached to the inactive enzymes by only one combining site. This leaves free the second combining site of the antibody. The membrane is then incubated in a solution containing native enzyme, which is captured by the bound antibody at the location of the SDS-inactivated enzyme. Zymography can then be performed as if the separation and blotting conditions had been non-destructive.

Using this antibody capture technique, enzymatic activity has been demonstrated for polymerase, α-1-esterase, and phosphodiesterase following SDS-electrophoresis (49, 54, 83).

## Acknowledgement

We are very grateful to Dr E. J. Zapolski, National Institutes of Health, Bethesda, MD 20892, for his valuable input and critical reading of the manuscript.

## References

1. Gabriel, O. (1971). In: *Methods in Enzymology*, (ed. W B. Jakoby), Vol. 22, p. 578. Academic Press, New York and London.
2. Heeb, M. J. and Gabriel, O. (1984). In: *Methods in Enzymology*, (ed. W. B. Jakoby), Vol. 104, p. 416. Academic Press, New York and London.
3. Hames, B. D. and Rickwood, D. (1981). *Gel Electrophoresis of Proteins. A Practical Approach*. IRL Press, Oxford.
4. Righetti, P. G. and Dysdale, J. W. (1976). In: *Laboratory Techniques in Biochemistry and Molecular Biology*, (ed. T. S. Work), Vol. 5, p. 424. North-Holland Publishing Company, Amsterdam, London.
5. Righetti, P. G. (1981). In: *Electrophoresis '81*, (ed. R. C. Allen and P. Arnaud), p. 3. Walter de Gruyter, Berlin.
6. *The Agarose Monograph* (1988). (4th edn). FMC Bioproducts, FMC Corp., Rockland, Maine, USA.
7. Baldo, B. A. and Tovey, E. R. (ed.) (1989). *Protein Blotting*. Karger, Basel.
8. Davis, B. J. (1964). *Ann. N.Y. Acad. Sci.*, **121,** 404.
9. Ornstein, L. (1964). *Ann. N.Y. Acad. Sci.*, **121,** 321.
10. Laemmli, U. K. (1970). *Nature*, **227,** 680.
11. Chrambach, A. (1985). *The Practice of Quantitative Gel Electrophoresis*. VCH Publishers, Weinheim.
12. Gersten, D. M., MacGregor, C. H., McElhaney, G. E., and Ledley, R. S. (1982). *Electrophoresis*, **3,** 231.
13. Allen, R. C., Popp, R. A., and Moore, D. J. (1965). *J. Histochem. Cytochem.*, **13,** 249.
14. Gabriel, O. (1971). In: *Methods in Enzymology*, (ed. W. B. Jakoby), Vol. 22, p. 565. Academic Press, New York and London.
15. Brewer, J. M. (1967). *Science*, **156,** 256.
16. Allen, R. C., Budowle, B., Saravis, C. A., and Lack, P. M. (1986). *Acta Histochem. Cytochem.*, **19,** 637.
17. Radola, B. J. (1989). *Electrophoresis*, **1,** 43.
18. Gombocz, E. and Chrambach, A. (1989). *Applied and Theoretical Electrophoresis*, **1,** 109.
19. Prestidge, R. L. and Hearn, M. T. W. (1979). *Analyt. Biochem.*, **97,** 95.
20. Gorg, A., Fawcett, J. S., and Chrambach, A. (1988). In: *Advances in Electrophoresis*, (ed. A. Chrambach, M. J. Dunn, and B. J. Radola), Vol. 2, p. 1. VCH Press, Weinheim.

21. Frey, M. D., Atla, B. J., and Radola, B. J. (1984). In: *Electrophoresis '84*, (ed. V. Neuhoff), p. 122. Weinheim; Dearfield Beach, Fl.
22. Saravis, C. A. and Zamchek, N. (1979). *J. Immunol. Meth.*, **29**, 91.
23. Vesterberg, O. (1973). *Ann. N.Y. Acad. Sci.*, **209**, 23.
24. Nguyen, N. Y. and Chrambach, A. (1977). *Analyt. Biochem*, **82**, 226.
25. Gianazza, E., Astrua-Testori, S., and Righetti, P. G. (1985). *Electrophoresis*, **6**, 113.
26. Guevara, J., Johnston, D. A., Ramagli, L. S., Martin, B. A., Capitello, S., and Rodriguez, L. V. (1986). *Electrophoresis*, **3**, 197.
27. Beisiegel, U. (1986). *Electrophoresis*, **7**, 1.
28. Southern, E. M. (1975). *J. Mol. Biol.*, **98**, 503.
29. Renart, J., Reiser, J., and Stark, G. R. (1979). *Proc. Natl. Acad. Sci. (USA)*, **76**, 3116.
30. Montelaro, R. C. (1987). *Electrophoresis*, **8**, 432.
31. Peferoen, M., Huybrechts, A., and DeLoff, A. (1982). *FEBS Lett.*, **145**, 369.
32. Towbin, H., Staehlin, T., and Gordon, J. (1979). *Proc. Natl. Acad. Sci. (USA)*, **76**, 4350.
33. Ohlsson, B. G., Westrom. B. R., and Karlson, B. W. (1987). *Electrophoresis*, **8**, 415.
34. Hagar, D. A. and Burgess, R. R. (1980), *Analyt. Biochem.*, **109**, 76.
35. Karsnas, P. and Roos, R. (1977). *Analyt. Biochem.*, **77**, 168.
36. Djondjurov, L. and Holzer, H. (1979). *Analyt. Biochem.*, **94**, 274.
37. Lewis, U. J. and Clark, M. O. (1963). *Analyt. Biochem.*, **6**, 303.
38. Stephens, R. E. (1975). *Analyt. Biochem.*, **65**, 369.
39. Nguyen, N. Y., DiFonzo, J., and Chrambach, A. (1980). *Analyt. Biochem.*, **106**, 78.
40. Bhown, A. G., Mole, J. E., Hunter, F., and Bennett, J. C. (1980). *Analyt. Biochem.*, **103**, 184.
41. Tuszynski, G. P., Damsky, C. G., Fuhrer, J. P., and Warren, L. (1977). *Analyt. Biochem.*, **83**, 119.
42. Ziola, B. R. and Scraba, D. G., (1976). *Analyt. Biochem.*, **72**, 366.
43. Stralfors, P. and Belfrage, P. (1983). *Analyt. Biochem.*, **127**, 7.
44. Thelu, J. (1988). *Analyt. Biochem.*, **172**, 124.
45. Mozhaev, V. V., Berezin, I. V., and Martinek, K. (1987). In: *Methods in Enzymology*, (ed. K. Mosbach), Vol. 135, p. 586. Academic Press, New York, London.
46. Sambrook, J., Fritsch, E. F., and Maniatis, T. (1989). *Molecular Cloning, A Laboratory Manual*, (2nd edn). Cold Spring Harbor Laboratory Press, Cold Spring Harbor, NY
47. Cleveland, D. W., Fisher, S. G., Kirschner, M. W., and Laemmli, U. K. (1977). *J. Biol. Chem.*, **252**, 1102.
48. Thomas, O. R. T. and White, G. F. (1989). *Biotechnol. Appl. Biochem.*, **11**, 318.
49. Heegaard, P. M. H. (1988). In: *Handbook of Immunoblotting of Proteins*, (ed. O. J. Bjerran and N. H. H. Heegaard), Vol. 1, p. 221. CRC Press, Boca Raton, FL.
50. Bjerrum, O. J., Selmer, J. C., and Lihme, A. (1987). *Electrophoresis*, **8**, 388.
51. Hodes, M. E., Crisp, M., and Gelb, E. (1977). *Analyt. Biochem.*, **80**, 239.
52. Anfinsen, C. B. (1986). In: *Protein Engineering*, (ed. M. Inouye, and R. Sarma). Academic Press, New York.

53. Lacks, S. A. and Springhorn, S. S. (1980). *J. Biol. Chem.*, **255,** 7467.
54. Muilerman, H. G., Ter Hart, H. G. J., and Van Dijk, W. (1982). *Analyt. Biochem.*, **120,** 46.
55. Weber, K. and Kuter, J. (1971). *J. Biol. Chem*, **246,** 4504.
56. Chang, L. M. S., Plevani, P., and Bollum, F. J. (1982). *Proc. Natl. Acad. Sci. (USA)*, **79,** 758.
57. Givol, D., DeLorenzo, F. E., Goldberger, R. R., and Anfinsen, C. B. (1965). *Proc. Natl. Acad. Sci. (USA)*, **53,** 676.
58. Pigiet, V. P. and Schuster, B. J. (1986). *Proc. Natl. Acad. Sci. (USA)*, **83,** 7643.
59. Szewczyk, B. and Summers, D. F. (1988). *Analyt. Biochem.*, **168,** 48.
60. Pearse, A. G. E. (1975). *Histochemistry. Theoretical and Applied.* Churchill Livingstone, Edinburgh.
61. Troyer, H. (1980). *Principles and Techniques of Histochemistry.* Little, Brown and Co., New York.
62. Seymour, J. L., Lazarus, R. L., and May, J. W. (1989). *Analyt. Biochem.*, **178,** 243.
63. Marshall, J. H., Bridge, P. D., and May, J. W. (1984). *Analyt. Biochem.*, **139,** 359.
64. Liebenguth, F. (1974). *Biochem. Genet.*, **13,** 263.
65. Hearing, V. J. (1987). In: *Methods in Enzymology*, (ed. S. Kaufman), Vol. 142, p. 154. Academic Press, New York and London.
66. Holstein, T. J. (1967). *Proc. Soc. Exp. Biol. Med.*, **126,** 415.
67. Gersten, D. M., Zapolski, E. J., and Ledley, R. S. (1985). *Electrophoresis*, **6,** 191.
68. Nimmo, H. G. and Nimmo, G. A. (1982). *Analyt. Biochem.*, **121,** 17.
69. Greyson, J. E., Jon, R. J., and Butterworth, P. J. (1979). *Biochem. J.*, **183,** 239.
70. Berg, W., Gutschker, G. G., and Schauer, R. (1985). *Analyt. Biochem.*, **145,** 339.
71. Hodson, A. W. and Skillen, A. W. (1988). *Analyt. Biochem.*, **169,** 253.
72. Every, D. (1981). *Analyt. Biochem.*, **116,** 519.
73. Hoeffelmann, M., Kittensteiner-Eberle, R., and Schrier, P. (1983). *Analyt. Biochem*, **128,** 217.
74. Nelson, R. L., Povey, S., Hopkinson, D. A., and Harris, H. (1977). *Biochem. Genet.*, **15,** 1023.
75. Gabriel. O. and Wang, S. F. (1969). *Analyt. Biochem.*, **27,** 545.
76. Kinzkofer, A. and Radola, B. J. (1983). *Electrophoresis*, **3,** 408.
77. Scrutton, M. C. and Fatabene, F. (1975). *Analyt. Biochem.*, **69,** 247.
78. Yasuda, T., Mizuta, K., Ikehara, Y., and Kishi, K. (1989). *Analyt., Biochem.*, **183,** 84.
79. Bertazzoni, U., Scovassi, A. I., Mezzina, M., Sarasin, A., Franchi, F., and Izzo, R. (1986). *Trends in Genetics*, **23,** 67.
80. Spanos, A. and Huebscher, U. (1983). In: *Methods in Enzymology*, (ed. C. H. W. Hirs and S. N. Timasheff), Vol. 91, p. 263. Academic Press, New York, London.
81. Spanos, A., Sedgwick, S. G., Yarranton, G. T., Huebscher, U., and Banks, G. R. (1981). *Nucl. Acid Res.*, **9,** 1825.
82. Parchment, R. E. and Shaper, J. H. (1987). *Electrophoresis*, **8,** 421.
83. Van der Meer, J., Dorssers, L., and Zabel, P. (1983). *EMBO J.*, **2,** 233b.

# 9

# Techniques for enzyme extraction

N. C. PRICE

## 1. Introduction: scope of chapter

This chapter discusses the techniques used to extract enzymes from cells. The principal aim of such procedures is to obtain the enzyme in as high a yield as possible, consistent with the retention of maximal catalytic activity. The various procedures involved in subsequent purification of enzymes to homogeneity, which is necessary for detailed studies of structural and kinetic properties, will not be discussed here; full details can be found elsewhere (1, 2, 3).

It is not always necessary to break open cells in order to obtain enzymes; many enzymes are secreted from cells or tissues and these can be purified directly from a culture filtrate or supernatant. A different type of problem has come to light in attempts to express genes coding for eukaryotic proteins in *E. coli*, either directly or as fusion proteins, where these expressed proteins often appear in an insoluble form as inclusion bodies within the cells (4). Recovery of the proteins in a useful form requires breakage of the cells and solubilization of the protein in a strong denaturing agent such as guanidinium chloride. It is then necessary to devise conditions under which the denatured protein can be re-folded to yield a biologically-active product; this latter stage is often the most difficult part of the entire process of production of the recombinant protein (4).

Section 2 deals with the question of the choice of tissue and methods for disruption of that tissue, separation of cells, and rupture of cells. In addition to general procedures, mention is made of specific procedures for plant cells and microorganisms (fungi, bacteria, and so on) which pose special problems. Plants and bacteria possess tough cell walls which must be broken in order to liberate the cell contents. Fungi possess vacuoles which contain large quantities of proteolytic enzymes which might damage the enzymes which are being extracted. Plant cells often contain large quantities of phenolic compounds which can interfere with extraction of enzymes.

Section 3 is concerned with the protection of enzyme activity during and after disruption of cells. Damage to enzymes can result from a number of causes, such as proteolysis, oxidation of thiol groups, and so on. Although a number of protocols are available to minimize such damage, it must be emphasized that each extraction should be investigated in preliminary experiments in order to establish optimum conditions.

Section 4 deals with assays of enzyme activities in crude (unfractionated) cell extracts. The general principles of enzyme assays have already been discussed in Chapter 1 of this volume; this section will outline a number of special considerations which apply in the case of crude extracts.

In Section 5 the separation of the various subcellular fractions in a cell extract will be considered. Characterization of such fractions is crucially dependent on assays of enzyme activity; a number of 'marker' enzymes serve as indicators of the presence of particular subcellular components.

Finally, in Section 6, the technique of permeabilization of cells is mentioned; this allows assays of enzyme activity to be carried out *in situ* without the need for extraction.

# 2. Disruption of tissues and cells

## 2.1 Choice of tissue

The choice of tissue for extraction of enzymes depends on a number of factors. In many cases, the choice will be made on the grounds of availability, cost, or abundance of enzyme. Thus, heart muscle is an excellent source of the enzymes of the tricarboxylic acid cycle because of the high number of mitochondria in this tissue. Large quantities of heart, brain, or liver tissue are available from meat animals. In other cases it may be important to choose a tissue so that information is obtained which can be compared with that from a previously-studied tissue in the same or another species. For some purposes it might be necessary to choose plants or microorganisms either because the enzymes of interest are unique to such organisms or perhaps because recombinant proteins have been expressed in micro-organisms. The amounts of many enzymes in micro-organisms can be regulated by appropriate choice of components in the growth medium. On account of its economic importance in the baking and brewing industries, yeast (*Saccharomyces cerevisiae*) is available in large quantities (either as a compressed cake or in a dried form) and serves as an excellent source of many enzymes, especially those of the glycolytic pathway.

Unless proteinases are themselves the object of interest, it may well be possible to avoid some potential problems (see Section 3.3) in enzyme extraction by a suitable choice of source. Thus, certain animal tissues (e.g. liver, spleen, kidney, and macrophages) are rich in lysosomal proteinases (notably cathepsins) and this should be borne in mind when these tissues are used as sources. In the case of micro-organisms, it may be possible to select or construct mutant strains which are deficient in certain proteinases. This approach has been successfully employed in yeast and *E. coli* (5).

## 2.2 Disruption of tissue and separation of cells

In some cases it may be desirable to disrupt tissues and prepare homogeneous populations of intact cells, prior to disruption of these cells. Many types of cells from complex multicellular organisms can also be grown under defined conditions in culture. The preparation of isolated cells offers a number of advantages over intact tissues in terms of the study of, for example, transport properties and response to hormones. It may also be useful to separate the different types of cells from a tissue before performing extractions on these types, to allow comparative information to be obtained which may not be available if the whole tissue is studied.

Cell suspensions can be prepared from tissues by mechanical or enzymatic methods, or a combination of the two (6). Mechanical methods such as shaking or loose homogenization often damage the integrity of the cells, so that enzymatic methods are preferred. In these it is normal to include EDTA to chelate $Ca^{2+}$ ions which are often involved in cell adhesion; similarly, addition of bovine serum albumin is beneficial, possibly by complexing with free fatty acids which might otherwise damage cell membranes. The enzyme which is most frequently employed (6) is collagenase from *Clostridium histolyticum* at a concentration of 0.01–0.1% (w/v) for periods from 15 min–1 h; trypsin, elastase, and pronase have also found application. During the incubation, the tissue is seen to disintegrate and the isolated cells go into suspension.

The cells so obtained can be separated on the basis of a number of properties such as charge and antigenic properties, but most commonly separation is performed on the basis of cell size and density by centrifugation. Full details of the methods are given in a companion volume in this series (7). The media used for centrifugal density gradient separations must fulfil certain conditions, i.e. they must be non-toxic and non-permeable to cells and form iso-osmotic gradients of the appropriate density; solutions of Percoll, Ficoll, and metrizamide have been widely used.

In the migrating-slug stage of the cellular slime mould *Dictyostelium discoideum*, the precursor of the mature spore and stalk cells can be separated by brief (3 min) centrifugation on a pre-formed Percoll density gradient. Identification of cell-types could be made on the basis of enzyme assays (e.g. UDP-galactose polysaccharide transferase only occurs in the pre-spore cells) and of immunological detection of pre-spore vacuole contents (8). The separation of the cell types allows their individual protein compositions and developmental characteristics to be determined. Full details of the methods of isolation of homogeneous cell preparations from different tissues are given in a number of sources (9, 10).

## 2.3 Disruption of cells

A wide variety of methods is available to bring about disruption of cells; some of the principal procedures are listed in *Table 1*. Classification of the methods

**Table 1.** Methods for disruption of cells.

| **Method** | **Underlying principle** |
|---|---|
| *Gentle* | |
| Cell lysis | Osmotic disruption of cell membrane. |
| Enzyme digestion | Digestion of cell wall; contents released by osmotic disruption. |
| Potter–Elvehjem homogenizer | Cells forced through narrow (0.05–0.6 mm) gap between pestle and glass vessel. Cell membranes removed by shear forces. |
| *Moderately harsh* | |
| Waring blender | Cells broken and sheared by rotating blades. |
| Grinding with sand or alumina, or glass beads | Cell walls removed by abrasive action of sand or alumina particles. |
| *Vigorous* | |
| French press | Cells forced through small orifice at very high pressure; shear forces disrupt cells. |
| Explosive decompression | Cells equilibrated with inert gas at high pressure. On release of the contents into atmospheric pressure disruption occurs and the contents are released. |
| Bead mill | Rapid vibrations with glass beads lead to removal of the cell wall. |
| Ultrasonication | High-pressure sound waves cause cell breakage by cavitation and shear forces. |

For further details see reference (3).

is broadly in terms of their harshness. Generally, it is advisable to use a method which is as gentle as possible, consistent with extraction of the enzyme of interest, so as to avoid damage to the enzyme or the release of degradative enzymes from subcellular organelles such as vacuoles or lysosomes. Full details of the uses of the various methods can be found in references (3, 7, 11); some of the more important points are described below in connection with particular types of tissue.

### 2.3.1 Mammalian tissue

Cut the tissue into small pieces and remove as much fat and connective tissue as possible. Soft tissue such as liver can be homogenized in a Potter–Elvehjem homogenizer in which a rotating pestle (Teflon piston attached to a metal shaft) fits into an outer glass vessel. The clearance varies from about 0.05 mm to about 0.6 mm in different types of homogenizers; too tight a fitting can lead to rupture of organelles. For tougher tissues such as skeletal or heart muscle, it is advisable to mince the chopped tissue prior to homogenization in a Waring Blender. (Three or four bursts, each of 15 s, are normally sufficient to give a smooth extract.) Stir the extract for about 30 min to allow further extraction of enzymes, and then centrifuge (10 000 *g* for 20 min) to give a clear extract.

The solution used for homogenization will depend on the nature of the extract required. If it is important to isolate subcellular organelles, iso-osmotic sucrose or mannitol (0.25 M) lightly buffered with Tris, Hepes, or Tes (5–20 mM) at pH 7.4 is generally used. For certain purposes, such as the isolation of mitochondria, it is best to avoid ionic solutions since these can remove peripheral proteins from the membranes. EGTA (1 mM) can be added to remove $Ca^{2+}$ which would lead to uncoupling (7). When it is not important to isolate the intact organelles, the solution used should be chosen to give a good yield of the desired enzyme(s). Thus most soluble enzymes, such as creatine kinase, are extracted from muscle using solutions of low ionic strength (0.01 M KCl). Myosin can be selectively extracted from muscle in solutions of high ionic strength (0.3 M KCl, 0.15 M potassium phosphate.

### 2.3.2 Plant tissues

Plant tissues pose special problems when extracting enzymes, not only because of the presence of the tough cellulose cell wall, but also because of the presence of vacuoles which occupy a large proportion of the total cell volume. Disruption of the vacuoles would lead to the release of proteinases and a lowering of the pH of the extract if it is not adequately buffered. An additional complication is caused by the presence of phenolic compounds in the plant cells, which in the presence of oxygen are converted to polymeric pigments by the action of phenol oxidases. These pigments can adsorb and inactivate enzymes in the extract. In order to minimize these effects it is usual to add a reducing agent such as 2-mercaptoethanol (see Section 3.4) to inhibit the phenol oxidases, and a polymer such as polyvinylpolypyrollidone to adsorb the phenolic polymer.

The extraction of ribulose bisphosphate carboxylase/oxygenase from spinach leaves (12) involves homogenizing the leaves with 2 volumes of a buffer at pH 8 containing 50 mM bicine, 1 mM EDTA, 10 mM 2-mercaptoethanol and 2% (w/v) polyvinylpolypyrollidone in a Waring blender for 40 s at low speed. The resulting extract is filtered through cheesecloth prior to centrifugation at 23 000 *g* for 45 min.

### 2.3.3 Yeasts

Apart from problems caused by the presence of a tough cell wall, yeasts and other fungi contain large amounts of proteinases which could damage enzymes during extraction. As mentioned earlier (Section 2.1), it may be possible to select or construct mutants which are deficient in proteinase production, or to repress the synthesis (or secretion) of proteinases by growth on media which do not contain protein substrates. A number of methods have been used to extract enzymes from yeasts; some of these are listed below:

(a) *Autolysis with toluene.* Treat fresh yeast cake with a small amount of toluene [6% (v/w)] and 2-mercaptoethanol [0.2% (v/w)] and incubate the mixture at 37°C for about 1 h until the yeast forms a smooth liquid due to extraction of the cell wall components. Add a solution (10 times the volume of toluene) of EDTA (15 mM) and 2-mercaptoethanol (5 mM), adjusted to pH 7.0, and stir the mixture at room temperature overnight to allow degradation of the cell wall by the action of endogenous enzymes (autolysis). The extract can be clarified by centrifugation at 15 000 *g* for 30 min (13). This method has two disadvantages; the high temperature and presence of toluene may lead to inactivation of some enzymes, and the use of toluene may be unacceptable on safety grounds. Ethyl acetate can be used in place of toluene, but the extraction is not as successful with certain strains of yeasts (3).

(b) *Lysis of sphaeroplasts.* Sphaeroplasts are formed by enzymatic removal of the cell walls. Treatment of yeast with snail gut digestive juice (which contains 3-glucanase) in the presence of a high concentration (0.7 M) of a non-permeable substance such as mannitol or sucrose leads to digestion of the cell wall while maintaining the integrity of the plasma membrane and subcellular organelles. Glucanases (such as zymolase) from microorganisms can also be used in this procedure, and sphaeroplasts can be prepared from plant tissues by use of cellulase preparations (11). Subsequent lysis of the sphaeroplasts can be achieved by incubation with DEAE–dextran under iso-osmotic conditions, resulting in the release of vacuoles and mitochondria with negligible contamination by proteinases (14).

(c) *Shaking with glass beads.* This technique involves shaking a suspension of yeast cells with small glass beads (1 mm diameter); subsequent centrifugation gives a clear extract (15). In the small-scale procedure, shake 1 ml of cell suspension at 2500 r.p.m. with 2.5 g glass beads. Maintain the temperature at 10°C during this process. Maximum degrees of extraction are obtained after 20 min shaking. Of the 27 strains of yeast tested, >80% breakdown was found in 25, and >95% in 20.

Other methods which have been used for disruption of yeasts include treatment for 18 h with 1 M ammonia solution (16). Although this method is effective in causing disruption of the cells, it is likely that many enzymes are inactivated under these alkaline conditions (pH 10). A freeze–thaw method has also been used successfully and the low temperatures involved have been claimed to be an important factor in minimizing damage caused by endogenous proteinases (17).

### 2.3.4 Bacterial cells

Bacteria possess very tough cell walls and vigorous mechanical methods are usually necessary to break these down. Such methods include the French press, explosive decompression, ultrasonication, grinding with alumina, or bead milling. Apart from damage which might be done to the cellular contents, it is not always easy to scale-up these treatments to deal with large amounts of cells. A gentler method of disruption involves the enzymatic breakdown of the cell wall. Gram-positive species (e.g. *Bacillus*, *Micrococcus*, *Streptococcus*) are readily susceptible to the action of lysozyme. Typical conditions involve incubation with hen egg-white lysozyme (0.2 mg/ml) at 37°C for 15 min (3).

Gram-negative bacteria (e.g. *E. coli*, *Klebsiella* spp., *Pseudomonas* spp) are much less susceptible to the action of lysozyme in the absence of additional treatments. A detailed study (18) showed that the digestion by lysozyme could be made much more effective by incorporating firstly, a preliminary washing of the cells in dilute detergent [0.1% (v/v) *N*-lauroyl-sarcosine), and secondly, a mild osmotic shock in which a cell suspension in sucrose (0.7 M), Tris (0.2 M), EDTA (0.04 M) is diluted with 4 volumes of distilled water. The first step may alter the permeability of the outer membrane and the second step involves a destabilization of the lipopolysaccharide-containing cytoplasmic membrane by the high concentration of Tris and EDTA. On dilution, lysozyme molecules are drawn osmotically into the murein layer of the cell wall, thus promoting digestion.

The release of DNA on cell lysis makes the resulting extract highly viscous and this can cause severe problems in subsequent purification steps. Deoxyribonuclease I (10 μg/ml) can be added to degrade the DNA; alternatively, nucleic acids can be precipitated by the addition of protamine (3). Solutions of this highly basic protein should be neutralized before addition to the extract.

### 2.3.5 The degree of cell breakage

Whatever method of disruption is used, it is important to have an estimate of the degree of cell breakage, so that the effectiveness of the procedure can be evaluated and the minimum degree of harshness required can be employed. For suspensions of single-celled organisms (e.g. yeasts, bacteria), an estimate of the degree of cell-breakage can be easily obtained by analysis of the extract using a haemocytometer (15). A quantitative estimate of the release of cell constituents can be obtained by measurement of the protein that is not sedimented by centrifugation (30 000 *g* min) relative to the total protein in the organism (11).

### 2.3.6 Membrane-bound enzymes

Many enzymes occur within cells physically associated with membranes. The association can range from the relatively weak (primarily electrostatic) interactions characteristic of peripheral proteins to the strong (predominantly hydrophobic) interactions of integral proteins (19). The methods used to extract different enzymes from membranes will depend on the mode and strength of the interactions involved.

Peripheral enzymes (e.g. glyceraldehyde 3-phosphate dehydrogenase or aldolase) can be extracted from erythrocyte membranes by treatment with EDTA (0.1 M) or KCl (0.7 M) plus NaCl (0.14 M) (19). The extraction of integral proteins requires more drastic treatments which disrupt the membrane structure (20), such as use of organic solvents (butanol), chaotropic agents ($NaClO_4$, urea), detergents (Triton X-100 or sodium deoxycholate) or enzymes (phospholipases or proteinases). Some of these treatments may cause loss of enzyme activity; many enzymes are readily denatured by the anionic detergent sodium lauryl (dodecyl) sulphate even at 0.1% (w/v). By contrast, non-ionic detergents such as Triton X-100 can normally be tolerated up to at least 2 or 3% (v/v).

An example of the extraction of an integral membrane protein is provided by the $NAD^+$-dependent cytochrome $b_5$ reductase from calf liver microsomal membranes (21). Extraction of the enzyme with Triton X-100 leads to a form of the enzyme of $M_r$ 43 000, whereas extraction by treatment of the microsomes with lysosomal proteinases (cathepsins) gives a smaller form of the enzyme ($M_r$ 33 000). In the latter case, the smaller size is due to release of a hydrophobic portion of polypeptide chain which is responsible for anchoring the enzyme to the membrane.

After extraction with a detergent such as Triton X-100, it may be necessary to remove excess detergent, since this could interfere with subsequent steps; for example, addition of $(NH_4)_2SO_4$ would cause Triton X-100 to separate as a layer on top of the aqueous phase, and this might contain some of the protein of interest (3). Various methods for removal of detergent, including hydrophobic adsorption chromatography and ion-exchange chromatography, have been reviewed by Furth (22), who also lists some of the important properties of a number of detergents. It should be noted that attempted removal of all detergent from some extracted enzymes (e.g. the $Ca^{2+}$-dependent ATPase from sarcoplasmic reticulum) leads to loss of activity and formation of aggregates (23), presumably as a result of association of hydrophobic areas of the protein which would otherwise be in contact with hydrophobic regions of the detergent (or membrane).

## 3. Protection of enzyme activity

During the process of tissue and cell disruption or during subsequent treatment such as subcellular fractionation or chromatography, enzyme activity can be lost for a variety of reasons. It is therefore essential to consider strategies for protection of the activity. In this section some of the more important factors will be considered. [It is important, however, to ensure that the measures taken to protect activity do not interfere with the extraction of the enzyme(s) or its (their) subsequent assay.]

## 3.1 Control of pH

Many enzymes are only active within a fairly narrow range of pH and exposure to pH values outside this range can led to irreversible loss of activity. [Sometimes the marked stability of a particular enzyme can be used to aid purification; thus adenylate kinase is stable at low pH, and incubation of a muscle extract at pH 2 can be used to denature and precipitate unwanted proteins (24)]. It is thus advisable to ensure that a suitable buffer is used during the extraction process. The pH within certain subcellular organelles can differ markedly from neutral pH (e.g. the pH in the interior of vacuoles and lysosomes is estimated to be approximately pH 5) and the buffering capacity used must be sufficient to account for this if these organelles are ruptured. In addition, metabolic processes could continue within an extract affecting the pH. Thus the breakdown of glycogen in muscle extracts could lead to a decrease in pH of one unit due to the accumulation of lactate and pyruvate over a period of 1 h (3). The following factors should be taken into account when the choice of buffer is made; a further discussion on buffers is given in Chapter 11 in this volume.

(a) Over what range of pH is buffering required? (Buffers should not, as a rule, be used outside a range of $\pm 1$ pH unit from the appropriate p*K*a.)

(b) What ionic strength is required to provide adequate buffering capacity and optimum extraction of enzyme?

(c) Would the buffer have any effect on the activity of the enzyme of interest? As an example, polyanionic buffers such as citrate or phosphate could act as chelators of metal ions which might be essential for enzyme action.

(d) Would the presence of the buffer interfere with any subsequent procedures after extraction? High concentrations of polyvalent buffers could interfere with ion-exchange chromatography by competing with the protein for charged sites on the ion exchanger.

(e) Does the pH of the buffer depend markedly on temperature or ionic strength? These effects, detailed in Chapter 11, are often ignored but can be substantial (3).

## 3.2 Control of temperature

During cell disruption, especially using the harsher methods listed in *Table 1*, the temperature can rise considerably (by up to 30°C or greater). In order to avoid such excessive rises in temperature it is advisable to use pre-cooled solutions (and apparatus) and, if necessary, take steps to dissipate heat generated during the extraction.

It is almost axiomatic that the temperature should be kept low (near 4°C) during extraction in order to minimize the rate of denaturation of enzymes and reduce the activity of proteinases (17). However, it should be remembered that there are a number of well-documented examples of enzymes (usually oligomeric) where exposure to low temperatures leads to inactivation and dissociation. This effect could be due to a shift in p*K*a with temperature of an ionizing side-chain involved in the association/dissociation process, or to a weakening of hydrophobic forces at the low temperature (25). It is thus important to check the effect of temperature on the particular enzyme of interest.

## 3.3 Control of proteolysis

The control of the degradation of enzymes by endogenous proteinases during or after extraction represents one of the most difficult challenges in this type of work. Fuller discussions of the problems have been given (3, 5). Indications that proteolysis is a problem include:

(a) the isolation of a particular enzyme or protein in poor yield
(b) loss of enzyme activity on incubation
(c) poor resolution of proteins on SDS–PAGE, reflecting heterogeneity in $M_r$ values
(d) discrepancies between reported and observed properties of proteins

A number of strategies are available to minimize or suppress unwanted proteolysis. These include lowering the temperature so as to slow the action of proteinases, and the use of proteinase-deficient strains or tissues if possible (see Section 2.1). However, the most commonly employed method involves the addition of proteinase inhibitors during extraction and subsequent steps. The major types of proteinases in various tissues and the inhibitor 'cocktails' which can be used to inhibit them are listed in *Table 2* (5).

Some particular points deserve comment:

(a) *Safety*. PMSF is highly toxic and should be handled with care. DCI could be used as an alternative to PMSF (see footnote to *Table 2*). Gloves should be worn when handling both the solid and solutions. DMSO should be handled with care, as it is very easily absorbed through the skin.
(b) *Solubility*. Several of the inhibitors listed in *Table 2* are of only limited solubility in aqueous solvents, and are thus prepared as stock solutions in organic solvents. The volume of this stock solution added during extraction should be kept to a minimum to avoid damage to the enzyme(s) of interest. The maximum solubility of PMSF in aqueous solutions is about

**Table 2.** Inhibitors used to control proteolysis.

| Type of tissue | Major types of proteinases | Inhibitors added | Stock solution (aqueous solution unless otherwise indicated) |
|---|---|---|---|
| Animal tissues | Serine | PMSF (1 mM)* | 0.2 M in methanol |
| | Metallo | EDTA (1 mM) | 0.1 M |
| | Aspartic | Benzamidine (1 mM) | 0.1 M |
| | | Leupeptin (10 μg/ml) | 1 mg/ml |
| | | Pepstatin (10 μg/ml) | 5 mg/ml in methanol |
| | | Aprotinin (1 μg/ml) | 0.1 mg/ml |
| | | Antipain (0.1 mM) | 10 mM |
| Plant tissues | Serine | PMSF (1 mM)* | 0.2 M in methanol |
| | Cysteine | Chymostatin (20 μg/ml) | 1 mg/ml in DMSO |
| | | EDTA (1 mM) | 0.1 M |
| | | E64 (10 μg/ml) | 1 mg/ml |
| Yeasts, fungi | Serine | PMSF (1 mM)* | 0.2 M in methanol |
| | Aspartic | Pepstatin (15 μg/ml) | 5 mg/ml in methanol |
| | Metallo (possibly) | Phenanthroline (5 mM) | 1 M in ethanol |
| Bacteria | Serine | PMSF (1 mM)* | 0.1 M in methanol |
| | Metallo | EDTA (1 mM) | 0.1 M |

* If the use of PMSF is considered undesirable on safety grounds, 3,4-dichloroisocoumarin (DCI) can be used instead. This compound is less toxic than PMSF and is more reactive towards many serine proteinases (50). DCI is, however, much more expensive than PMSF. The stock solution of DCI (10 mM) is prepared in DMSO; the final concentration in the extraction medium should be 0.1 mM. DCI is relatively unstable in aqueous solutions; the half-life at near neutral pH is 20–30 min (50).

Abbreviations: DMSO dimethylsulphoxide
PMSF phenylmethanesulphonylfluoride
E64 L-*trans*-epoxysuccinyl leucylamido (4-guanidino)butane

For further details see reference (5).

2 mM and it is important to note that this decreases markedly as the ionic strength of the solution increases.

(c) *Stability*. PMSF is unstable in aqueous solution with a half-life at 25°C, pH 7.0, of about 30 min (5). Repeated additions of the stock solution of the inhibitor might be advantageous.

Using the data shown in *Table 2*, it should be possible to avoid many of the problems caused by proteolysis, although it must be emphasized that the 'cocktails' represent only general guidelines and should be tested by preliminary experiments in each case.

## 3.4 Protection of thiol groups

The thiol groups of the cysteine side-chains of proteins can be damaged during extraction. Within a cell, the prevailing reducing environment maintains cysteine side-chains in the reduced (–SH) form; however, on cell rupture and exposure to oxygen, there is a tendency for the side-chains to form either disulphide bonds or oxidized species such as sulphinic acids. Traces of heavy metals can catalyse this process by forming complexes with the otherwise rather unreactive oxygen molecule. Protection against such oxidative damage is normally provided by inclusion of a reagent containing a thiol group, such as 2-mercaptoethanol or dithiothreitol. It would also be advisable to add EDTA at a low concentration (e.g. 0.1 mM) to remove any heavy metal ions.

2-Mercaptoethanol is a dense liquid (density 1.12 g/ml) with a most disagreeable odour, and is toxic. It is usually necessary to add it to a final concentration of 10–20 mM to provide protection for thiol groups in proteins for up to 24 h (3). Dithiothreitol, on the other hand, because of its greater reducing power, can provide protection at lower concentrations (1 mM). [The standard redox potential of dithiothreitol at pH 7.0 is quoted as −0.33 V, some 0.12 V more negative than that for cysteine (26).] Dithiothreitol is also much more convenient to handle; it is a white solid with little odour. The principal disadvantage of dithiothreitol is its cost; it is approximately 20 times more expensive to make a solution of 1 mM dithiothreitol than the equivalent volume of 20 mM 2-mercaptoethanol. In a study of the stability of solutions of dithiothreitol, 2-mercaptoethanol, and other reagents containing thiol groups, it was found that the reagents were less stable at higher pH, higher temperatures, in aerated water compared with nitrogen-purged distilled water and in the presence of low concentrations of $Cu^{2+}$ ions. Inclusion of EDTA increased the stability of the solutions (27). These points should be borne in mind when solutions of the reagents are made up or when the reagents are to be used for 'long term' storage of extracts.

## 3.5 Protection against heavy metals

Heavy metals (such as Cu, Pb, Hg, or Zn) can inhibit enzymes, usually by reacting with cysteine side-chains. These metals can arise from the tissue used for extraction, the glassware or distilled water used, or can occur as contaminants in the reagents employed. Inclusion of EDTA ($\leq$1 mM) in the extraction medium will minimize any effects of these heavy metals; however, it is important to check that the EDTA does not remove any essential metal ions which may be required for the activity of a given enzyme (e.g. $Zn^{2+}$ for alcohol dehydrogenase). In the latter case, it may be necessary either to add a lower concentration of EDTA, or to supplement the extraction medium with the specific metal ion required. The supplement should, of course, be of as high a purity as possible, so as to avoid further potential contamination.

## 3.6 Control of free radical formation

Cell extracts prepared by ultrasonic disintegration are susceptible to damage by free radicals, which are thought to originate from the breakdown of $H_2O$ molecules caused by local high temperatures in the solution. From a study of the effects of ultrasound on enzymes (28) it was concluded that damage during extraction could be minimized by sonication of high concentrations of cells, and in the presence of media components such as sugars which could act as radical scavengers. Alternatively, certain gases could provide protection, e.g. $N_2O$ acts as a radical scavenger (29), while $H_2$ influences bubble behaviour, thereby affecting the degree of sonochemical modification of biological molecules (11).

## 3.7 Control of mechanical stress

During cell disruption by harsh techniques such as the French press or sonication, the cell contents are subjected to high pressure, which can lead to inactivation of some enzymes. Jaenicke (30) has shown that the effects of high pressure on enzymes can be complex. For many oligomeric enzymes such as lactate dehydrogenase or glyceraldehyde-3-phosphate dehydrogenase, the effect of moderate pressure (up to about 2 kbar) is to cause reversible dissociation to inactive monomers. At higher pressures the monomers can then aggregate to form a denatured polymer, a process which is normally irreversible.

In practice, therefore, it is important to control the period of time and pressure applied during vigorous disruption procedures in order to minimize potential damage to enzymes. The chosen conditions should however be consistent with the need to obtain adequate degrees of extraction.

## 3.8 Effects of dilution

When a tissue or cell is extracted, there can be a high degree of dilution of the enzymes and proteins within the cell. In some cellular compartments (e.g. the mitochondrial matrix) the protein concentration is estimated to be as high as 500 mg/ml. The concentration might be reduced to 5 mg/ml in a cell extract, and to 5 μg/ml in a solution of pure enzyme used for assays of activity.

In practice, it has been found that many enzymes lose activity fairly rapidly on storage in dilute solution. This effect can often be overcome by inclusion of an 'inert' protein such as bovine serum albumin at a concentration of 1–10 mg/ml in the solution. It is possible that the added protein may help to prevent loss of enzyme by adsorption on the surface of the vessel, or it may act as a 'sacrificial' substrate for proteinases, thereby protecting the enzyme of interest (3).

Alternative protective agents for enzymes include polyols such as glycerol, glucose, or sucrose. The mode of action of these compounds has been investigated in detail by Arakawa and Timasheff (31). In the aqueous medium, the polyol is preferentially excluded from the domain of the protein, resulting in preferential hydration of the protein. The net effect is to make the native state of the protein more stable in the presence of the polyol than in its absence. Glycerol at high concentrations [50% (w/v)] lowers the melting point of aqueous solutions below the normal temperature of most laboratory freezers (−20 °C). Such solutions are suitable for long term storage of proteins, since freezing (which can be damaging to many enzymes) is avoided (3). However, 50% (w/v) glycerol solutions are very viscous and unsuitable for chromatographic procedures. Lowering the glycerol concentration to 20% (w/v) represents a suitable compromise between the need for viscosity to be reduced and for a degree of protection to be maintained. Concentrated solutions of sorbitol may also offer a practical alternative since they provide protection but are much less viscous than solutions of glycerol (32).

An additional consequence of the dilution of cell contents on extraction can be the dissociation of cofactor (e.g. pyridoxal-5′-phosphate for aminotransferases). Apart from the requirement to add the cofactor during assays of enzyme activity, it is possible that the apoenzyme may be less stable than the holoenzyme. This appears to be the case for pyridoxal-5′-phosphate-dependent enzymes where the apoenzyme is more susceptible than the holoenzyme to the action of intracellular proteinases (33). If such an effect is suspected, it would be advisable to include the cofactor in the buffer used for extraction.

## 4. Assays of enzymes in unfractionated cell-extracts

The general principles involved in assaying enzymes have been described in earlier chapters in this volume. The present section will highlight the particular considerations which should be borne in mind when performing assays on crude cell-extracts. Such assays are useful in providing data on the fluxes through metabolic pathways and in helping to formulate and test theories of the regulation of these pathways (34). Thus, data on the $V_{max}$ values of the various enzymes in a pathway can help to pinpoint those enzymes which are present in low amounts and which therefore could act as control points. In addition, a comparison of the kinetic properties of an enzyme in a cell-extract with those of the corresponding purified enzyme can indicate any effects on the enzyme during the purification procedure. This type of comparison is important in order to assess the validity of conclusions about the enzyme in the cell which have been reached from studies of the purified enzyme. Some of the problems which arise in assays of crude extracts are listed below.

### 4.1 The presence of endogenous inhibitors

A crude extract may contain an inhibitor of the enzyme of interest, so that only a low rate is observed during the assay; for example, the concentration of AMP in muscle extracts is sufficient to cause significant inhibition of fructose bisphosphatase (34). Inhibitors of low $M_r$ can be removed by dialysis or gel filtration prior to assay. (Although the latter method is quicker, which can be an advantage if the enzyme of interest is unstable, it does lead to dilution of the extract which has to be taken into account in subsequent calculations. Also, the extracts may have too great a volume to make gel filtration a practical possibility.) Inhibitors of high $M_r$ are often much more difficult to remove and some considerable further purification of the extract may be required; for example, in the purification of the neutral proteinases from mycelial extracts of *Aspergillus nidulans*, an endogenous inhibitor is removed after a heat-treatment step or on prolonged storage of the extract at room temperature (35). The loss of the inhibitor leads to a considerable (8-fold) increase in the total amount of activity compared with the crude extract.

## 4.2 Interference from other reactions

Other enzyme-catalysed reactions taking place in the extract could complicate the assays of the enzyme of interest. In such cases, it may be possible to estimate the 'blank rate' in the absence of the specific substrate and then subtract this rate from the rate measured in the presence of this substrate. Thus, assays of glycogen phosphorylase in muscle extracts are usually performed by measuring the phosphate released in the reaction shown below (in the presence of the activator AMP):

$$(\text{glycogen})_n + \text{glucose-1-phosphate} \rightarrow (\text{glycogen})_{n+1} + \text{Pi}$$

There could be interference, however, from the action of non-specific phosphatases present in the extract. In order to overcome this, the buffer used for extraction contains 35 mM glycerol-2-phosphate. The contribution of the phosphatases can then be assessed by measuring the rate of production of phosphate in the presence of glycogen and AMP, but in the absence of glucose-1-phosphate. By subtracting this rate from that observed in the presence of glucose-1-phosphate, the rate of the phosphorylase-catalysed reaction can be measured (36).

An alternative approach is to inhibit the interfering reaction. Assays of $NAD^+$-dependent dehydrogenases, such as lactate dehydrogenase in muscle extracts, can be interfered with by the activity of the electron transport chain which leads to oxidation of NADH. The latter activity can be eliminated by the addition of suitable electron transport inhibitors, such as 1 mM potassium cyanide (*Care*: *poison*!) (34, 37).

## 4.3 Removal of substrate

The presence of a competing reaction in an extract could reduce the concentration of the substrate available for the enzyme of interest so that the measured activity of that enzyme is reduced. If it is not feasible to inhibit the interfering activity, it may be possible to add a substrate-regenerating system so as to maintain the concentration. In assays of hexokinase in muscle extracts there is possible interference from the presence of ATPases which would lower the concentration of ATP. Phosphocreatine and creatine kinase can be added to the assay system in order to replenish ATP (37). Assays of hexokinase are initiated by addition of glucose and control assays from which glucose is omitted are run in parallel. (See Chapter 1, Section 2.1.1 for a fuller discussion of the effects of substrate depletion.)

### 4.4 Turbidity of extract

In any spectrophotometric assay, problems can arise if the extract to be assayed is turbid or contains high concentrations of an interfering absorbing species. It may be possible to overcome the first of these problems by clarifying the extract by centrifugation, assuming that the enzyme of interest is not sedimented by this procedure. Absorbing species of low $M_r$ can be removed by gel filtration or dialysis, but addition of the extract to the assay solution could still give rise to large background absorbance or turbidity which requires the spectrophotometer to be 'backed off' so that subsequent changes in absorbance due to the enzyme-catalysed reaction are brought on to scale. It is important to check that under these conditions the instrument is still capable of giving accurate readings and that Beer's Law is obeyed (see Chapter 2). For practical purposes, it is often more convenient to arrange to initiate the reaction by addition of a non-absorbing substrate to the solution already containing the extract, so that there will only be a minimal change in absorbance as the reaction is started. Problems with high background absorbance are more acute in assays performed in the far UV where more species are likely to interfere. Thus, for example, assays of enolase (or coupled assays of phosphoglycerate mutase involving enolase), which rely on the increase in absorbance at 240 nm on formation of phosphoenolpyruvate from 2-phosphoglycerate, are prone to interference from absorbance due to large quantities of proteins in an extract.

## 5. Subcellular fractionation

Following extraction of cell contents, it is often desirable to fractionate the extract so as to obtain purified preparations of the various subcellular structures and organelles. This fractionation allows a detailed study of the properties of each type of organelle and the relationships between the different parts of a complex eukaryotic cell in terms of movements of intermediates of metabolism, macromolecules, and so on. In addition, purified preparations of organelles are of considerable value in reconstitution experiments, where, for example, the pathways of translocation and processing of targetted proteins might be studied. The separation of the different types of organelle might also be of value as a first purification step if it is required to purify different isoenzymes which occur in distinct cellular locations.

In order not to destroy the integrity of subcellular organelles it is necessary to ensure that the method of cell disruption (Section 2) is as gentle as possible and that osmotic shock is avoided by the appropriate choice of extraction medium.

Subcellular fractionation is almost always performed by centrifugation, exploiting the differences in sedimentation characteristics or buoyant density of the different fractions. The techniques employed have been described in detail in a companion volume in this series (7) and in several other books and articles (32, 38, 39); these give the necessary details concerning the different types of centrifugation experiments for the separation of subcellular fractions from a number of different cell types.

Characterization of the various fractions obtained by centrifugation is undertaken by a combination of morphological and analytical techniques (32, 40). The former usually involves an assessment of homogeneity and integrity by electron microscopy, whereas the latter involves an analysis of the content of enzymes or other macromolecules, such as DNA. Central to the analysis is the idea of 'marker' enzymes, i.e. that certain enzymes are found exclusively in particular subcellular locations. A list of some of these marker enzymes is given in *Table 3*, together with references for the methods of assay. The assays of these enzymes in any fraction obtained by centrifugation allows a quantitative assessment of the degree of purity of that fraction.

Extraction of enzymes from the preparations of subcellular organelles involves the techniques discussed in Section 2, except that in most cases only gentle methods, such as osmotic shock or mild treatment with detergent, are required to release the contents. In an organelle such as the mitochondrion, different conditions are required to release enzymes from the various locations. Thus, the non-ionic detergent digitonin, at a concentration of 1 mg/ 10 mg protein, releases all the adenylate kinase from the intermembrane space, whereas a concentration approximately 50% higher is required to release monoamine oxidase from the outer membrane. Under these latter conditions, inner-membrane or matrix enzymes, such as cytochrome *c* oxidase or malate dehydrogenase respectively, are only released to a very small extent (41). Release of these matrix and inner-membrane enzymes from mitochondria can be achieved by extraction with Triton X-100, with the former group requiring a lower concentration of the detergent than the latter.

## 6. *In situ* assays using permeabilization techniques

The assay of enzymes within cells or within subcellular organelles is normally prevented by the lack of permeability of the membrane towards the substrate(s) acted on or the products(s) formed. However, there can be many advantages in studying the activity of enzymes in a permeabilized system rather than an extract:

(a) the need for lengthy or damaging extraction procedures is avoided

**Table 3.** Marker enzymes for subcellular components.

| Subcellular component | Marker enzyme(s) | Method of assay reference |
|---|---|---|
| Nucleus | (DNA) | (46) |
| | DNA-dependent RNA polymerase | (47) |
| | $NAD^+$ pyrophosphorylase | (47) |
| Chloroplast | Ribulose bisphosphate carboxylase/oxygenase | (12) |
| Mitochondrion | (outer membrane) Monoamine oxidase | (48) |
| | (inner membrane) Cytochrome *c* oxidase | (48) |
| | (matrix) Glutamate dehydrogenase | (49) |
| | (intermembrane space) Adenylate kinase | (41) |
| Lysosome, vacuole | Acid phosphatase | (46) |
| | Aryl sulphatase | (46) |
| Peroxisome | Catalase | (46) |
| Endoplasmic reticulum | (smooth) Glucose-6-phosphatase | (46) |
| | (rough) Glucose-6-phosphatase plus membrane-bound ribosomes | (46) |
| | (Golgi) Galactosyl transferase | (46) |
| Plasma membrane | 5′-Nucleotidase | (46) |
| Cytosol | Lactate dehydrogenase | (37) |

(b) the enzymes can be studied in a system which more closely resembles their natural state in terms of cellular organisation

(c) dilution of the cell contents, which occurs on extraction, is avoided (see Section 3.8)

Various methods have been described for rendering different types of cells permeable to substrates and products. Thus, a number of Gram-negative bacteria (e.g. *Pseudomonas aeruginosa*, *Klebsiella aerogenes*) can be permeabilized by treatment with 5% (v/v) toluene at 37°C for 10 min (42). The structural detail within the cells is preserved, as is the activity of all the enzymes tested. In addition, the inhibition of citrate synthase activity by NADH is the same in permeabilized cells as in sonicated extracts, allowing conclusions to be extrapolated from studies *in vitro* to the situation *in vivo* (42).

Yeast (*Saccharomyces cerevisiae*) cells can be permeabilized by treatment with nystatin (a polyene fungicide produced by *Streptomyces noursei*). Low $M_r$ substances but not macromolecules can then pass in and out of the cells. The high concentrations of enzymes remaining within the cells favours the association between the enzymes arginase and ornithine carbamoyltransferase which serves to inhibit the activity of the latter. This association does not occur in crude cell extracts, presumably because the protein concentration is too low (43).

Animal cells in culture can also be rendered permeable to small molecules such as nucleotide and cyclic nucleotides. When baby hamster kidney cells are treated with a hypertonic cell-culture medium (Eagles medium, without serum, but containing 0.7 M NaCl), they retain gross morphology, the subcellular organelles, 100% DNA and 91% protein, but become permeable to small molecules. The process can be reversed by exposure to the complete Eagles cell-culture medium (44).

It is also possible to make a subcellular organelle such as the mitochondrion permeable to small molecules, allowing a direct assay of matrix enzymes such as those of the tricarboxylic acid cycle (e.g. citrate synthase, glutamate dehydrogenase, isocitrate dehydrogenase). Permeabilization can be brought about by treatment with toluene [2% (v/v)] for 2 min in a buffered iso-osmotic medium, but it is also necessary to include polyethylene glycol [8.5% (w/v); $M_r$ 6000] to prevent release of enzymes during toluene treatment and subsequent assay (45).

## 7. Concluding remarks

In this chapter, some of the factors involved in extracting enzymes from tissues and cells have been discussed. Rather than presenting a series of recipes, the points to be borne in mind when devising experiments have been emphasized. As mentioned in Section 1, preparation of a cell extract is often only the first step towards the ultimate purification of an enzyme for structural or kinetic characterization. It is vital in this type of work to ensure that as much activity as possible is extracted in as intact and stable a state as possible (i.e. resembling the presumed native state). The considerations described in this chapter should help the investigator to achieve this goal.

## References

1. Jakoby, W. B. (ed.) (1984). *Methods in Enzymology*, **104**.
2. Suelter, C. H. (1985). *A Practical Guide to Enzymology*, Wiley, New York.
3. Scopes, R. K. (1987). *Protein Purification: Principles and Practice*, (2nd edn). Springer, New York.
4. Marston, F. A. O. (1986). *Biochem. J.*, **240,** 1.
5. North, M. J. (1989). In: *Proteolytic Enzymes: A Practical Approach*, (ed. R. J. Beynon and J. S. Bond), pp. 105–124. IRL Press, Oxford.
6. Elliott, K. R. F. (1979). In *Techniques in Metabolic Research*, Vol. B204, pp. 1–20. Elsevier/North Holland, Amsterdam.
7. Graham, J. (1984). In: *Centrifugation: A Practical Approach*, (2nd edn), (ed. D. Rickwood), pp. 161–182. IRL Press, Oxford.
8. Ratner, D. and Borth, W. (1983). *Exp. Cell Research*, **143,** 1.
9. Fleischer, S. and Packer, L. (ed.) (1974). *Methods in Enzymology*, **32**.
10. Hardman, J. G. and O'Malley, B. W. (ed.) (1975). *Methods in Enzymology*, **39**.

11. Lloyd, D. and Coakley, W. T. (1979). *Techniques in Metabolic Research*, Vol. B201, pp. 1–18. Elsevier/North Holland, Amsterdam.
12. Hall, N. P. and Tolbert, N. E. (1978). *FEBS Lett.*, **96,** 167.
13. Yun, S-L., Aust, A. E., and Suelter, C. H. (1976). *J. Biol. Chem.*, **251,** 124.
14. Schwenke, J., Canut, H., and Flores, A. (1983). *FEBS Lett.*, **156,** 274.
15. Naganuma, T., Uzuka, Y., and Tanaka, K. (1984). *Anal. Biochem.*, **141,** 74.
16. de la Morena, E., Santos, I., and Grisolia, S. (1968). *Biochim. Biophys. Acta*, **151,** 526.
17. Fell, D. A., Liddle, P. F., Peacocke, A. R., and Dwek, R. A. (1974). *Biochem. J.*, **139,** 665.
18. Schwinghamer, E. A. (1980). *FEMS Microbiology Letters*, **7,** 157.
19. Singer, S. J. (1974). *Ann. Rev. Biochem.*, **43,** 805.
20. Penefsky, H. S. and Tzagaloff, A. (1971). *Methods in Enzymology*, **22,** 204.
21. Spatz, L. and Strittmatter, P. (1973). *J. Biol. Chem.*, **248,** 793.
22. Furth, A. J. (1980). *Anal. Biochem.*, **109,** 207.
23. Le Marne, M. Lind, K. E., Jørgensen, K. E., Røigaard, H., and Møller, J. V. (1978). *J. Biol. Chem.*, **253,** 7051.
24. Heil, A., Müller, G., Noda, L., Pinder, T., Schirmer, H., Schirmer, I., and von Zabern, I. (1974). *Eur. J. Biochem.*, **43,** 131.
25. Bock, P. E. and Frieden, C. (1978). *Trends in Biochem. Sciences*, **3,** 100.
26. Cleland, W. W. (1964). *Biochemistry*, **3,** 480.
27. Stevens, R., Stevens, L., and Price, N. C. (1983). *Biochem. Educ.*, **11,** 70.
28. Coakley, W. T., Brown, R. C., James, C. J., and Gould, R. K. (1973). *Arch. Biochem. Biophys*, **159,** 722.
29. McKee, J. R., Christman, C. L., O'Brien, W. D. Jr., and Wang, S. I. (1977). *Biochemistry*, **16,** 4651.
30. Jaenicke, R. (1981). *Ann. Rev. Biophys. Bioeng.*, **10,** 1.
31. Arakawa, T. and Timasheff, S. N. (1982). *Biochemistry*, **21,** 6536.
32. Lloyd, D. and Poole, R. K. (1979). In *Techniques in Metabolic Research*, Vol. B202, pp. 1–27. Elsevier/North-Holland, Amsterdam.
33. Katunuma, N., Kominami, E., Banno, Y., Kito, K., Aoki, Y., and Urata, G. (1976). *Adv. Enzyme Regulation*, **14,** 325.
34. Crabtree, B., Leech, A. R., and Newsholme, E. A. (1979). In: *Techniques in Metabolic Research*, Vol. B211, pp. 1–37. Elsevier/North Holland, Amsterdam.
35. Ansari, H. and Stevens, L. (1983). *J. Gen. Microbiol.*, **129,** 1637.
36. Crabtree, B. and Newsholme, E. A. (1972). *Biochem. J.*, **126,** 49.
37. Zammit, V. A. and Newsholme, E. A. (1976). *Biochem. J.*, **160,** 447.
38. Fleischer, S. and Packer, L. (ed.) (1974). *Methods in Enzymology*, **31**.
39. Birnie, G. D. (1982). *Subcellular Components. Preparation and Fractionation* (2nd edn.). Butterworths, London.
40. Baudhuin, P. (1974). *Methods in Enzymology*, **32,** 3.
41. Schnaitman, C. and Greenawalt, J. W. (1968). *J. Cell Biol.*, **38,** 158.
42. Weitzman, P. D. J. (1973). *FEBS Lett.*, **32,** 247.
43. Wiame, J. M. (1971). *Curr. Top. Cell Regul.*, **4,** 1.
44. Castellot, J. J. Jr., Miller, M. R., and Pardee, A. M. (1978). *Proc. Natl. Acad. Sci. USA*, **75,** 351.
45. Matlib, M. A., Shannon, W. A. Jr., and Srere, P. A. (1979). *Methods in Enzymology*, **56,** 544.

46. Rickwood, D. (1984). *Centrifugation: A Practical Approach* (2nd edn), Appendix V. IRL Press, Oxford.
47. Tata, J. R. (1974). *Methods in Enzymology*, **31,** 253.
48. Comte, J. and Gautheron, D. C. (1979). *Methods in Enzymology*, **55,** 98.
49. Bendall, D. S. and de Duve, C. (1960). *Biochem. J.*, **74,** 444.
50. Harper, J. W., Hemmi, K., and Powers, J. C. (1985). *Biochemistry*, **24,** 1831.

# 10

# Statistical analysis of enzyme kinetic data

PETER J. F. HENDERSON

## 1. Introduction

Characterization of an enzyme usually includes determination of maximum reaction velocity, $V_{max}$, and of a 'Michaelis constant', $K_m$, for each substrate. Knowledge of $V_{max}$ and $K_m$ is useful for a number of biochemical purposes: for the estimation of intracellular reaction rates, detection of metabolic control points, comparison of isoenzymes from different tissues or organisms, determination of the molecular events of catalysis, determination of turnover numbers, quantitative comparison of alternative substrates, and for definition of the potency of inhibitors or activators. The problem for the novice is the choice of a method for calculating these valuable parameters, given the variety of methods available and the apparent debate about which is best. One might think that valid statistical arguments could resolve the debate, but these are not always widely understood. As an example, as described below, there are three straight-line graphs, transformations of the Michaelis–Menten equation, that can be used for estimating $K_m$ and $V_{max}$. The debate about which is the best was settled in 1961 by Johansen and Lumry (1) [see also Wilkinson, (2)], with a statistical argument describing the correct weighting factors for each plot. While many kineticists appreciated and applied the correctly-weighted analyses, the literature continues to contain estimates of $K_m$ and $V_{max}$ based on unweighted fits to the Lineweaver–Burk double reciprocal plot—the least appropriate one to use!

Statistical criteria may settle some problems, but they can raise others. As an example, since 1963 kinetic data have increasingly been analysed by computerized least-squares fitting techniques, particularly those developed by Cleland (3, 4, 5). However, arguments have been advanced (6, 7, 8) that the least-squares criterion of best fit may not be safe, unless certain assumptions are justified, e.g. replicated velocity values should follow a normal distribution, and any systematic variations in such distributions with velocity should be compensated by appropriate weighting. Such justifications require hundreds of measurements, and for reasons of expense or fragility of materials are often impractical. Without the support of such tests it would be better to use non-parametric criteria of best fit, which rely on fewer assumptions (reviewed by Cornish-Bowden, 8), but even these techniques have their own disadvantages and limitations (see Section 4.3).

Fortunately, experience has shown that valid and useful conclusions can be made from determinations of $K_m$ and $V_{max}$ values, despite the lack of absolute confidence in the methods of determination. The important criterion is that the interpretation of an experiment involving calculations of $K_m$ and $V_{max}$ should be independent of the method by which the values are calculated, and independent of the magnitude of their standard deviations.

This chapter is limited to describing the various methods for the determination of $K_m$, $V_{max}$, $K_i$ (defined below) and their standard deviations, with the emphasis on the advantages and disadvantages of each. The approach is designed to answer the questions often asked by research workers who are familiar only with the Lineweaver–Burk plot and whose aim is to determine reliable values of $K_m$ and $V_{max}$ without too great a commitment of time, money, or mathematical expertise. This chapter will not attempt to describe the principles of enzyme kinetics, which are well covered in a number of textbooks and articles (e.g. 8, 9, 10, 11, 12, 13).

# 2. Preliminary considerations

## 2.1 Assumption that the Michaelis–Menten equation applies

It is initially assumed that the velocity ($v$) of an enzyme reaction is related to substrate concentration ($s$) by the Michaelis–Menten equation (Equation 1).

$$v = \frac{V_{max}\, s}{K_m + s} \tag{1}$$

$K_m$ and $V_{max}$ are the two parameters that define a rectangular hyperbola relating $v$ to $s$ (*Figure 1*). The derivation of the equation relies on three assumptions: the rate of the reverse reaction is negligible during the measurement period, the measurement is made in the steady state when the concentration(s) of enzyme–substrate complex(es) is unchanging, and the formation of enzyme–substrate complex(es) does not significantly deplete the concentration of free substrate. Tests for the validity of these assumptions are described in Section 3.1.

## 2.2 Definition of $K_m$ and $V_{max}$—'true' and 'apparent' values

Strictly, $K_m$ is the ratio of two sets of rate constants, the number and arrangement of which depend upon the number and order of steps in the mechanism of an enzyme (e.g. 14, 15). Similarly, $V_{max}$ comprises rate constant(s) and enzyme concentration.

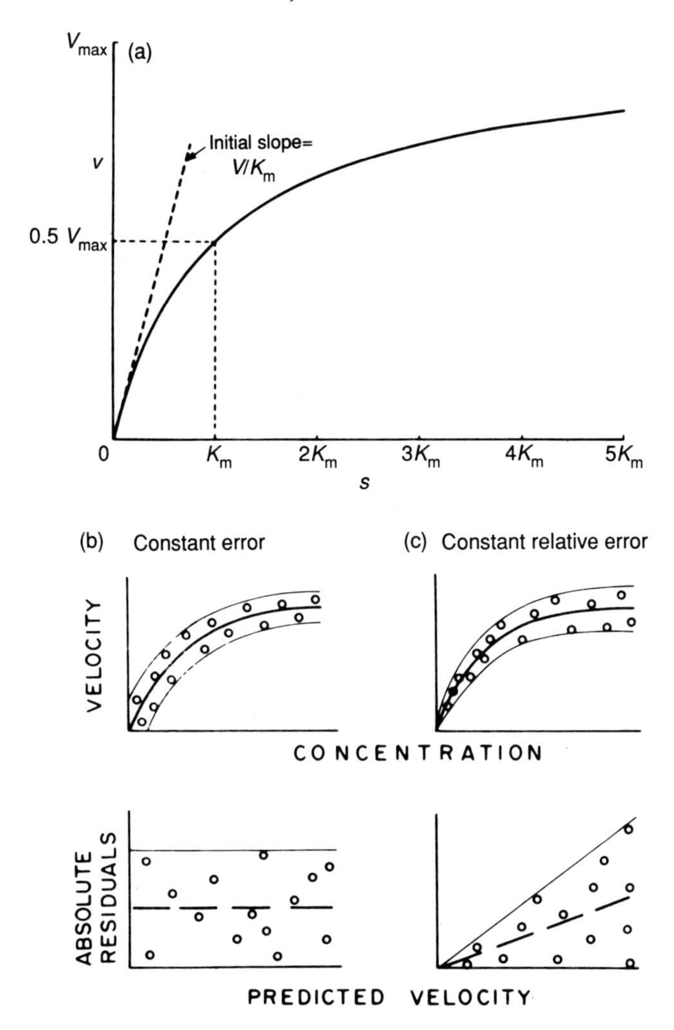


**Figure 1.** (a) Plot of initial velocity, $v$, against substrate concentration, $s$, for a reaction obeying Equation 1 (7, 8); illustration of constant absolute errors (b), or constant relative errors (c), in $v$, and the associated plots of residuals (24).

However, one virtue of these parameters is the validity of conceptually simpler working definitions. Thus, true $K_m$ is the concentration of substrate required to achieve half the maximum velocity of the enzyme, provided that all other substrates are saturating. True $V_{max}$ is the velocity of the enzyme reaction when all substrates are at saturating concentrations.

Even if one substrate of an enzyme-catalysed multi-substrate reaction is maintained at a constant, below-saturating concentration, the relationship of

$v$ to $s$ for other substrate(s) usually still follows Equation 1. However, the measured $V_{max}$ value is then not the 'true', but a lower 'apparent' value. Similarly, $K_m$ is altered to an 'apparent' value, which may be higher or lower (depending on the steady-state mechanism) than the true value measured when all substrates are saturating, a situation that may be impossible to realize in practice. One of the appealing features of Cleland's strategy for analysing multi-substrate enzyme reactions is the logical way it proceeds from measurement of apparent $K_m$ and $V_{max}$ values to determination of the true values.

## 2.3 Optimization of the assay conditions

Investigations prior to $K_m$, $V_{max}$ determination have usually established optimal assay conditions of pH, temperature, ionic strength, cation concentration ($Mg^{2+}$ or $K^+$ may be particularly important), coupling enzyme concentration, and so on. The aim is generally to maximize enzyme velocity at the minimal enzyme concentration required for satisfactory detection of reaction rate. The first criterion important for our purposes is that the enzyme must be stable over the measurement period, as discussed in Chapter 1. The other criteria are described in Section 3.1.

## 2.4 The value of graphs

### 2.4.1 Visual confirmation that experimental conditions are satisfactory and that the correct rate equation has been chosen

This is probably the most important function of graphical display of experimental data. If the data points deviate systematically from an expected line they can reveal crucial information: inadequacy of the assay method (see also Chapter 1), inappropriateness of the Michaelis–Menten equation, allosteric interactions, substrate inhibition, the presence of impurities in substrate or enzyme.

If a systematic deviation is suspected but not obvious, a re-plot of residuals (differences between points and expected line) may be clearer (see *Figure 1* and Section 5.4). Alternatively, more measurements could be made.

A graph will also reveal if the scatter of the data points is too great for sensible conclusions to be drawn. In this case, it will probably be necessary to improve the assay method before further progress is possible.

### 2.4.2 Communication of results

The clearest way to communicate the precision of one's results to a reader is by a graph showing the data points and a best-fit line that follows a stated equation. In enzyme kinetics the Lineweaver–Burk (16) reciprocal plot is popularly used for this 'display' purpose because of its familiarity and its linearity; also, the intercepts indicate reciprocal values of $K_m$ and $V_{max}$

directly. It has the disadvantages that one must adjust to reciprocals—the 'lowest' points represent the highest rates and substrate concentrations, and vice versa—and the high concentration points tend to be compressed towards the $1/v$ axis; this problem can be offset by obtaining additional intermediate points.

#### 2.4.3 Calculation of $K_m$ and $V_{max}$ from graphs

The Lineweaver–Burk plot ($1/v$ against $1/s$) or the Eadie–Hofstee plot ($v$ against $v/s$) should never be used to calculate $K_m$, $V_{max}$ values for publication. The Hanes plot ($s/v$ against $s$) is less dangerous, but, even so, for statistical reasons (Section 4) these three plots should not be used for anything other than display purposes.

There is only one statistically sound method for the determination of the best fit $K_m$, $V_{max}$ values by visual inspection of a graph. This is the direct linear plot (7, 8, 17, 18). The method uses a non-parametric criterion of best fit, in which the median values of a series of estimates of $K_m$ and $V_{max}$ are shown to be the best (Section 4.3).

The usual objection to any graphical method of calculating parameter values is that confidence limits or standard deviations are not also determined. The direct linear plot does not suffer from this disadvantage (8, 19, 20). Even so, it is more practicable to determine the confidence limits using a computer (see below) than by drawing graphs.

### 2.5 Is a computer necessary?

The direct linear plot is the only method by which best-fit values of $K_m$, $V_{max}$ can be satisfactorily determined without the aid of a computer. Even in this case, a computerized procedure (21, 22) shortens the calculation time and facilitates the estimation of confidence limits. For the best least-squares analysis, the weighted least-squares fit to a hyperbola [Section 3.4.2, (4)], or more complex equations, a computer is the only practicable way to handle the intricacies of the calculations.

Nevertheless, it must be emphasized that some type of graph should always be plotted. Visual inspection of the plots is an essential method of detecting abnormalities, e.g. non-linearity, that might invalidate the assumptions underlying the analysis.

## 3. General strategy for determination of $K_m$ and $V_{max}$

### 3.1 Validation of the assumptions of the Michaelis–Menten relationship

The measured initial velocity should be linear with time. This is to ensure that the steady-state is operative and that the reverse reaction is negligible, two of the assumptions underlying the Michaelis–Menten relationship (Equation 1).

If the reaction gives a continuous curve it is customary to calculate the initial rate from a tangent drawn to the first part. This procedure may be inaccurate, and methods are available for checking that the true initial rate is measured (8, pp. 34–37; 10, Chapter 2; 18). If reaction occurs in the absence of any of the components, care must be taken to make a proper correction.

Velocity should be linearly proportional to enzyme concentration. This is both an implicit requirement of the Michaelis–Menten relationship and a check of the assumption that combination with enzyme does not significantly deplete substrate concentration. Departures from linearity could occur if $s$ rapidly became depleted (possible at high enzyme concentrations), if the enzyme underwent polymerization–depolymerization as a result of co-operative interactions, if a tight-binding inhibitor or activator were present, or for a number of other reasons [see Chapter 1 and (23)].

## 3.2 Measurement of velocity over a range of substrate concentrations

### 3.2.1 The trial run

A rough estimate of the $K_m$ can be obtained from a knowledge of the physiological substrate concentration, or by a direct linear plot (Section 4.3) of measurements of velocities at only two substrate concentrations, a high and low value giving the clearest intersection point. A wide range of substrate concentrations should then be used in a trial run; practical limits are the sensitivity of the assay at low substrate concentration, and substrate solubility at high substrate concentration. These data must be inspected for abnormalities (Section 3.3), but first some features of the experimental design should be considered.

### 3.2.2 What range and spacing of substrate concentrations should be used?

Usually the experimenter wishes to confirm the validity of Equation 1, as well as to obtain reasonably precise values of $K_m$ and $V_{max}$. The statistical problem is to define the $v$ against $s$ curve (*Figure 1*) as precisely as possible. For this purpose the substrate concentration range should be about $K_m/2$–$10K_m$, spaced more closely at low $s$ values and with at least one high concentration to approach the $V_{max}$ value (4, 8, 24). A convenient method for spacing the $s$ concentrations that gives approximately even spacing of $v$ values has been described (25).

### 3.2.3 How many points are necessary?

There is no dogmatic answer to this question. Cleland (4) suggested that reasonably precise estimates or $K_m$ and $V_{max}$ would have standard deviations of less than 10% of the mean values. The more accurately the velocity values are measured, the fewer points will be required to obtain this empirical level of precision (assuming that Equation 1 applies, of course). For a straightforward single substrate enzyme, as few as five velocity measurements covering the range $K_m/2$–$8K_m$ may be sufficient, but using more points will often offset the perturbation produced by a single outlier (Section 8.6.5.). Twelve measurements are always adequate in our experience.

In the case of a two-substrate enzyme, many more combinations of substrate concentrations will be necessary, often 25 ($5 \times 5$) measurements or more. The stability of the enzyme and/or the time required for each assay may then limit the number of measurements.

### 3.2.4 The value of replicate measurements

The above experimental design is not the most efficient for all circumstances. The objective may be the precise determination of $K_m$ and $V_{max}$ or the determination of the effect of a perturbing agent (inhibitor, activator, and so on), when it is already established that Equation 1 satisfactorily relates $v$ to $s$. In these cases, the data can be collected at only two values of $s$; a value well above the $K_m$ (limited by solubility), and a value at or below $K_m$ (limited by the precision of the assay technique). At least four measurements should be made at each value of $s$. This experimental design gives more precise estimates than an equal number of observations at intermediate $s$ values (19, 26).

If an alternative model to Equation 1 is being considered, Porter and Trager (19) advocate that only three values of $s$ are necessary—a low and a high value (as above), and a value at which the alternative model is expected to produce a maximum deviation from Equation 1; replicate measurements should be made at each concentration.

Replication of velocity measurements at one $s$ concentration provides an estimate of the standard deviation of $v$ [note that at least five replicates are necessary for such a standard deviation to be reliable (27), so there is little advantage in the common practice of making duplicate observations]. Comparison of the standard deviation of $v$ obtained at low $s$ concentration with that obtained at a high $s$ concentration (and intermediate values if practicable) is one way of deciding the correct weighting scheme to be used (27, 28, 29, 30; Sections 5.1–5.5), an important consideration in the design of experiments (8, 13, 24, 26, 31, 32, 33, 34). However, the weighting scheme can also be determined by consideration of the distribution of residuals (22, 32, 34, 35), so replicate measurements are by no means essential.

The procedure recommended next (Section 3.3 and Section 3.4) is based on three considerations. The applicability of Equation 1 should be tested, so a wide range of substrate concentrations is needed. Precise values of $K_m$ and $V_{max}$ are required, for which a geometric spacing of substrate concentrations is reasonably ideal. A weighting scheme may be needed; if a plot of residuals is inadequate to choose a scheme, then replicate measurements may be helpful.

## 3.3 Visual inspection of the data

The data are plotted as $1/v$ against $1/s$ or, much better, $v$ against $v/s$ (Section 4). The following three questions are examined:

(a) Is there systematic departure from linearity, indicating that Equation 1 is inadequate?

(b) Is the scatter of data too great, indicating that the assay method should be improved?

(c) Do the substrate concentrations used cover at least the range $K_m/2$–$5K_m$?

If the answers to (a) and (b) are not obvious, it is wise to consider improvements in the assay method before embarking on sophisticated statistical methods of detecting abnormalities.

## 3.4 Determination of the best-fit $K_m$ and $V_{max}$ values

Provided that the answers to questions (a)–(c) above are satisfactory, the best-fit values can be obtained by one of the methods in Section 3.4.1 or Section 3.4.2. These methods are more reliable than linear transformations of the Michaelis–Menten equation, for reasons to be described in Section 4.

### 3.4.1 Direct linear plot

Pairs of ($v$, $s$) values are obtained in the usual way. Then a $v$ value is plotted onto a $V_{max}$ vertical axis (*Figure 2*) and the corresponding negative $s$ value is plotted onto a $K_m$ horizontal axis. The two points are joined and the line is extrapolated into '$V_{max}$, $K_m$ parameter space'. The process is repeated for each pair of ($v$, $s$) values. Hence $n$ lines are obtained for $n$ points. With data that fit Equation 1 exactly, all the lines intersect at the coordinates of the best fit $K_m$, $V_{max}$ values (*Figure 2*). For the practical case when the data contain measuring errors, a number $[n(n-1)/2]$ of different intersections are obtained (*Figure 3*). The coordinates of each intersection provide estimates of $K_m$ and $V_{max}$. Cornish-Bowden and Eisenthal (6) showed that the median values of these estimates are the best-fit values $K_m$ and $V_{max}$. The points of intersection are ranked from left to right to obtain the $K_m$ values, and from bottom to top to obtain $V_{max}$ values. In determining the median, one only requires numerical values for the middle points; the median is the middle one of an odd-numbered set, or the mean of the middle pair in an even-numbered set.

Replicate observations can be included in the direct linear plot, provided an intersection given by a pair of duplicates is ignored when finding the median or when calculating the probability distribution (19, 36). Note that replicates must not be averaged to include one mean velocity and substrate concentration in the plots. An intersection may appear in the second quadrant (top left) of the plot if substantial experimental errors occur in two $s$

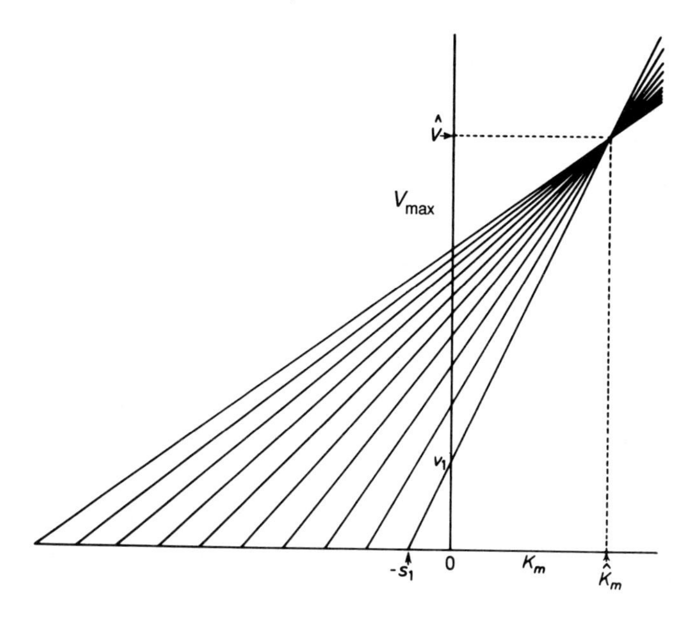


**Figure 2.** The direct linear plot of data that fits Equation 1 exactly (7, 8, 17).

values much higher than $K_m$; these intersections are simply included in the overall selection of the median values (8). An intersection may appear in the third quadrant (bottom left) as a result of one or both $s$ values being small compared with $K_m$; the negative estimates of $K_m$ and $V_{max}$ should then be treated as if they had large positive values (8). An alternative plot avoiding the need for special rules is described by Cornish-Bowden and Eisenthal (36).

Porter and Trager (19) and Cornish-Bowden *et al.* (20) showed how the confidence limits of $K_m$ and $V_{max}$ can be determined from the direct linear plot.

Computer software is available (21, 22) that eliminates the need to draw the plot, and provides confidence limits easily. The method has proved to be reliable and is recommended equally with the least-squares fit to the hyperbola (Section 3.4.2); the advantages/disadvantages of each are discussed in Section 4.3 and Section 5.7.

### 3.4.2 Least-squares fit to a hyperbola

If available, the computer should also be used to calculate the least-squares fit of the data points to a $v$ against $s$ hyperbola [Section 5.6; (4, 5, 21)]. A weighting scheme can be included if it is known. This method calculates best-fit values of $K_m$, $V_{max}$, $K_m/V_{max}$, $1/V_{max}$ and their standard deviations. The best-fit values should not differ significantly from those determined by the

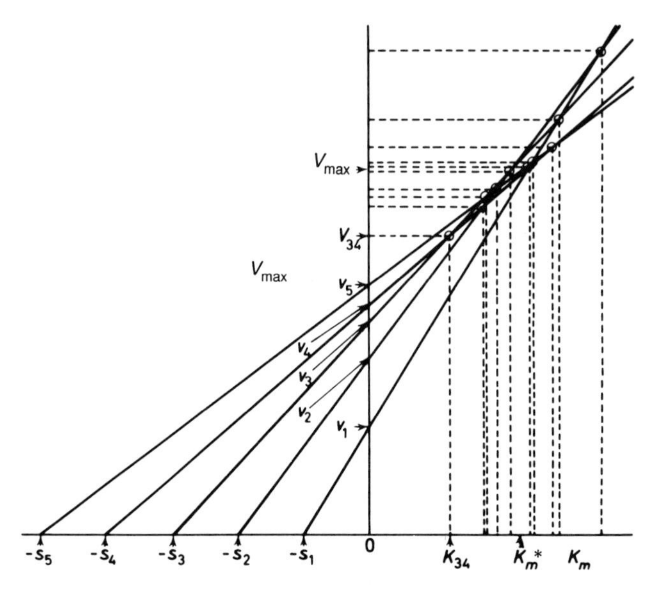


**Figure 3.** The direct linear plot of data containing errors. $K_m$* and $V_{max}$* are the median, best-fit, values (7, 8, 17).

direct linear plot. If they do, or if the standard deviations/confidence limits are unsatisfactory, there may be a departure from Equation 1 too subtle to be detected by visual inspection (Section 3.3). These problems are examined further in Section 8.5 and Section 8.6.

# 4. Calculation of $K_m$ and $V_{max}$ using graphs

$K_m$ and $V_{max}$ are commonly calculated by measuring velocities at a series of substrate concentrations and then drawing a straight line through the $1/v$ against $1/s$ plot; $K_m$ and $V_{max}$ are determined from the intercepts (16, 17). It is well known that this is not the only graphical method, since the Michaelis–Menton Equation 1 can be rearranged in several ways (Equations 2–4) to give straight-line relationships (16, 37, 38, 39, 40).

$$\frac{1}{v} = \left(\frac{K_m}{V_{max}} \times \frac{1}{s}\right) + \frac{1}{V_{max}} \qquad \text{(see } \textit{Figure 4}\text{)} \quad (2)$$

$$\frac{s}{v} = \left(\frac{1}{V_{max}} \times s\right) + \frac{K_m}{V_{max}} \qquad \text{(see } \textit{Figure 5}\text{)} \quad (3)$$

$$v = \left(-K_m \times \frac{v}{s}\right) + V_{max} \qquad \text{(see } \textit{Figure 6}\text{)} \quad (4)$$

The methods of estimating $K_m$ and $V_{max}$ from the slopes and intercepts are illustrated in *Figures 4–6*.

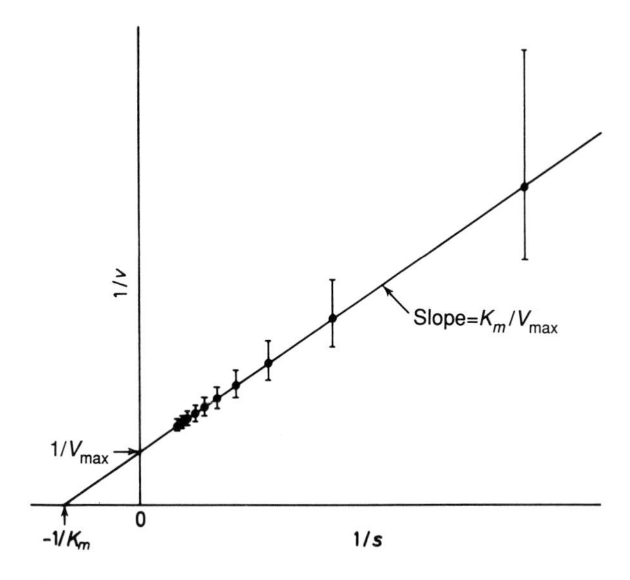


**Figure 4.** Plot of $1/v$ against $1/s$ from *Equation 2*, showing the effect of the reciprocals on errors of $\pm 0.05\ V_{max}$ (7, 8).

Which of the three plots is the 'best' to use? If the data fitted Equation 1 perfectly, each plot would yield the same $K_m$, $V_{max}$ values. In practice, because there are errors in the measured variable which are distorted differently in each plot (*Figures 4–6*), each gives a different answer. Cornish-Bowden (8) has neatly illustrated the source of the problem by modifying the Michaelis–Menten Equation 1 to include the measurement error (Equation 5)

$$v = [(V_{max}s)/(K_m + s)] + E \tag{5}$$

This realistic equation cannot be rearranged to any of the conveniently linear transformations (Equations 2–4).

Clearly, it is impossible to compensate accurately for the asymmetrical error distributions (*Figures 4–6*) if a straight line is drawn through the experimental points by eye. However, it can be done by making a correctly weighted least-squares fit using a computer (Section 5.1). Since fitting reciprocal plots by eye is condoned by some textbooks and many journals, the reasons why it is not acceptable practice are now considered in detail.

## 4.1 Why the Lineweaver–Burk double reciprocal plot is not acceptable

A technique of computerized simulation can be used to decide which plot is most or least likely to reveal the correct answer. A set of artificial data which exactly fits a hyperbola of known $K_m$ and $V_{max}$ is calculated by a computer

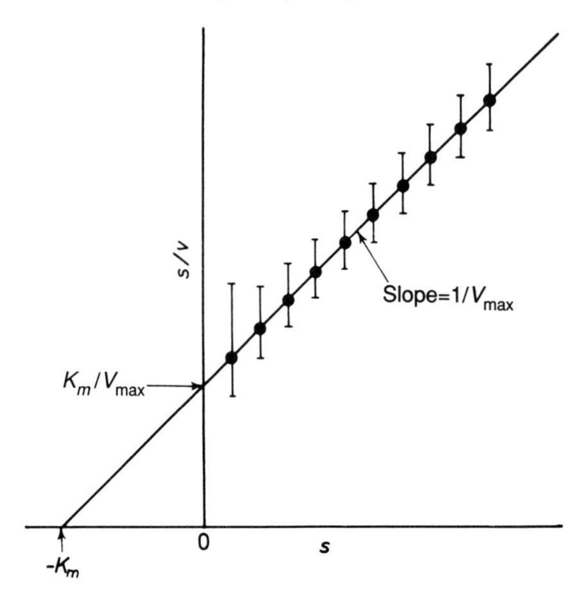


**Figure 5.** Plot of $s/v$ against $s$ from Equation 3, with error bars of $\pm 0.05\ V_{max}$ (7, 8).

program. The data are then modified by a random number generator so that random errors, of a magnitude comparable with experimental errors, are imposed. The data are then plotted in each of the three ways. The best straight line for each may be drawn by eye or by any of the available statistical methods, and the $K_m$ and $V_{max}$ calculated from the intercepts and slopes. These calculated values can then be compared with the 'true' values defined by the original 'perfect' data.

Dowd and Riggs (41) used this procedure to show that the Lineweaver–Burk reciprocal plot (*Figure 4*) gave the best-looking straight line when fitted by eye. However, this was nearly always associated with worse estimates of $K_m$ and $V_{max}$ than the other two plots, because the range of distortions of the experimental errors was greatest with the reciprocal plot; compare the error bars in *Figure 4* with those in *Figures 1*, *5*, and *6* (1, 2, 8, 42). The fact that this plot actually conceals a poor fit between the data and a straight line (41) may be the reason for its popularity!

## 4.2 Why the $s/v$ against $s$ plot is preferred

From their computerized simulation study Dowd and Riggs (41) concluded that the $v$ against $v/s$ plot is marginally better than the $s/v$ against $s$ plot for determining the kinetic constants by eye. In particular, it best reveals the goodness-of-fit between the data and the line, and so is recommended for checking that Equation 1 applies (Section 3.3). Unlike the double reciprocal plot (Section 4.1) the $v$ against $v/s$ plot actually encourages good experimental

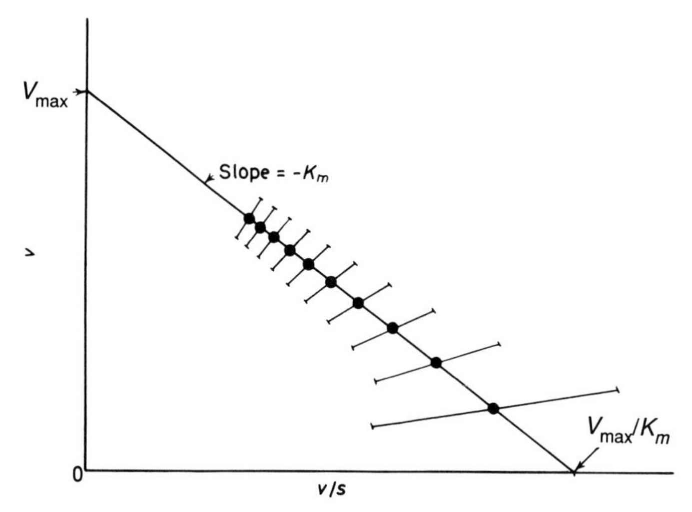


**Figure 6.** Plot of $v$ against $v/s$ from Equation 4, with error bars of $\pm 0.05\ V_{max}$ (7, 8).

design because it shows everything; you cannot launder it by choosing a suitable scale (personal communication from E. A. Boeker to A. Cornish-Bowden).

However, the dependent variable $v$ occurs in both the $y$ and $x$ axis of this plot. This leads to an over-riding disadvantage pointed out by Johansen and Lumry (1). The errors in the $y$ values are not independent of errors in the $x$ values, i.e. an error in $v$ gives a corresponding directional error in $v/s$. This gives an angular distortion of the error bars (*Figure 6*) which might make fitting by eye more difficult, but, more importantly, the usual methods of least-squares fit cannot be applied under such conditions (Section 5, introduction).

By a process of elimination, the $s/v$ against $s$ plot is found to be the safest to use, a conclusion supported by the simulation study of Endrenyi and Kwong (24). Comparison of *Figure 5* with *Figure 4* and *Figure 6* shows how distortion of the error bars is minimal in this type of plot.

## 4.3 How the direct linear plot avoids the difficulties

### 4.3.1 Advantages of the direct linear plot

(a) The data can be plotted easily and quickly, since the $v$ and $s$ values are used directly without transformation to a reciprocal or other arithmetical combination.

(b) The best estimates of the $K_m$ and $V_{max}$ values can be found by eye.

(c) The only assumption underlying the technique is that the error in any observation is as likely to be positive as negative; this is a considerable advantage over the least-squares fit method, which has several assumptions (Section 5, introduction) that are difficult to verify.

(d) The method is much less sensitive to outliers (Section 8.6.5) than the least-squares fit method.

(e) Weighting is not relevant.

(f) Confidence limits can be calculated for the $K_m$ and $V_{max}$ values.

### 4.3.2 Disadvantages of the direct linear plot

(a) It is not suitable for the visual display of data, only small numbers of measurements can be accommodated on one graph.

(b) If the data depart from a fit to Equation 1, i.e. are non-hyperbolic, this does not become readily apparent in the direct linear plot.

(c) The method cannot be extended directly to fitting equations more complex than Equation 1, i.e. containing more than two parameters as do the equations for multi-substrate reactions or inhibition.

(d) A trivial disadvantage is that the better the data the closer together come the intersections, and the more difficult it is to distinguish them—larger scales or computerized methods overcome this problem.

### 4.3.3 Comparison of the direct linear plot with other statistical methods

Clearly, the direct linear plot avoids the need to validate the normal distribution of errors and to determine the correct weighting scheme. Cornish-Bowden and Eisenthal (6) undertook a computer simulation study to compare the parameter values obtained by a direct linear plot with those obtained by a least-squares fit to a hyperbola (Section 5.6), the most acceptable of the other methods. When the assumptions underlying the least-squares-fit method (Section 5, introduction) are satisfied, the least-squares-fit method gives marginally better estimates of $K_m$ and $V_{max}$ than the direct linear plot. When the assumptions are not valid, a situation that would usually go unrecognized, then the direct linear plot is the superior method (6, 20).

A more extensive comparison of seven different methods of estimating $K_m$ and $V_{max}$ was described by Atkins and Nimmo (43). Their conclusion was the same: 'Unless the error is definitely known to be normally distributed and of constant magnitude, Eisenthal and Cornish-Bowden's method (17) is the one to use.'

### 4.3.4 Other uses of the direct linear plot

The pattern of intersecting lines of a direct linear plot is changed in the presence of a reversible inhibitor, because the apparent $K_m$, apparent $V_{max}$, or both, are altered. The direction in which the median intersection moves as the concentration of inhibitor is changed can be used to diagnose whether inhibition is competitive, non-competitive, mixed, or uncompetitive (8, 17).

This method is just as useful as the more familiar inhibition patterns of reciprocal plots (see e.g. 9, 12). However, determination of the $K_i$ value(s) and standard deviation(s) is probably more conveniently performed by computerized least-squares fit methods (Section 7.2) (22).

Similarly, the pattern of intersections is changed by altering the concentration of a second substrate (B), i.e. $K_m^{app}$ and $V_{max}^{app}$ change. A plot of $V_{max}^{app}$ against $K_m^{app}$ can be used directly to distinguish steady-state mechanisms (17). This plot has two disadvantages compared with the $K_m^{app}/V_{max}^{app}$ against 1/B replots advocated by Cleland (Section 6): it is less familiar and less versatile. Also, as mentioned above, the direct linear plot cannot be used to determine parameter values (and standard deviations) for multi-substrate rate equations with more than two parameters, whereas computerized least-squares fit methods can be (Section 6) (4, 5, 21, 22).

A modified plot in which lines joining $1/v$ and $s/v$ values in $1/V_{max}$ against $K_m/V_{max}$ 'parameter space' has been suggested (36). This has the advantages that most of the intersection points appear without extrapolation of the lines representing the observations, and that complexities due to negative estimates of $K_m$ and $V_{max}$ are eliminated. Also, the median values of $1/V_{max}$ and $K_m/V_{max}$ can be employed directly in Cleland-type 'intercept' and 'slope' replots against inhibitor or reciprocal substrate concentration (Section 6).

A second variation of the direct linear plot provides a good method of checking that the true initial velocity is being measured (8, 18, 44), especially when the reaction rate is markedly non-linear or when a discontinuous assay must be used (e.g. 45).

## 5. Calculation of $K_m$ and $V_{max}$ using computerized least-squares-fit methods

The general principles of fitting data to an equation by the least-squares criterion have been described elsewhere (e.g. 7, 46, 47). The following assumptions are implicit in the least-squares-fit methods used in the statistical analysis of enzyme kinetic data (1, 2, 48).

(a) Random errors in replicate values of the measured velocity, $v$, follow a normal distribution.

(b) There is no error in the substrate concentration, $s$.

(c) The correct weights are known, i.e. if there is systematic variation of the standard deviations between low and high values of $v$, this variation must be compensated by weighting the points appropriately (Sections 5.2–5.4).

(d) In general, and this applies particularly to transformations of the Michaelis–Menten relationship (Equations 2–4), fluctuations in the $y$ values must be independent of any fluctuations in the $x$ values. This is the reason for eliminating error in $s$.

It is obviously desirable to justify these assumptions before using least-squares fitting methods, but tests for normality and for distribution of errors require hundreds of points. Such repetition of biological experiments is often limited by fragility of the material and by expense. Accordingly, it is rarely possible to check the validity of the above assumptions, and this uncertainty undermines conclusions drawn from any least-squares estimations of $K_m$ and $V_{max}$. Occasionally, the assumptions have been checked (e.g. 27, 29, 49), and the conclusions drawn are discussed in Sections 5.2–5.4.

Despite these disadvantages, the computerized least-squares-fit methods devised by Cleland (3, 4, 5) provided a major advance in analyses of steady-state enzyme kinetics. His system for writing rate equations (14, 15, 50) is easy to understand and use, the programs provide estimates of $K_m$ and $V_{max}$ and their standard deviations, and the method extends easily to multi-parameter rate equations, i.e. for more than one substrate or for inhibition. This last point is a most important advantage, and is the main reason least-squares fits have not been superseded by the statistically more robust direct linear plot.

## 5.1 Weighted least-squares fit to a straight line

Consider *Figures 1* and *4*. *Figure 1* shows a $v$ against $s$ plot with the best-fit hyperbola. Suppose that at each substrate concentration the velocity measurement was repeated sufficient times to obtain a standard deviation of the measurement. Two possibilities are commonly considered. The standard deviations might be the same whatever the value of $v$ (*Figure 1b*), or they might be proportional to $v$ (*Figure 1c*). The former case is described as constant absolute error, the latter case as constant relative error. *Figure 4* shows what happens when the homogenous errors of *Figure 1b* are transformed into reciprocals; even though the errors were the same throughout the hyperbola, they systematically increase along the straight-line re-plot. A less extensive, but nevertheless similar, effect occurs with constant relative errors (e.g. 4).

From *Figure 4* it is clear that the relative reliability of the points at low $s$ (high $1/s$) is decreased by the act of taking reciprocals. Hence, in fitting the data points, less weight should be applied to the points at high $1/s$ values than to those at low $1/s$ values. Yet if one attempted to draw the best line by eye, it is impossible to avoid putting more weight on these points. The problem is not nearly so acute when the $s/v$ against $s$ plot is used, as illustrated in *Figure 5*.

Cleland (4) and Cornish-Bowden (7, 51) discussed the correct weighting factors to be used in a least-squares analysis of an $s/v$ against $s$ or $1/v$ against $1/s$ plot. The theoretical considerations were considered by Burk (48), Johansen and Lumry (1), and Wilkinson (2), and the general procedure is described briefly here.

If a measurement of $v$ is made many times so that its standard deviation ($\sigma$)

can be precisely calculated, then the correct weight, $w$, for that point in fitting to an equation is

$$w = 1/\sigma^2 \tag{6}$$

Note that this formula applies only if $s$, the controlled variable, is known precisely, a usual assumption in enzyme kinetics studies (Section 5). The weighted least-squares-fit is then equivalent to an unweighted fit to an enlarged set of points, each point of the original set being repeated the appropriate $w$ times. Formulae for calculating the slope and intercept of such a weighted fit to a straight line are given in (1, 21); the calculations are tedious by hand but they can easily be processed on a computer (21).

Now we consider how these general considerations apply to the particular case of a $1/v$ against $1/s$ plot, assuming that standard deviations ($\sigma$) are constant throughout the original hyperbola as illustrated in *Figure 1a*. It can be shown (1, 2, 47) that the standard error of $1/v$ is related to $\sigma$ by the approximation.

$$\sigma^2(1/v) \approx \sigma^2/\hat{v}^2v^2 \approx \sigma^2/v^4 \tag{7}$$

$\hat{v}$ is the calculated value of $v$, but for most applications its use is an unnecessary refinement. Hence, if $\sigma$ is constant through the range of measured $v$ values, a reasonable first approximation of the weight to be used is

$$w = v^4 \tag{8}$$

Similarly, for the $s/v$ against $s$ re-plot of data with constant absolute error, a reasonable approximation of the weight is

$$w = v^4/s^2 \tag{9}$$

When the appropriate weighting factors are applied, a least-squares fit to each type of straight line should give the same $V_{max}$ and $K_m$. However, precise agreement will not be achieved unless the data are iteratively re-processed with refined weights $v\hat{v}^3$ or $v\hat{v}^3/s^2$, respectively (22, 51).

It is startling that appropriate weighting factors are often omitted in fitting data to many other types of transformed equations. Such weighting, necessary because the equations have been transformed, must be distinguished carefully from weighting required because of inherent variability in the experimental measurements, a problem considered next.

## 5.2 Are measuring errors in $v$ homogeneous?

The vast majority of published $K_m$ and $V_{max}$ determinations assumed that experimental errors in measuring $v$ were constant, i.e. homogeneous as illustrated in *Figure 1b*. However, evidence is accumulating that the measurement errors are more likely to increase with the value of $v$, i.e. are heterogeneous, as illustrated in *Figure 1c* (e.g. 27, 28, 29, 30, 45). If the measurement errors are simply proportional to velocity, i.e. constant relative error, then the weighting to be applied to a $1/v$ against $1/s$ plot is:

$$w = v^2 \tag{10}$$

in contrast to $w = v^4$ for homogeneous errors (Section 5.1) (1, 4). Similarly, for constant relative error the correct weighting for a $s/v$ against $s$ plot is (1, 4)

$$w = v^2/s^2 \tag{11}$$

So the available evidence shows that errors in measuring $v$ are likely to be homogeneous. In at least one case (30, 34) the measuring errors were not uniquely defined by $v$, but also by both substrate and inhibitor concentrations in a systematic way.

## 5.3 Is it necessary to establish the correct weighting scheme?

If $K_m$ and $V_{max}$ are determined by the direct linear plot, it is not necessary to establish a weighting scheme.

If a conventional least-square fitting method is chosen, then ideally the weighting scheme should be established. However, to do so requires more measurements of velocity at different substrate concentrations. The number of velocity values which can be determined in a given period of time is often limited by fragility of the biological material or expense. How necessary is this extra effort?

To answer this we must return to the principle stated in the Introduction. If, for example, the question posed is 'Are the values of $K_m$ of each of two isoenzymes significantly different?', then the experimenter must find whether a change in the weighting scheme would change the apparent answer to this question. The easy way is to use the computerized least-squares fit methods of Cleland (4, 5), into which chosen weighting factors can be inserted easily. The fit may then be repeated with the same data but different hypothesized weighting schemes, and the effects on the final answer determined. If the type of weighting scheme significantly alters the conclusion from the experiment, then it is incumbent on the experimenter to perform more measurements and find which scheme is the correct one to use.

## 5.4 How is the weighting scheme established?

A plot of the residuals $(v - \hat{v})$ against predicted velocity $(\hat{v})$ should reflect only experimental measuring errors provided that the correct rate equation has been chosen (8, 31, 32, 34). Examples are shown in *Figure 1*. Such plots are probably the simplest way to establish a weighting scheme, but they do require many measurements (about 50 or more) for any trend to become apparent through the random scatter.

The novel least-squares fitting method of Cornish-Bowden and Endrenyi

(35) obviates the necessity for a separate plot. Any information about weighting in the scatter of the data is accommodated in the fitting procedure. A computer program for the method is now available (22). This will probably supersede plots of residuals and the method of Mannervik and coworkers, who suggested using the relationship:

$$\sigma^2(v) = K \times v\alpha \tag{12}$$

where $K$ and $\alpha$ are empirical constants (28, 29, 30, 34). The relative weighting factor for each $v$ value is then

$$w = v^{-\alpha} \tag{13}$$

'Values of $\alpha = 0$ and $\alpha = 2$ correspond to the classical error structures of constant absolute error and constant relative error, but experience . . . shows that any value of $0 < \alpha < 2$ (or even higher) can be found' (34). Once these weight factors are established they are most easily applied using the least-squares fit to an hyperbola (Section 5.6; references 4, 21).

It might be thought that experimental estimates of $\sigma$, obtained by a few replicate observations, could be used directly for weighting the least-squares analysis. Storer *et al.* (27) tested this possibility by a computerized simulation study. They found that 'With weights calculated from . . . duplicate observations the results are very bad, almost as bad as those from the double reciprocal plot!' In fact, 'the method is never significantly better than methods based on sensible (through not necessarily correct) assumptions about weights', a conclusion supported by experiments of Mannervik and coworkers (29, 34). So replicate observations enable one to choose the best weighting scheme, and it is the scheme, not the observations themselves, that would be used in the actual analysis.

## 5.5 Do replicate values of *v* follow a normal distribution?

Mabood *et al.* (49) measured the velocity of acetylcholinesterase at four substrate concentrations, repeating each observation 75 times. The values of $\sigma$ were very similar for three concentrations of $s$, but higher at one of the intermediate concentrations. This differed from the systematic heterogeneity of $v$ reported for glucokinase and chymotrypsin (27). However, for all these enzymes there appeared to be significant deviation of the replicated values from a normal distribution. This would support the general use of the direct linear plot, rather than least-squares fitting techniques, for calculating $K_m$ and $V_{max}$, since the former is insensitive to the way in which the error is distributed (Section 4.3).

Two other interesting observations on the error of measuring $v$ were reported by Mabood *et al.* (49). The shapes of the distributions were apparently related to the angle of the line recording the spectrophotometric change used to calculate $v$. Consequently, measurements made on recorders should have the chart speed adjusted to a relatively constant angle in all measured lines. Secondly, as might be expected, one experimenter produced consistently different standard deviations from another experimenter.

The distribution of errors in $v$ does not have to deviate greatly from the normal curve for the least-squares estimates of $K_m$ and $V_{max}$ to become less accurate than those determined by the direct linear plot method (6). However, the standard deviations of $K_m$ and $V_{max}$, as calculated by the least-squares method, are relatively robust against departures of the $v$ distribution from normality, but they are sensitive to incorrect weighting (20).

## 5.6 Least-squares fit to a hyperbola

The mathematical techniques for fitting data to a straight line by the least-squares criterion have been available for some time. However, techniques for making a least-squares fit to a hyperbola were probably first described by Wilkinson (2). The method requires initial estimates of $K_m$ and $V_{max}$ (provided by one of the linear transformations). These estimates are then refined by an iterative technique until they converge on the best-fit values by the least-squares criterion. Such calculations are tedious to perform without the aid of a computer, and suitable programs have been described by Cleland (3, 4, 5, 21) and by Leatherbarrow (52).

The sequence of operations in the Cleland program may be summarized as follows:

(a) initialize certain variables to fixed values
(b) read pairs of $s$ and $v$ values with weights, if any
(c) use a matrix method to obtain initial estimates of $K_m$ and $V_{max}$ from a weighted least-squares fit to a reciprocal plot
(d) use matrix methods and initial estimates to fit data points to a hyperbola and obtain fresh estimates of $K_m$ and $V_{max}$
(e) repeat (d) until estimates do not change significantly
(f) calculate standard deviations of estimates
(g) print out the results.

A modified version (5) fits data to the equation

$$\log v = \log [(V_{max}\, s)/(K_m + s)] \tag{14}$$

This form is appropriate when standard deviations of $v$ are directly proportional to $v$ (5).

## 5.7 Comparison of the least-squares fit to a hyperbola with other statistical methods

There are at least two advantages of fitting the ($v$, $s$) values directly to the hyperbola (*Figure 1*), rather than transforming the values to a linear relationship (Equations 2–4). Firstly, weighting is unnecessary provided the errors are homogeneous. Secondly, the computer programs are arranged so that it is simple to 'test' the influence of different types of hypothesized weighting schemes on the final values obtained for $K_m$ and $V_{max}$.

The disadvantages of the least-squares fit to a hyperbola are the assumption of normal distribution of error, the need to know the correct weighting scheme, and its sensitivity to outliers (Section 5, Section 8.6.5). In the absence of information about error distribution and weights it is safer to calculate $K_m$, $V_{max}$, and their standard deviations by the direct linear plot (Section 3.4, Section 4.3). More recently, Cornish-Bowden and Endrenyi (35) have proposed a method based on the least-squares criterion that does not require prior knowledge of the weighting scheme and is insensitive to outliers. The method works well with simulated data, and only awaits testing with experimental results before adoption on a wider scale (22). Least-squares methods, including this novel one (22, 35), have one important advantage over the direct linear plot: they are easily extended to fitting more complex rate equations, e.g. for multi-substrate reactions or inhibitions (3, 4, 5) (Section 6 and Section 7).

### 5.8 Conclusions

In practice, with good data that fit Equation 1, the answers given by each method may not be significantly different. The availabilty of computerized analyses (5, 21, 22, 52) makes it easy to perform several types of analysis on one data set, with varied weighting schemes if needed, and see whether the range of $K_m$ and $V_{max}$ obtained affects the conclusions to be drawn from the experiment.

## 6. Multi-substrate reactions

### 6.1 Introduction

Segal *et al.* (53) were among the first to recognize that steady-state kinetic mechanisms could be distinguished by visual inspection of initial-rate plots. By kinetic mechanism is meant the existence of discrete chemical steps in the enzyme-catalysed reaction which may, or may not, impose an obligatory order of substrate entry and product exit. Subsequently, Alberty (54), Dalziel (55), and Cleland (14) devised systematic ways of distinguishing between different possible mechanisms (reviewed in 8, 9, 10, 11, 12, 15). Numerical analysis of such data permitted determination of $V_{max}$ and $K_m$ for each substrate and, if appropriate, other parameters associated with the particular reaction mechanism.

The complex rate equations that relate velocity to concentrations of several substrates can be written in various ways, but the simplest nomenclature to use is probably that of Cleland (4, 5, 14, 15, 50). His strategy for finding the correct rate equation and measuring kinetic constants is now illustrated, but the reader should consult the more comprehensive reviews listed above before undertaking such an analysis.

## 6.2 Determination of kinetic constants for a two-substrate, ordered reaction

Consider the reaction (Equation 15), in which substrate A combines with an enzyme obligatorily before substrate B, followed by departure of the products in the order P then Q (Equation 16).

$$A + B \rightleftharpoons P + Q \tag{15}$$

$$E \underset{}{\overset{A \quad\quad B \quad\quad\quad\quad\quad P \quad\quad Q}{\overline{\downarrow \;EA\; \downarrow \;EAB, EPQ \uparrow\; EQ\; \uparrow}}} E \tag{16}$$

If the concentration of B is held constant in the absence of P and Q, and then velocities are measured at different concentrations of A, Equation 1 is often followed so that 'apparent' values (Section 2.2) of $K_m^A$ and $V_{max}$ can be determined by the methods described above. These values only hold at the defined concentration of B. Repetition of the measurements at a different fixed concentration of B yields different apparent $K_m^A$ and $V_{max}$ values, and a pattern of reciprocal plots as illustrated in *Figure 7*. The crucial point is that this intersecting pattern is reasonably diagnostic of the mechanism (Equation 16), for which the overall rate equation is

$$v = \frac{V_{max}AB}{K_m^A B + K_m^B A + AB + K_{iA}K_m^B} \tag{17}$$

In reciprocal form this equation leads to the following

$$\frac{1}{v} = \left[\frac{K_m^A}{V_{max}}\left(1 + \frac{K_{iA}K_m^B}{K_m^A B}\right)\right]\frac{1}{A} + \frac{1}{V_{max}}\left(1 + \frac{K_m^B}{B}\right) \tag{18}$$

$$\text{Intercept} = \left(\frac{K_m^B}{V_{max}} \times \frac{1}{B}\right) + \frac{1}{V_{max}} \tag{19}$$

$$\text{Slope} = \left(\frac{K_{iA}K_m^B}{V_{max}} \times \frac{1}{B}\right) + \frac{K_m^A}{V_{max}} \tag{20}$$

The reader should derive Equations 18–20 from Equation 17 to be satisfied that the relationships are linear provided [B] is kept constant.

So, if intercept values ($1/V_{max}^{app}$) are plotted against 1/B the relationship should be linear and the true values of $K_m^B$ and $V_{max}$ can be deduced from the intercept and slope of the re-plot. A second linear re-plot of slope values

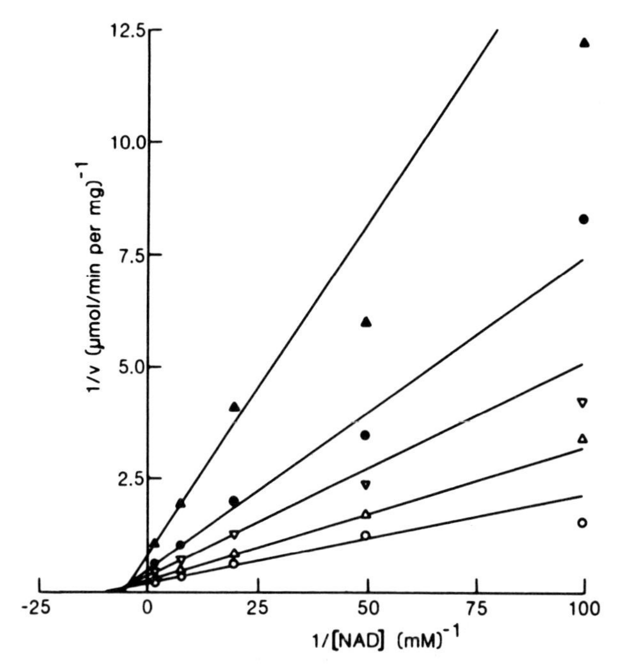


**Figure 7.** 'Primary' double reciprocal plots obtained with the two substrate enzyme, galactose dehydrogenase. NAD concentration was varied between 0.01–0.5 mM, while galactose was at different fixed concentrations: (o) 10 mM; (△) 1 mM; (▽) 0.4 mM; (●) 0.2 mM; (▲) 0.1 mM. Each line was obtained by an unweighted least-squares fit to a hyperbola (Section 5.6). The experiments were performed by Stephen Hopkins (University of Cambridge, 1979).

($K^{app}_m/V^{app}_{max}$) against 1/B yields $K^A_m$ and $K_{iA}$ from its intercept and slope, given the values of $K^B_m$ and $V_{max}$ from the first re-plot (see *Figure 8*).

This process of making primary reciprocal plots and secondary slope and intercept replots is the essence of the Cleland strategy for deducing enzyme mechanisms and the true values of the kinetic constants. The statistical methods used for each stage are summarized below.

### 6.2.1 Data acquisition

Choose concentrations of substrate that, ideally, span the range of $K_m/2$–$8\ K_m$ (i.e. $K^{app}_m$) (Section 3.2). Vary [A] and measure steady-state reaction velocities at a fixed [B]. Repeat the measurements at several different fixed [B]; the range of concentrations of B should similarly cover $K_m/2$–$8\ K_m$. At least five different concentrations of each substrate will be necessary.

### 6.2.2 Data processing (Section 3.4, Section 4.3, and Section 5.6)

Using the direct linear plot or the least-squares fit to a hyperbola (appropriately weighted) the apparent $K_m$, $V_{max}$, $K_m/V_{max}$ (slope), $1/V_{max}$ (intercept)

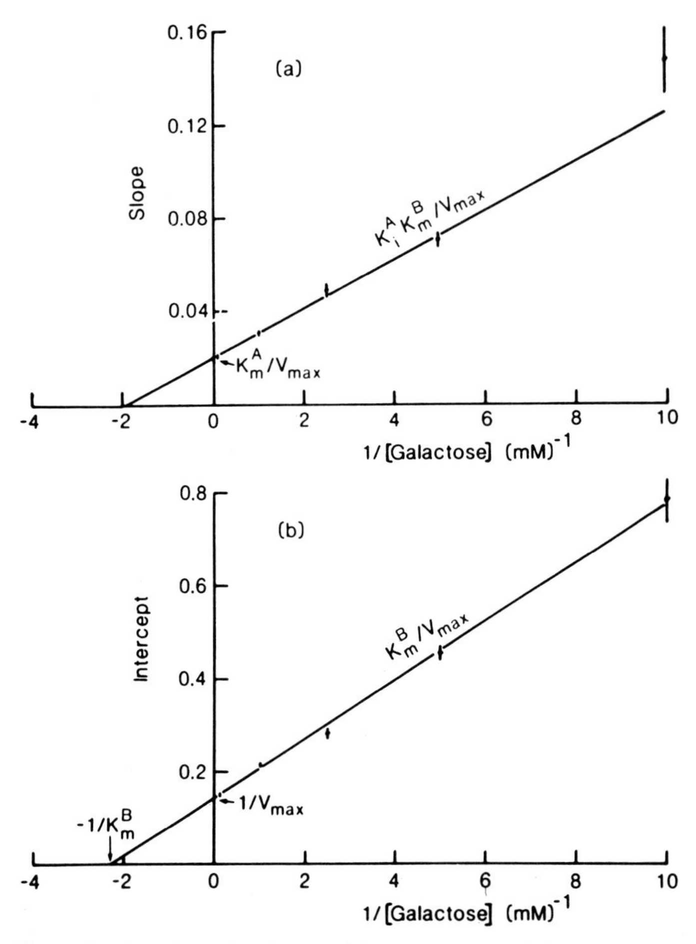


**Figure 8.** 'Secondary' replots of the slopes (a) and intercepts (b) of the lines in *Figure 7* against 1/[galactose]. The vertical bars represent standard deviations calculated by each unweighted least-squares hyperbolic fit. The positions of the lines were determined by an unweighted least-squares fit of all data points to Equation 17. The reaction parameters were found to be $V_{max} = 6.72 \pm 0.10\,\mu$mol/min per mg, $K_{NAD} = 0.133 \pm 0.006$ mM, $K_{galactose} = 0.407 \pm 0.029$ mM, $K_{iNAD} = 0.186 \pm 0.026$ mM.

values, and their standard deviations are determined for substrate A at each fixed concentration of substrate B.

### 6.2.3 Data display

The direct linear plot becomes rather complex with two substrates (but see 17) and any one of the linear transformations in Section 4 is probably more convenient. Cleland routinely used the double-reciprocal plots, because their slopes and intercepts are used directly in subsequent replots. However, for assessing the fit of the data to a particular equation, the $v$ against $v/s$ is plot is better. Note that the transformations are used only for display and assessment, not for quantitative analysis. Chapter 9 of Segel (9) shows the patterns obtained where graphs other than the reciprocal plot are used for the primary analysis. Check carefully that the plots are reasonably linear, indicating that Equation 1 is valid. If they are non-linear, refer to (9, 15, 23).

The vital second stage of this assessment is that the calculated $K^{app}_{m}/V^{app}_{max}$ (slope) or $1/V^{app}_{max}$ (intercept) ratios are plotted against the reciprocal concentration of the second substrate. Standard deviations should be displayed in such re-plots. Check carefully whether each re-plot is reasonably linear. If so, a least-squares fit, weighted according to the reciprocal of the square of each standard deviation, can be used to obtain true $K_m$ and $V_{max}$ values from the slopes and intercepts (Section 5.1; *Figure* 7; 21). Alternatively, the 'true' parameters may be obtained from the hyperbolic relationships of $V^{app}_{max}$ and $V^{app}_{max}/K^{app}_{m}$ to the second (fixed) substrate concentration. A better procedure is to fit all the data directly to the overall rate equation (Section 6.2.4). If the re-plots are non-linear appropriate programs to carry out the least-squares analysis are available (3, 4, 5, 14).

### 6.2.4 Selection of, and fitting data to, the overall equation

The pattern and shape of the primary reciprocal plots (intersecting, parallel, linear, non-linear) combined with the linearity/non-linearity of the secondary re-plots often enables the selection of a rate equation that satisfactorily describes the relationship between reaction velocity and concentrations of all substrates (3, 4, 5, 9, 14, 15).

Sometimes technical difficuties, e.g. inhibition by substrate (e.g. 56), insensitivity of an assay method, expense of a substrate, make it difficult to obtain a complete kinetic pattern. Often sufficient information may nevertheless be acquired by assaying the enzyme in the reverse direction.

Once the appropriate equation has been chosen all the data points can be fitted to it, and all the kinetic parameters with their standard deviations calculated. Cleland devised least-squares-fit procedures for the commonly encountered equations, and his computer programs are readily available (3, 4, 5, 21). The methods rely on the validity of the assumptions described in Section 5, and more robust methods are now available (22, 35). Other computerized techniques for fitting data directly to rate equations of any order of complexity have been described (13, 34, 57, 58, 59, 60). These more general methods are felt to be beyond the scope of this chapter.

The two-substrate enzyme was used as an example here, but the same strategy extends to three-substrate reactions. More plots have to be made and analysed, and the overall rate equation is more complex (reviewed in 9).

Fitting to the complete rate equation is a more satisfactory way of determining all the $K_m$ values, $V_{max}$, and their standard deviations, than the weighted least-squares fit to re-plots described in Section 5.1.

# 7. Inhibition and the determination of the inhibitor constant ($K_i$ values)

## 7.1 Types of inhibitor and the significance of $K_i$

### 7.1.1 'Dead-end' inhibitors

Some compounds combine reversibly with the enzyme and slow the reaction rate, without participating in the catalysed reaction; they are known as 'dead-end' inhibitors (8, 9, 12, 50, 61). Provided such compounds give linear re-plots (Section 7.3), the measured $K_i$ value is the dissociation constant of the enzyme–inhibitor or (enzyme–substrate)–inhibitor complex. This can be used to calculate $\Delta G°$ ($-RT\ln K_i$). A comparison of $\Delta G°$ values for a range of structurally different 'substrate analogue' inhibitors can illuminate the topology of the binding site (e.g. 62).

$K_i$ values with standard deviations provide a quantitative measure of inhibitor potency, which is useful for comparing inhibitors that interact with a single receptor, especially one of pharmacological interest. Conversely, $K_i$ values determined for a single inhibitor reacting with different isoenzymes may be used for comparison of the latter. Note that the concentration of inhibitor required to give 50% inhibition is not necessarily equal to the dissociation constant, though '$I_{50}$' or $K_i^{app}$ can be a useful parameter.

A dead-end inhibitor of the competitive type (Section 7.2) may be particularly useful for distinguishing between different steady-state mechanisms of multi-substrate enzyme reactions (see 63 for review).

### 7.1.2 Product inhibitors

The presence of one or more of the normal products will slow the velocity of the forward reaction. Such 'product inhibition' usually follows competitive, simple non-competitive, mixed non-competitive, or uncompetitive (Section 7.2) patterns and is invaluable for discriminating between different mechanisms of multi-substrate enzyme reactions (8, 9, 12, 50, 64). These and other references describe the derivations of the appropriate rate equations and the predicted graphical patterns, and include very useful tables of the inhibition patterns expected for different mechanisms (9, 65). Note that the apparent $K_i$ value determined for product inhibitors is not necessarily the true dissociation constant, but is a ratio of velocity constants, the composition of which depends on the reaction mechanism (9, 50).

### 7.1.3 Alternative substrate inhibitors

The velocity of enzyme-catalysed conversion of one substrate is reduced if an alternative substrate is introduced. Again, the inhibition patterns observed can help reveal the kinetic mechanism (9, pp. 790–810; 66). The technique is facilitated by the assay of a reaction with radioisotope-labelled substrates, e.g. in studies of transport, where inhibition by an unlabelled alternative substrate is easy to measure. Kinetic constants for both the labelled and unlabelled compound can be determined (9).

The apparent $K_i$ value is not necessarily the same as the $K_m$ for the alternative substrate. The correct interpretation is not usually difficult (9, 50), but it is unfortunately common to equate the apparent $K_i$ with the apparent $K_m$ in studies of inhibition by alternative substrates.

### 7.1.4 Substrate inhibitors

Sometimes higher, usually non-physiological, concentrations of substrate inhibit activity of an enzyme. The possible reasons are discussed by Segel (9), Cornish-Bowden (8), and Cleland (15, 56). Such 'substrate inhibition' may be diagnosed by the occurrence of non-linear reciprocal plots, or by unconventional patterns of reciprocal plots when they are repeated at different fixed concentrations of a second inhibiting substrate (see above references). Substrate inhibition may appear to be a nuisance in the statistical analysis of data, because measurements must be made at more substrate concentrations. Nevertheless, the effect can be analysed quantitatively (56) and may be very useful in discriminating between different steady-state mechanisms of enzyme action (5, 8, 9, 15, 55, 56).

If the substrate simply combines as a dead-end inhibitor with an enzyme form with which it is not supposed to react, then the $K_i$ value determined will be the true dissociation constant.

### 7.1.5 Irreversible chemical modification (inactivation)

A reagent that chemically modifies an enzyme may inhibit its catalytic activity. Such a case cannot be analysed by the steady-state methods to be described below, which apply only to reversible inhibitors.

### 7.1.6 Tight-binding inhibitors

Some inhibitors, e.g. transition state analogues, antibiotics, and antibodies, bind very tightly to an enzyme, but without chemical modification. Often, their slow interaction with the enzyme precludes a steady-state analysis. Provided the steady-state is achieved and the ratio $[E]/K_i^{app}$ is less than 0.01, the analysis described in Section 7.2 is valid. If $[E]/K_i^{app}$ is greater than 0.01, a different steady-state approach is possible (67, 68), but non-steady-state methods may be easier (69, 70, 71).

## 7.2 Choice between competitive, simple non-competitive, mixed non-competitive, or uncompetitive mechanisms, and the statistical determination of $K_i$ value(s)

Dead-end, product, alternative substrate, and substrate inhibition can all be examined by the approach described in this section. The strategy now described for graphical/statistical analysis is analogous to that given in

Section 6 for a two-substrate reaction and was first described in detail by Cleland (50).

### 7.2.1 Data acquisition

As before (Section 3.2) the substrate(s) and inhibitor concentrations must be chosen to define hyperbolic saturation curves. Thus, concentrations from $K^{app}_{m}/2$–$8K^{app}_{m}$ (or higher) should be used for substrate(s) and $K_i/2$–$8K_i$ (or higher) for inhibitor (8). With this in mind, the following strategy is followed using again the two substrate reaction (Equation 15, 16) as an example.

Vary [A] at fixed concentrations of B and I, and measure steady-state reaction velocities; repeat this at several different fixed [I]. Vary [B] at fixed concentrations of A and I; repeat this at several different fixed [I].

### 7.2.2 Data processing

Using the direct linear plot or the least-squares fit to a hyperbola (appropriately weighted), the apparent $K_m$, $V_{max}$, $K_m/V_{max}$ (slope), $1/V_{max}$ (intercept) values and their standard deviations are determined for each substrate at each fixed inhibitor concentration.

### 7.2.3 Data display

Any one of the linear transformations in Section 4 is probably more convenient for plotting inhibition data than the direct linear plot (but see 17). Cleland routinely used double reciprocal plots, because their slopes and intercepts are used directly in subsequent re-plots. In our two-substrate reaction example $1/v$ is plotted against 1/[A] at different fixed [I] and $1/v$ is plotted against 1/[B] at different fixed [I]. Other methods of plotting inhibition data are illustrated in Chapters 3 and 4 of Segel (9). Check carefully that the plots are reasonably linear indicating that Equation 1 is valid. If they are non-linear, refer to Cleland (15, 50) or Segel (9).

The vital second stage is that for each 'primary' reciprocal plot the apparent $K_m/V_{max}$ (slope) and the apparent $1/V_{max}$ (intercept) values are re-plotted against [I]. Standard deviations should be displayed in such 'secondary' plots. Check carefully whether each replot is reasonably linear. If so, a least-squares fit, weighted according to the reciprocal of the square of each standard deviation, can be used to obtain estimates of the $K_i$ value(s) from the intercept(s) on the [I] axis. A much better procedure is to fit all the data directly to the overall rate equation (Section 7.2.4). Non-linearity of the re-plots can result from a number of circumstances, including partial inhibition, binding of more than one inhibitor molecule, inhibition by an alternative product, or allosteric effects; these possibilities can often be analysed by appropriate programs (3, 4, 5, 50).

### 7.2.4 Selection of, and fitting data to, the overall rate equation

Inhibitors can be divided into three classes, competitive, mixed (non-competitive), or uncompetitive, on the basis of their influence on the slopes, the intercepts, or both slopes and intercepts of the primary $1/v$ against $1/s$ plots (*Figure 9*; see also 8, 9, 10, 12, 25). The competitive (intersecting on the $1/v$ axis), non-competitive (intersecting to the left of the $1/v$ axis), or uncompetitive (parallel) pattern and shape (linear or non-linear) of the primary plots (*Figure 9*), combined with the linearity/non-linearity of the secondary re-plots, is diagnostic of the relationship between reaction velocity, concentration of substrate, and concentration of inhibitor (3, 4, 5, 15, 50). The appropriate equation can then be selected and all the data fitted to it by a computerized least-squares-fit method (3, 4, 5). This is the most satisfactory way of obtaining the $K_i$ value(s), $K_m$ value(s), $V_{max}$, and their standard deviations.

Particular carc must bc exercised when the inhibition is thought to be competitive or uncompetitive. In the former, $1/V_{max}$ (intercept), and in the latter $K_m/V_{max}$ (slope), should remain constant. However, a small, but systematic, increase in either parameter with increasing [I] may be difficult to detect. Re-plots which include standard deviations must be examined rigorously. Better, the data can be fitted to the equation for non-competitive inhibition as well as the equation for the more obvious mechanism. If the non-competitive fit gives a significant second $K_i$ value (e.g. 5) then a simple competitive or uncompetitive mechanism is contra-indicated.

### 7.3 Use of the Dixon plot

Dixon (72) described how patterns of $1/v$ against [I] plots obtained at different fixed [$s$] can be used to diagnose and display mechanisms of inhibition. Such plots have been widely used but they do not on their own distinguish between the competitive and mixed non-competitive cases. Such a distinction can be made when they are combined with $s/v$ against [I] plots (73).

$K_i$ value(s) can be deduced from visual inspection of such plots, but a much more reliable value can be obtained from a least-squares fit to the appropriate overall rate equation, as described in Section 7.2.4.

## 8. Uses for the standard deviations

### 8.1 A warning—the estimate of $V_{max}$ is not independent of the estimate of $K_m$

Cleland (4, 5) pointed out that the estimates of $V_{max}$ and $K_m$ are not statistically independent of each other. Hence there will be a strong positive correlation between the standard deviation calculated for $V_{max}$ and that calculated for $K_m$. If, for example, the purpose of calculating the parameters is to compare two isoenzymes, it would be most unwise to compare just two calculated $K_m$ values, or just two calculated $V_{max}$ values, with their standard deivations. Instead, the ($K_m$ and $V_{max}$) pair for one enzyme should be

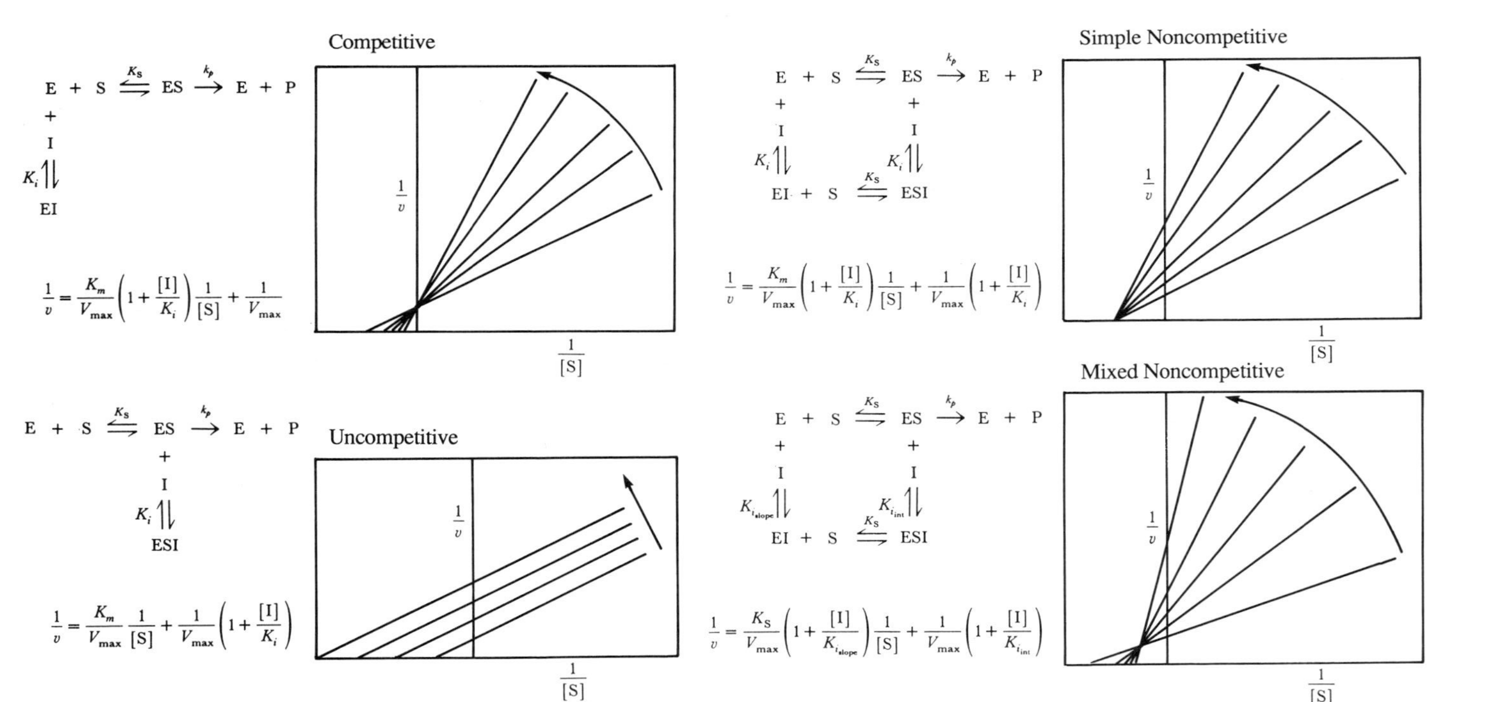


**Figure 9.** Illustrations of competitive, uncompetitive, simple non-competitive, and mixed non-competitive inhibition. Double reciprocal (Lineweaver–Burk) plots are shown for each case. The direction of the arrow indicates increasing concentrations of inhibitor; the lines are calculated for [I] = O, [I] = 0.5 $K_i$, [I] = $K_i$, [I] = 2 $K_i$, [I] = 3 $K_i$, except for the uncompetitive case, where the second line is omitted. For a single-substrate reaction the different mechanisms are consequences of the inhibitor combining only with free enzyme (competitive), only with ES complex (uncompetitive), or with both (non-competitive); in the non-competitive case if the dissociation constants for combination with both free enzyme and ES complex are the same, then the family of reciprocal plots intersects on the $1/s$ axis (simple non-competitive inhibition), whereas if they are different the lines intersect above or below the $1/s$ axis, depending on the ratio of $K_{ii}/K_{is}$ values (mixed non-competitive inhibition). For multi-substrate reactions the algebraic origins of the plots are more complex, but essentially the same patterns arise and can be analysed to yield true $K_i$ values (9, 10, 50, 65). This figure is derived from data in Segel (1975).

compared with the ($K_m$ and $V_{max}$) pair for the other. The relative error (i.e. standard deviation/mean) of $K_m/V_{max}$ (calculated by the appropriate computer programs described above) is invariably smaller than that of $K_m$ and usually smaller than that of $V_{max}$; hence putative errors in $K_m$ or $V_{max}$ should be consistent with the standard deviation of $K_m/V_{max}$ (7).

## 8.2 A second warning—the reliability of standard deviations depends upon the validity of the weighting scheme used

Standard deviations calculated from least-squares fit equations are unlikely to be correct if the wrong weighting scheme is used (20, 24). If there is uncertainty about the weighting, standard deviations calculated by non-parametric methods are safer (20).

## 8.3 Tests for the significance of the difference between two estimates

It may be that $K_m$ or $V_{max}$ determined on one day appears to be different from that determined on another. Or it may be that the $V_{max}$ values for two putative isoenzymes are clearly the same but it is not clear whether the $K_m$ values are different. The experimenter may wish to perform a statistical test to find the level of significance of the difference. Given that the standard deviation of each estimate has been reliably calculated, Johansen and Lumry (1) proposed that the difference is tested as follows.

Suppose two $K_m$ values are being compared, represented by $K_1 \pm \sigma_1$ and $K_2 \pm \sigma_2$.

$$t = \frac{\text{difference between values}}{\text{standard error of the difference}} = \frac{K_1 - K_2}{\sqrt{(\sigma_1^2 + \sigma_2^2)}} \tag{21}$$

This ratio can be used in three ways (Section 8.3.1–8.3.3). The validity of each depends on the assumptions that can be made about the weighting scheme used in the original fit.

### 8.3.1 The correct weighting scheme is known

In this case $t$ is calculated for infinite degrees of freedom. If $t$ is less than 1.96 the probability, $P$, that $K_1$ and $K_2$ are identical is greater than 5%, i.e. they may be the same. If $t$ is between 1.96 and 2.58 then $P$ is between 1% and 5% and $K_1$ and $K_2$ are probably different. If $t$ is greater than 2.58 then $P$ is less than 1%; $K_1$ and $K_2$ are, therefore, almost certainly different.

### 8.3.2 The weighting scheme is uncertain, but the experimental techniques are comparable

The assumption here is that, although the weighting scheme is not known, the same scheme applies in both experiments. The results can then be compared by Student's $t$-test (47); $t$ is calculated as in Equation 21. From a table of the $t$ distribution (available in sets of statistical tables and many textbooks) the critical values of $t$ for $n_1 + n_2 = 4$ degrees of freedom at the 5% and 1% levels are obtained, and compared with the calculated value; $n_1$ and $n_2$ are the number of points in each experiment. If $n_1$ is very different from $n_2$ or $\sigma_1$ is very different from $\sigma_2$, a more rigorous procedure should be adopted (4, p. 20).

### 8.3.3 Weighting scheme unknown; incomparable experimental techniques

In this case, the $d$ test of Fisher (74) is used; $d$ is the same as $t$ (Equation 21), but a second quantity ($\theta$) is also calculated.

$$\theta = \tan^{-1}(\sigma_1/\sigma_2) \tag{22}$$

From Table VI or VI.1 of Fisher and Yates (75), the critical $d$ values of the 5% and 1% levels of significance ($P$) for $n_1 - 2$ and $n_2 - 2$ and the calculated angle $\theta$ are obtained. If one of the weighting schemes is known, the corresponding number of degrees of freedom is taken as infinity.

### 8.3.4 Real value of the statistical tests of difference

According to Johansen and Lumry (1) if the calculated statistic, $t$ or $d$, exceeds the critical value at the 5% (or 1%) level of $P$, the values in question are probably (or almost certainly) different. Note that two more assumptions apply in Section 8.2.1, Section 8.2.2, and Section 8.2.3; they are that the data actually fit Equation 1, and that replicate determinations of $K_1$ or $K_2$ would follow a normal distribution.

It is noteworthy, if not significant, that none of the wealth of textbooks on enzyme kinetics appearing after Johansen and Lumry's paper (1) have mentioned the tests described above. This may be because they are not necessarily superior to the many other tests available (e.g. 47), especially if non-parametric tests should be applied. However, it may also reflect that a statistical test rarely does more than confirm the experimenter's intuitive opinion! Common sense may be the best guide in deciding whether two values are the same, different, or not yet sufficiently different. Cornish-Bowden (7) rightly said, 'It is often safer to avoid statistical tests unless one thoroughly understands their theoretical basis: if one blindly applies a textbook recipe and reaches a conclusion that is obviously wrong, one is in danger of looking foolish.'

## 8.4 Averaging parameter estimates and their standard deviations

Suppose a kinetic analysis is repeated several times. Several estimates of $K_m$, $V_{max}$, or $K_i$ would be obtained, each with its calculated standard deviation.

How should the estimates be averaged, and what would be the standard deviation of the mean value? Since the standard deviations are likely to be different, the weighted mean must be calculated (1, 47).

$$\bar{x} = \Sigma w_i x_i / \Sigma w_i \tag{23}$$

So for several $K_m$ values:

$$\bar{K}_m = \left(\frac{K_1}{\sigma_1^2} + \frac{K_2}{\sigma_2^2} + \cdots \frac{K_n}{\sigma_n^2}\right) \Big/ \left(\frac{1}{\sigma_1^2} + \frac{1}{\sigma_2^2} \cdots \frac{1}{\sigma_n^2}\right) \tag{24}$$

Provided that the standard deviation of one determination is uncorrelated with that of another, the standard deviation of the mean $K_m$ ($\sigma_{\bar{K}_m}$) may be calculated from *Equation 25* (1, 47).

$$\sigma_{\text{mean}} = \sqrt{1/\Sigma(1/\sigma_i^2)} \tag{25}$$

$$\sigma_{\bar{K}_m} = 1 \Big/ \sqrt{\frac{1}{\sigma_1^2} + \frac{1}{\sigma_2^2} + \ldots \frac{1}{\sigma_n^2}} \tag{26}$$

The averaging of parameter values and their standard deviations may be unsafe. As an example, it is often seen that on one day precise values of $K_m$ and $V_{max}$, i.e. with low standard deviations, are obtained, but they are different from precise determinations made on another day. It is better to acknowledge the existence of uncontrolled experimental variations than to hide them in specious mean values.

## 8.5 Choice of the best model for the kinetic mechanism

In multi-substrate reactions (Section 6), $K_m$ and other parameters are determined in relation to a particular mathematical model of the enzyme mechanism. Often there is a choice between two or more models, and the problem is to decide which best fits the data. Reich and co-workers provided a computerized strategy for such selections, and illustrated types of defects that may occur (see *Figure 10*) (31, 59). In general, several criteria can be used to distinguish between models (e.g. 5, 7, 8, 13, 34, 42, 59, 76). The following are based on those of Bartfai and Mannervik (77, 78, 79):

(a) Success of a regression analysis. If regression methods converge to an optimized fit for one model, but fail to do so for another, the latter model would be rejected in favour of the former.

(b) Parameter values. Models which yield unreasonable (e.g. negative) parameters or unreliable (e.g. large standard deviation) parameters would be rejected.

(c) Distribution of residuals. Models which yield residual deviations (between predicted and measured values) that are non-randomly distributed or do not have zero mean would be rejected.

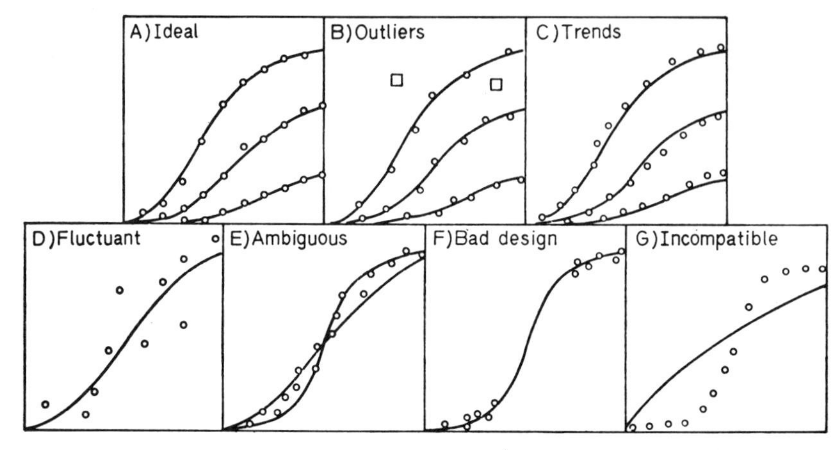


**Figure 10.** Possible defects of a parametric fit. All entries represent possible outcomes of the fitting procedure. The drawn curves are the model predictions, the points refer to (assumed) observations. Ordinate is *v*. Abscissa (A–F) is log substrate, (G) is substrate. (A) ideal fit with random deviations only; (B) outliers (squares) in the observations; (C) data trends (a series of observations above, another series below the predicted curve); (D) statistical uncertainty of the data; (E) ambiguous model (data do not uniquely determine the parameter values of the model); (F) bad experimental design (not the whole model curve has been experimentally tested, hence the half-saturation constant is uncertain), the situation is similar to (E); (G) incompatible model (the data obey a different rate law), this situation is reminiscent of (C). This figure is reproduced from Reich *et al.* (59).

(d) Sum of squares of residuals. The sum of squares of residual deviations should not be excessive relative to the variance of the experimental observations (7, p. 188).

(e) The smallest mean sum of squares. Of the models satisfying criteria (a)–(d), the one that yields a significantly smaller sum of squares would be the best.

## 8.6 Reasons for 'excessively large' standard deviations

It is clearly difficult to quantitate 'excessively large', though Cleland suggested that standard deviations should not be greater than 10% of the parameter values (4). However, it may be helpful to list a number of sources of error that, singly or in combination, contribute to imprecision in the measured estimates.

### 8.6.1 Excessive variability in velocity measurements

The reproducibility of measurement of $v$ should not vary beyond ±5%. This should usually be attainable with the laboratory methods currently in use, even at low $v$ values. Recorder chart speeds should be adjusted so that slopes of lines near 45° are measured. Note that pipetting should be accurate to

±1%, since it is a basic assumption that $s$ is known precisely (Section 5, introduction).

### 8.6.2 Insufficient number of data points

Calculation of standard deviations involves division by a term containing $n$, the number of data points (e.g. 47). So the higher the value of $n$ the smaller will be the standard deviations.

### 8.6.3 Choice of the incorrect model

See Section 8.5.

### 8.6.4 Inappropriate spread of data points throughout the chosen model

The distribution of substrate concentrations optimal for discriminating between models is not necessarily the best for determination of parameter values (13, 19, 24, 34). Furthermore, the best distribution of substrate concentrations is critically dependent upon the weighting scheme to be applied (13, 24, 32, 34). A relatively large standard deviation may simply reflect the choice of an inappropriate range of substrate concentrations, as illustrated by the following two examples.

It may be that a competitive inhibitor is investigated at only relative high concentrations of $s$ that substantially combat its potency. This will nearly always be reflected in a large standard deviation of the $K_i$ value determined by the methods in Section 7.1. In this case, the analysis should be repeated to include low substrate concentrations. A second and common example is the estimation of $K_m$ and $V_{max}$ from data determined at a range of $s$ below the $K_m$ value. This tends to give aesthetically pleasing linear plots, which is not surprising because a fit is being made to a small section of an hyperbola that tends to linearity anyway; $1/v$ against $1/s$ or $s/v$ against $s$ linear re-plots appear to pass very close to the origin (12). When the standard deviation of $V_{max}$ is determined, it is consequently found to be proportionately large. The error on $K_m$ is also high. In this case, the remedy is to include velocity measurements made at higher concentrations of $s$.

Sometimes a compromise is inevitable. Determination of velocity values near $V_{max}$ involves relatively large amounts of substrate, which may be impracticable because of a solubility limit or expense.

### 8.6.5 The inclusion of outliers

Sometimes a point(s) appear(s) to be totally different from the pattern displayed by the majority (*Figure 10*). Such apparently aberrant points are known as outliers (46). They may have a profound effect on the estimates of $K_m$ and $V_{max}$ and their standard deviations, particularly if the least-squares criterion of best fit has been used (an outlier makes a disproportionately large contribution to the sum of squares). One way of minimizing their effects is to use the direct linear plot (Section 4.3). Alternatively, they may be excluded from the analysis if they fulfil some criterion of rejection, e.g. that they are more than two standard deviations from the fitted best line. Note that there is no statistical justification for such rejection, it is an arbitrary choice of the experimenter.

The more points there are, the less will be the effect of an outlier. A possible criterion for rejection is that removal of the outlier(s) should substantially decrease the calculated standard deviation without significantly altering the mean parameter values.

## 8.7 Another warning—systematic errors

It may be that the standard deviations are satisfyingly small, but nevertheless $K_m$, $V_{max}$, or $K_i$ have been determined inaccurately because of some systematic error in the velocity measurements. Such errors could arise from using an incorrectly calibrated pipette, making an erroneous correction for a non-enzymic reaction rate, electrical/mechanical faults in a spectrophotometer, and so on. Day-to-day variability is often observed; on one day precise estimates are made, but the mean values are different from precise determinations made on a different day. Cornish-Bowden (7) quotes an example of Hwang *et al.* (80) in which $K_m$ values of bacterial nitrogenase showed marked day-to-day variation. Unless it is established that such fluctuations are normally distributed about a mean value (i.e. genuinely random variation), it is unsafe simply to average them (Section 8.4).

Clearly such systematic errors could be insidiously misleading, particularly in cases where sets of kinetic parameters are being compared. Repetition of the $K_m$ and $V_{max}$ determination on different days, different times, with different experiments, and so on, is one way to detect such errors, but the time and expense involved must be weighed against the type of conclusion originally required from the experiment (see Introduction).

# 9. The analysis of Hill plots

Equation 1 does not hold for some enzymes where the rate equation may be described by a polynomial function, and which exhibit a sigmoidal, rather than a hyperbolic, relationship between $v$ and $s$. As discussed by many authors (e.g. 7, 23, 42) a sigmoidal relationship may be described by the Hill equation (Equation 27), which can be rearranged to a linear form (Equation 28)

$$v = \frac{V_{max} s^h}{K + s^h} \tag{27}$$

$$\log \frac{v}{V_{max} - v} = h \log s - \log K \tag{28}$$

$K$ and $h$ (the 'Hill coefficient') are constants, the values of which can be useful for comparative purposes, or for speculating on the physiological role of the enzyme. The value of $h$ can be interpreted as a minimum estimate of the number of subunits in the enzyme, but great caution should be exercised in assigning any biological or physical meaning to $h$ or $K$.

If $V_{max}$ is known, $h$ or $K$ may be estimated from the slope and intercept of a plot of log $(v/V_{max} - v)$ against log $s$, the Hill plot. There are at least two problems with this. One is that $h$ often varies with $s$, so that a curve rather than a straight line is obtained; the steepest, and approximately linear, section of such a curve (usually when $s \simeq K$) can be used to obtain a minimum value of $h$. The second is the inherent unreliability of unweighted fits made by eye to a 'linear' transformation of the original equation (Section 4 and Section 5.1).

A safer analytical procedure is to estimate $V_{max}$, $h$, and $K$ by a computerized fit directly to Equation 27. Such programs that use a least-squares criterion of best fit have been described by Atkins (57, 81), Nimmo and Bauermeister (82), and Cleland (5); a more general method is that of Reich *et al.* (59). Note that calculation of $V_{max}$, $K$, and $h$, and their standard deviations by such statistically sound techniques does not improve their chances of having a real relationship to biological parameters, rate constants, and so on.

## Acknowledgements

I am indebted to Dr R. Eisenthal and Dr M. J. Danson for their very helpful advice during the preparation of this article. Professor W. W. Cleland, Dr A. Cornish-Bowden, Professor J. G. Reich, Dr L. Endrenyi, and Dr W. Trager very kindly gave permission to reproduce figures and use computer programs from their published and unpublished work. I thank Mrs R. Baxter for invaluable help in the preparation of the manuscript.

## References

1. Johansen, G. and Lumry, R. (1961). *C.R. Trav. Lab. Carlsberg*, **32,** 185.
2. Wilkinson, G. N. (1961). *Biochem. J.*, **80,** 324.
3. Cleland, W. W. (1963). *Nature*, **198,** 463.
4. Cleland, W. W. (1967). Adv. Enzymol., **29,** 1.
5. Cleland, W. W. (1979). In: *Methods in Enzymology*, Vol. 63, (ed. D. L. Purich), pp. 103–138. Academic Press, New York and London.
6. Cornish-Bowden, A. and Eisenthal, R. (1974). Biochem. J., **139,** 721.
7. Cornish-Bowden, A. (1976). *Principles of Enzyme Kinetics*. Butterworths, London.
8. Cornish-Bowden, A. (1979). *Fundamentals of Enzyme Kinetics*. Butterworths, London.
9. Segel, I. H. (1975). *Enzyme Kinetics*. Wiley, London.

10. Dixon, M., Webb, E. C., Thorne, C. J. R., and Tipton, K. F. (1979). *Enzymes*, (3rd edn). Longman, London.
11. Purich, D. L. (ed.) (1979). *Enzyme Kinetics and Mechanism, Methods in Enzymology*, Vol. 63. Academic Press, New York and London.
12. Engel, P. C. (1981). *Enzyme Kinetics*, (2nd edn), Chapman and Hall, London.
13. Endrenyi, L. (1981). In: *Kinetic Data Analysis*. (ed. L. Endrenyi), pp. 137–167. Plenum Press, New York and London.
14. Cleland, W. W. (1963). *Biochim. Biophys. Acta*, **67,** 104.
15. Cleland, W. W. (1970). In: *The Enzymes*, (3rd edn), Vol. 2, (ed. P. D. Boyer), pp. 1–65. Academic Press, New York and London.
16. Lineweaver, H. and Burk, D. (1934). *J. Am. Chem. Soc.*, **56,** 658.
17. Eisenthal, R. and Cornish–Bowden, A. (1974). *Biochem. J.*, **139,** 715.
18. Cornish–Bowden, A. (1975). *Biochem. J.*, **149,** 305.
19. Porter, W. R. and Trager, W. F. (1977). *Biochem. J.*, **161,** 293.
20. Cornish–Bowden, A., Porter, W. R., and Trager, W. F. (1978). *J. Theor. Biol.*, **74,** 163.
21. Henderson, P. J. F. (1985). In: *Techniques in the Life Sciences, Protein and Enzyme Biochemistry, BS114*, Vol. B1/II (ed. K. F. Tipton). Elsevier, Ireland.
22. Cornish–Bowden, A. (1985). *Techniques in the Life Sciences, Protein and Enzyme Biochemistry, BS115*. Vol. B1/II (ed. K. F. Tipton). Elsevier, Ireland.
23. Keleti, T. (1981). In: *Kinetic Data Analysis*. (ed. L. Endrenyi), pp. 353–373. Plenum Press, New York and London.
24. Endrenyi, L. and Kwong, F. H. F. (1972). In: *Analysis and Simulation of Biochemical Systems*, (ed. H. C. Hemker and B. Hess), pp. 219–237. North-Holland, Amsterdam.
25. Wharton, C. W. and Eisenthal, R. (1981). In: *Molecular Enzymology*, pp. 281–4. Blackie, Glasgow.
26. Duggleby, R. G. (1981). In: *Kinetic Data Analysis*, (ed. L. Endrenyi), pp. 169–179. Plenum Press, New York and London.
27. Storer, A. C., Darlison, M. G., and Cornish–Bowden, A. (1975). *Biochem. J.*, **151,** 361.
28. Siano, D. B., Zyskind, J. W., and Fromm, H. J. (1975). *Arch. Biochem. Biophys.*, **170,** 587.
29. Askelöf, P., Korsfeldt, M., and Mannervik, B. (1976). *Eur. J. Biochem.*, **69,** 61.
30. Mannervik, B., Jakobsen, I., and Warholm, M. (1979). *Biochim. Biophys. Acta*, **567,** 43.
31. Reich, J. G. (1970). *FEBS Lett.*, **9,** 245.
32. Endrenyi, L. and Kwong, F. H. F. (1981). In: *Kinetic Data Analysis*, (ed. L. Endrenyi), pp. 98–103. Plenum Press, New York, London.
33. Cornish–Bowden, A. (1981). In: *Kinetic Data Analysis*. (ed. L. Endrenyi), pp. 105–119. Plenum Press, New York and London.
34. Mannervik, B. (1981). In: *Kinetic Data Analysis*, (ed. L. Endrenyi), pp. 105–119. Plenum Press, New York and London.
35. Cornish–Bowden, A. and Endrenyi, L. (1981). *Biochem. J.*, **193,** 1005.
36. Cornish–Bowden, A. and Eisenthal, R. (1978). *Biochim. Biophys. Acta*, **523,** 268.
37. Woolf, B. (1932), cited by Haldane, J. B. S. and Stern, K. G. (1975). In: *Allgemeine Chemie der Enzyme*, pp. 119–120. Steinkopff, Dresden and Leipzig.

38. Hanes, C. S. (1932). *Biochem. J.*, **26,** 1406.
39. Eadie, G. S. (1942). *J. Biol. Chem.*, **146,** 85.
40. Hofstee, B. H. J. (1952). *J. Biol. Chem.*, **199,** 357.
41. Dowd, J. E. and Riggs, D. S. (1965). *J. Biol. Chem.*, **240,** 863.
42. Wong, J. F.-F. (1975). *Kinetics of Enzyme Mechanisms*. Academic Press, London.
43. Atkins, G. L. and Nimmo, I. A. (1975). *Biochem. J.*, **149,** 775.
44. Wharton, C. W. (1983). *Biochem. Soc. Trans.*, **11,** 817.
45. Nimmo, I. A. and Mabood, S. F. (1979). *Anal. Biochem.*, **94,** 265.
46. Draper, N. R. and Smith, H. (1966). *Applied Regression Analysis*, Wiley, New York.
47. Colquhoun, D. (1971). *Lectures on Biostatistics*, Clarendon, Oxford.
48. Burk, D. (1934). *Ergebrisse der Enzymforschung*, **3,** 23.
49. Mabood, S. F., Newman, P. F. J., and Nimmo, I. A. (1977). *Biochem. Soc. Trans.*, **5,** 1540.
50. Cleland, W. W. (1963). *Biochim. Biophys. Acta*, **67,** 173.
51. Cornish–Bowden, A. (1982). *J. Mol. Sci.* (Wuhar, China), **2,** 107.
52. Leatherbarrow, R. J. (1990). *TIBS*, **15,** 455.
53. Segal, H. L. K., Kachmar, J. F., and Boyer, P. D. (1952). *Enzymologia*, **15,** 187.
54. Alberty, R. A. (1953). *J. Am. Chem. Soc.*, **75,** 1928.
55. Dalziel, K. (1957). *Acta. Chem. Scand.*, **11,** 1706.
56. Cleland, W. W. (1979). In: *Methods in Enzymology*, Vol. 63, (ed. D. L. Purich), pp. 500–513. Academic Press, New York and London.
57. Atkins, G. L. (1971). *Biochim. Biophys. Acta*, **252,** 405.
58. Ottaway, J. H. and Apps, D. K. (1972). *Biochem. J.*, **130,** 861.
59. Reich, J. G., Wangermann, G., Falck, M., and Rohde, K. (1972). *Eur. J. Biochem.*, **26,** 368.
60. Garfinkel, L., Kohn, M. C., and Garfinkel, D. (1977). *CRC Crit. Rev. Bioeng.*, **2,** 329.
61. Todhunter, J. A. (1979). In: *Methods in Enzymology*, Vol. 63, (ed. D. L. Purich), pp. 383–411. Academic Press, New York and London.
62. Shindler, J. S. and Bardsley, W. G. (1976). *Biochem. Pharmacol.*, **25,** 2689.
63. Fromm, H. J. (1979). In: *Methods in Enzymology*, Vol. 63, (ed. D. L. Purich), pp. 467–486. Academic Press, New York and London.
64. Rudolph, F. B. (1979). In: *Methods in Enzymology*, Vol. 63, (ed. D. L. Purich), pp. 411–436. Academic Press, New York and London.
65. Plowman, K. M. (1972). *Enzyme Kinetics*. McGraw-Hill, London and New York.
66. Huang, C. Y. (1979). In: *Methods in Enzymology*, Vol. 63, (ed. D. L. Purich), pp. 486–500. Academic Press, New York and London.
67. Morrison, J. F. (1969). *Biochim. Biophys. Acta*, **185,** 269.
68. Henderson, P. J. F. (1972). *Biochem. J.*, **127,** 321.
69. Cha, S. (1975). *Biochem. Pharmacol.*, **24,** 2177.
70. Cha, S. (1975). *Biochem. Pharmacol.*, **25,** 2695.
71. Williams, J. W. and Morrison, J. F. (1979). In: *Methods in Enzymology*, Vol. 63, (ed. D. L. Purich), pp. 437–467. Academic Press, New York and London.
72. Dixon, M. (1953). *Biochem. J.*, **55,** 170.
73. Cornish–Bowden, A. (1974). *Biochem. J.*, **137,** 143.
74. Fischer, R. A. (1956). *Statistical Methods and Scientific Inference*. Oliver and Boyd, Edinburgh.

75. Fisher, R. A. and Yates, F. (1957). *Statistical Tables for Biological, Agricultural and Medical Research*. Oliver and Boyd, Edinburgh.
76. Ottaway, J. H. (ed.) (1969). *FEBS Lett.*, **2,** S1.
77. Bartfai, T. and Mannervik, B. (1972). *FEBS Lett.*, **26,** 252.
78. Bartfai, T. and Mannervik, B. (1973). *FEBS Lett.*, **32,** 174.
79. Mannervik, B. and Bartfai, T. (1973). *Acta Biol. Med. Germ.*, **31,** 203.
80. Hwang, J. C., Chen, C. H., and Burris, R. M. (1973). *Biochim. Biophys. Acta*, **292,** 256.
81. Atkins, G. L. (1973). *Eur. J. Biochem.*, **33,** 175.
82. Nimmo, I. A. and Bauermeister, A. (1977). *Anal. Biochem.*, **82,** 468.

# 11

# Buffers and the determination of protein concentrations

LEWIS STEVENS

## 1. Introduction

Probably the most frequently employed procedures in any biochemical laboratory are the use of buffers and the measurement of protein concentration. Perhaps because of their routine nature these procedures are not always given the attention that they deserve. As an example, it is probably the case today that many biochemists requiring a buffer in the region of neutral pH automatically opt for Tris; however, for many reasons this may not be the most suitable for the purpose. Nor is the choice of a method for protein determination without pitfalls; much reagent, time, expense, and valuable protein have been wasted in using a sensitive colorimetric method when UV absorption would have been perfectly adequate. The purpose of this chapter is to describe those features of buffers and methods of protein estimation that will allow the worker at the bench to make a rational choice.

## 2. Buffers and pH

A convenient definition of a buffer was first made by van Slyke in 1922 as 'a substance which by its presence in solution increases the amount of acid or alkali that must be added to cause unit change in pH'. The importance of controlling pH with buffers in the quantitative estimation of enzyme activity was recognized by the beginning of this century. Until the mid-1960s the range of buffers used in enzyme assays was limited to the old tried and tested such as phosphate, carbonate/bicarbonate, acetate/acetic acid, Tris/Tris–HCl, pyrophosphate, triethanolamine/triethanolamine–HCl, maleate/maleic acid, glycine/glycine–HCl. In 1966 Good *et al.* (1) made a systematic study of new buffers based mainly on organic amines and sulphonic acid derivatives which buffered in the range pH 6.15–8.35 and which met a number of criteria which made them suitable for assaying enzymes. These are discussed further in Section 2.2. Since that time a much wider range has been available.

## 2.1 Theoretical aspects of buffering

Buffer solutions comprise a weak acid [HA] and its conjugate base [$A^-$]. The dissociation of a weak acid is described by the equilibrium:

$$HA + H_2O = H_3O^+ + A^-$$

although this is more usually written for convenience omitting the hydration of the proton:

$$HA = H^+ + A^-$$

The dissociation constant $K_a$ is therefore equal to [$H^+$] [$A^-$]/[HA], and thus by taking logarithms of the equation it can be written in the form known as the Henderson–Hasselbach equation:

$$pH = pK_a + \log[A^-]/[HA] \quad (1)$$

In most buffer solutions HA and $A^-$ are either contributed to by a weak acid and its salt with a strong base, e.g. acetic acid and sodium acetate, or by the salt of a weak base with a strong acid and the weak base, e.g. triethanolamine–HCl and triethanolamine. The buffering capacity of any such solution is described as the 'buffer value', β, which is the amount of a strong base producing a resultant change in pH ($\beta = dB/d\,pH$). The equation relating the buffer value to the initial concentration of weak acid (i.e. the concentration, $C$, before any base has been added, and the degree of dissociation, α) is:

$$\beta = 2.3C\alpha(1 - \alpha) \quad (2)$$

Using this equation the buffer values at differing pH values can be calculated and, as shown in *Figure 1*, it can be seen that a maximum occurs when pH = p$K$. The effective range over which a buffer is normally used is p$K \pm 1$. It can be seen from Equations 1 and 2 that the buffering value is directly related to the concentration of the buffer, and also that a buffer would be expected to maintain its pH upon dilution, if both [$A^-$] and [HA] are reduced in equivalent

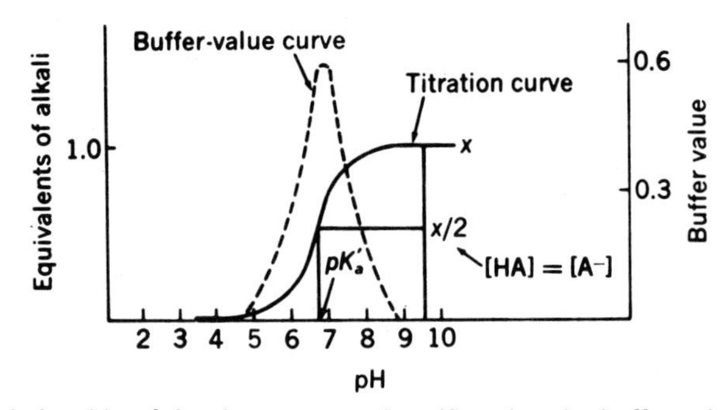


**Figure 1.** Relationship of titration curve to the p$K'_a$ and to the buffer value.

proportions. This is not strictly the case, although it is a useful approximation provided the dilution is not large. In Equation 1, $K_a$ is the thermodynamic equilibrium constant, and [HA] and [$A^-$] the activities of the undissociated acid and its conjugate base. The thermodynamic dissociation constant, $K_a$, is the value that applies at zero ionic strength, i.e. at infinite dilution. (This can be related to the practical dissociation constant $K_a^*$ for an acid $HA^{z+1}$ under any set of conditions by the equation

$$K_a^* = K_a \times f_{HA^{z+1}}/f_{A_z} \qquad (3)$$

where $f \times HA^{z+1}$ and $f \times A^z$ are the activity coefficients of $HA^{z+1}$ and $A^z$.) In dilute solutions, using the Debye–Huckel equation, it is possible to relate $K_a^*$ to $K_a$, the ionic strength ($I$), and the charge of the conjugate base of the buffer ($z$), as follows:

$$\mathrm{pK}_a^* = \mathrm{p}K_a + (2z + 1)[\{0.5I^{½}/(1 + I^{½})\} - 0.1I] \qquad (4)$$

This means that the Henderson–Hasselbach equation can be expressed in terms of the practical dissociation constant $K_a^*$ and the ionic strength of a buffer solution:

$$\mathrm{pH} = \mathrm{p}K_a^* + \log[(A^-)/(HA)] - (2z + 1)[\{0.5I^{½}/(1 + I^{½})\} - 0.1I] \qquad (5)$$

Note that ( ) are used to indicate concentrations, where [ ] are used for activities.

Equation 5 can thus be used to calculate the effect of dilution or change in ionic strength of a buffer on its pH arising from the changes in activity coefficients. Ionic strength is defined as: $-0.5\Sigma(c_i z_i^2)$, where $c_i$ is the concentration of each type of ion and $z_i$ is its charge. The changes in pH arising from the dilution of a buffer are generally small where the buffering ion is monovalent; e.g. dilution of 0.1 M buffer comprising equal amounts of HA and $A^-$ to 0.05 M causes a change of 0.024 pH units. However, if the buffer ions are polyvalent, e.g. phosphate or citrate, the change may be appreciable and large dilutions should be avoided.

The other important parameter which can affect the pH of a buffer is the temperature, and allowance must be made in calculating $\mathrm{p}K_a^*$. From the values of $\mathrm{d}\mathrm{p}K_a/\mathrm{d}t$ given in *Table 1* it can be seen that amine-containing buffers are most sensitive to changes in temperature, e.g. Tris–HCl adjusted to pH 8.0 at 25°C will have a pH of 8.78 at 0°C, whereas carboxylic acid buffers are least sensitive to changes in temperature; e.g. acetate buffer adjusted to pH 4.5 at 25°C will have a pH of 4.495 at 0°C. These differences are due to the differences in ΔH for ionization of the acids.

## 2.2 Practical considerations in the choice of buffer

When choosing a buffer in which to assay an enzyme, one of the primary aims will be to obtain maximum activity and stability of the enzyme at a particular

**Table 1.** Selected properties of selected buffers

| $pK_a^*$ | Compound† | $d(pK_a)/dt$ | Charge on conjugate base | Saturated solution at 0°C (M) | Price‡ |
|---|---|---|---|---|---|
| 4.64 | Acetic acid | 0.0002 | −1 | >10 | L |
| 5.28 | Succinic acid ($pK_2$) | 0 | −2 | 0.36 | L |
| 5.80 | Citric acid ($pK_3$) | 0 | −3 | >2 | L |
| 6.02 | Mes | −0.011 | −1 | 0.65 | M |
| 6.32 | Bis-Tris | −0.017 | 0 | — | M |
| 6.32 | Pyrophosphate ($pK_3$) | −0.01 | −3 | 0.1 | L |
| 6.62 | Ada | −0.011 | −2 | v. sol. | M |
| 6.67 | Aces | −0.020 | −1 | 0.22 | M |
| 6.77 | Mopso | −0.015 | −1 | 0.75 | M |
| 6.84 | Phosphoric acid ($pK_2$) | −0.0028 | −2 | 0.2 | L |
| 6.86 | Pipes | −0.0085 | −2 | v. sol. | H |
| 6.97 | Imidazole | −0.020 | 0 | — | L |
| 6.98 | Bes | −0.016 | −1 | 3.2 | M |
| 7.02 | Mops | −0.015 | −1 | 3.09 | M |
| 7.27 | Tes | −0.020 | −1 | 2.6 | H |
| 7.39 | Hepes | −0.014 | −1 | 2.25 | M |
| 7.42 | Dipso | −0.015 | −1 | 0.24 | H |
| 7.49 | Tapso | −0.018 | −1 | 1.0 | M |
| 7.77 | Heppso | −0.010 | −1 | 2.2 | H |
| 7.78 | Triethanolamine | −0.020 | 0 | v. sol. | L |
| 7.82 | Popso | −0.013 | −2 | v. sol. | H |
| 7.85 | Hepps | −0.015 | −1 | 1.58 | M |
| 7.92 | Tricine | −0.021 | −1 | 0.8 | L |
| 8.00 | Tris | −0.031 | 0 | 2.4 | L |
| 8.09 | Glycylglycine | −0.028 | −1 | 1.1 | M |
| 8.17 | Bicine | −0.018 | −1 | 1.1 | L |
| 8.19 | Taps | −0.018 | −1 | — | H |
| 8.88 | Diethanolamine | −0.024 | 0 | v. sol. | L |
| 9.05 | Pyrophosphate ($pK_4$) | −0.006 | −4 | 0.1 | L |
| 9.08 | Borate | −0.008 | −1 | 0.05 | L |
| 9.23 | Ches | −0.029 | −1 | 1.14 | M |
| 9.47 | Ethanolamine | −0.029 | 0 | v. sol. | L |
| 9.55 | Glycine | −0.025 | −1 | 4.0 | L |
| 9.96 | Carbonate ($pK_2$) | −0.009 | −2 | 0.8 | L |
| 10.05 | Caps | −0.032 | −1 | 0.47 | M |

Data compiled in this table are from (1, 3, 5, 6) and other sources including manufacturers catalogues.
† The structures of compounds with abbreviated names are given in *Figure 2*.
‡ The price categories are based on 1989 manufacturers catalogues. L = less than £20 per litre of M buffer solution, M = between £20–50 per litre of M buffer solution, and H = greater than £50 per litre of M buffer solution.

pH or range of pH values. There is now a wide choice of buffers throughout the normal range of pH values used (*Table 1*). Many buffers which were routinely used for many years have now been replaced by alternatives which have properties more suitable for biological systems. Many of these are zwitterionic buffers comprising substituted amines and sulphonic acids (1–4) (*Figure 2*). In choosing a buffer the first consideration is the pH range required. Generally, buffers are used within the pH range $pK \pm 1$ because of the reduction in buffer value outside that range (*Figure 1*). The range that is used may be varied if it is known that a buffer has only to counter the effects of acid or of base but not both; e.g. a dehydrogenase catalysed reaction in which NADH + $H^+$ is produced. The edge of the working range can be more satisfactorily used if the change in pH is towards the p*K*. Other considerations will be the stability of the buffer, whether it interacts with the substrates, cofactors, or metal ions, the temperature coefficients of its p*K*, the ionic strength at which it is used, its absorbance in the UV region of the spectrum, its cost, and its availability free from contaminants. Many of these aspects are summarized in *Table 1*. Other properties may also be important in certain instances; e.g. interference with methods of protein estimation, volatility, and its solubility for preparing stock solutions.

Most of the newer zwitterionic buffers do not appreciably bind divalent metal ions, are chemically stable, do not appreciably absorb light at wavelengths longer than 240 nm, and can be made up as concentrated stock solutions. Many of the buffers which have been in longer use have one or more disadvantages. They are generally cheaper, and if required in large quantities, e.g. for dialysis or column chromatography, may be used provided they have been tested or are known not to affect the enzyme in question. The following drawbacks associated with certain buffers should be borne in mind. Phosphate buffers tend to precipitate $Mg^{2+}$, $Ca^{2+}$, $Fe^{3+}$, and other polyvalent cations. Phosphate being an important metabolite is known to inhibit a number of enzymes; e.g. kinases, dehydrogenases, carboxypeptidase, fumarase, urease, aryl sulphatase, adenosine deaminase, phosphoglucomutase, and other enzymes involving phosphate esters. On the other hand, it may also stabilize enzymes, e.g. phosphoribosepyrophosphate synthase. Pyrophosphate buffer also precipitates polyvalent cations and has a tendency to form complexes. Like phosphate it is also a metabolite and may affect certain enzymes. Imidazole buffer has on occasions been used where phosphate is not acceptable, since both have similar buffering ranges. It is not generally a good alternative, since it is reactive, unstable, and also complexes with divalent metal ions. Mops or Bes might be a better alternative. Borate has the disadvantage of complexing with *vic* diols, which include many carbohydrates and the ribose moiety of nucleotides. It would best be avoided when assaying enzymes using NAD(P) or other nucleotides.

Tris/maleate has been used to cover a wide span of pH from 5.8–8.6, but it does not buffer uniformly between the two p*K*s, and maleate absorbs in the

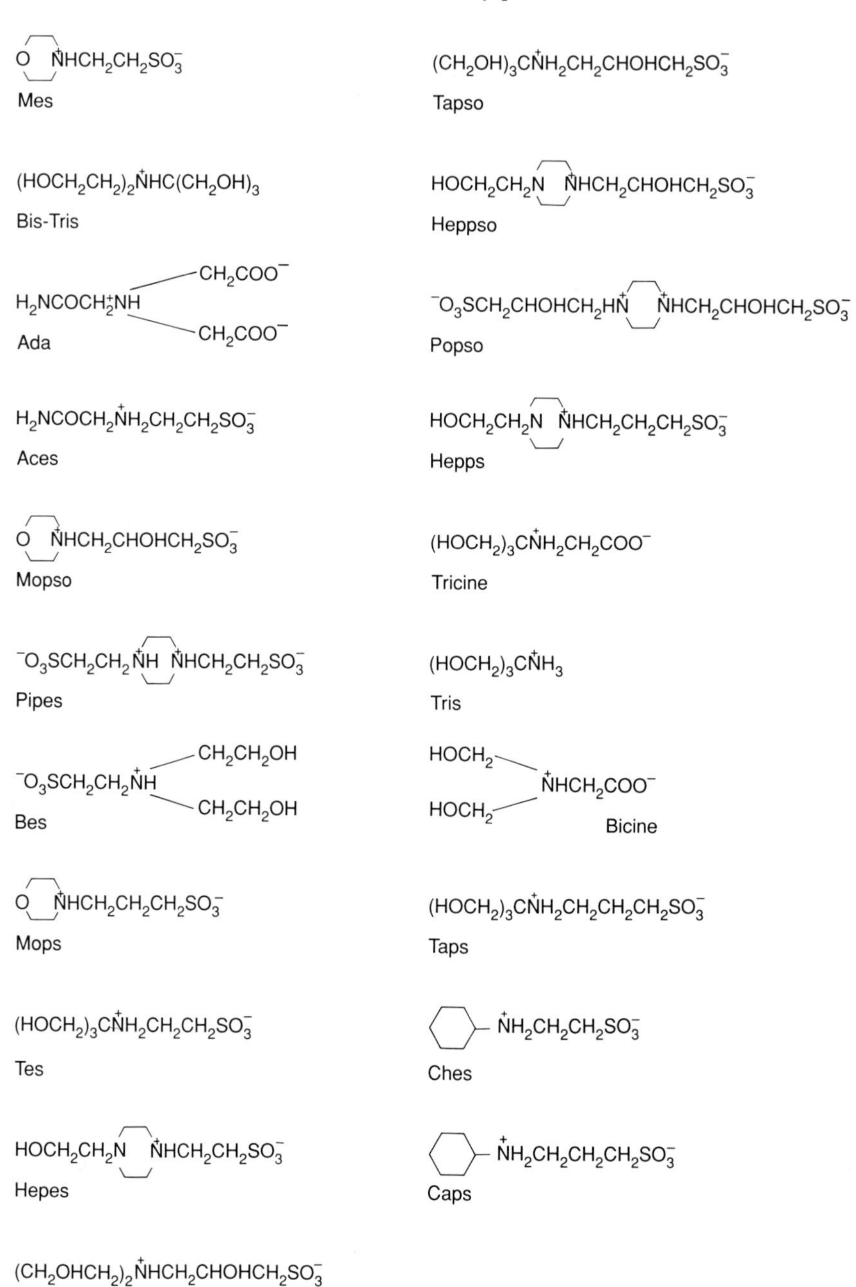


**Figure 2.** Structures of buffer components referred to in *Table 1*.

UV region of the spectrum. Barbituric acid buffer used to be used extensively in the range pH 7–9. Apart from being poisonous and inhibiting oxidative phosphorylation, it has the disadvantages of being unstable and of absorbing in the UV region. Citric acid binds some proteins and also complexes many metal ions. Its $pK_a$ values are influenced by the ionic strength of the buffer more than most (see below). Tris, although widely used, has the disadvantage common to other primary amine buffers of being reactive and forming Schiff bases with aldchyde and ketones. Glycylglycine binds certain metal ions; e.g. $Mn^{2+}$ and $Cu^{2+}$; it may be degraded by peptidases, it is expensive, and generally has no advantages over the cheaper Tricine. Bicarbonate buffer has the disadvantage of having to be used in a closed system to avoid the loss of carbonic acid as $CO_2$.

Apart from the specific disadvantages mentioned in the previous paragraphs, two other points, namely ionic strength and temperature, require attention. In general it is best to use buffers at the lowest concentration necessary to give adequate buffering capacity required for the reaction being studied. For a given buffering capacity the number of ions in solution will vary with the buffer, those of polybasic acids such as phosphoric acid contributing more than those of monobasic acids such as acetic acid. When making up a buffer of a given molarity at different pH values, the ionic strength will vary with the pH. This is important when measuring the pH optimum of an enzyme, or when making more detailed kinetic studies at differing pH. It is important not to attribute a change in activity to a change in pH when it may be in part due to an accompanying change in ionic strength. It is then important to use the buffer at a constant ionic strength. *Table 2* shows the change in ionic strength of a 0.05 M buffer adjusted from $pH = pK - 1$ to $pH = pK + 1$. It can be seen that the smallest changes in ionic strength occur with buffers for which the conjugate base has a change $(z) = 0$ or $-1$.

**Table 2.** The variation of ionic strength with pH for single component buffers with different charges on their conjugate bases (ionic strength contribution of buffering species only).

| **Charge on conjugate base‡** | **Example** | **Ionic Strength†** | | | **Change in ionic strength from p*K* − 1 to p*K* + 1** |
|---|---|---|---|---|---|
| | | **pH = p*K* + 1** | **pH = p*K*** | **pH = p*K* − 1** | |
| 0 | $TrisH^+$/Tris | 0.009 | 0.05 | 0.09 | −0.081 |
| −1 | Acetic acid/Acetate | 0.09 | 0.05 | 0.009 | +0.081 |
| −2 | $H_2PO_4^-/HPO_4^2$ | 0.28 | 0.2 | 0.118 | +0.162 |
| −3 | $HCitrate^{2-}/Citrate^{3-}$ | 0.568 | 0.45 | 0.324 | +0.244 |

‡ $HA^{z+1} = H^+ + A^z$
† The ionic strength is calculated in each case for 0.1 M buffer solutions

To achieve a constant ionic strength at different pH values the addition of varying amounts of a salt can be used. Alternatively, it is possible by varying the concentrations of the buffer components to make buffers which have a constant ionic strength. For many studies on the effects of pH, a wider range than 2 pH units is required. This can best be achieved by using suitable mixtures of more than one buffer. If one of the buffers is a weak base and the other a weak acid then less acid or alkali has to be added to this mixture to adjust pH. It is possible to obtain a buffer mixture giving a constant ionic strength throughout its buffering range. The theory of this is explained in (5) where a list of useful three-component buffer systems to achieve an constant ionic strength is given.

The second factor is that the p*K*s of a number of buffers have a significant temperature coefficient. This is important, for example, when a buffer is adjusted to give a required pH at room temperature and then used at a different temperature. Most frequently this occurs when buffers are used in some purification procedure at 4°C, but have been adjusted at room temperature. It can also occur when enzyme assays are carried out at 37°C and, although much less commonly, when an enzyme is assayed at a high temperature; e.g. in the case of an enzyme from a thermophilic organism. One of the highest temperature coefficients of a buffer is that of Tris, whereas some of the lowest are those of carboxylic acids. Most Good's buffers have low temperature coefficients (1–3). In the case of Tris when adjusted at 20°C ($\simeq$ room temperature) and then used at 4°C or 37°C the change in pH would be +0.5 and −0.52, whereas in the case of acetate buffer the same changes in temperature only cause changes in pH of −0.003 and +0.003 respectively.

Once a particular buffer has been chosen as suitable on the basis of the properties indicated in *Table 1* and any other consideration, there are a number of points to bear in mind in making it up. A buffer can be made up by mixing the components in the correct proportions as shown in *Table 3*. It will be necessary to check the pH, since many of the components are hydrated to a variable degree. Alternatively, the buffer may be made up by weighing the total amount of the proton donor and adding sufficient alkali to adjust to the required pH; e.g. for 0.1 M sodium phosphate buffer, pH 7.2, weigh out the amount of $NaH_2PO_4$ to give 0.1 M phosphate and then add NaOH to adjust to pH 7.2. In the case of a buffer such as Tris–HCl, the proton acceptor (Tris base) can be weighed out and adjusted with HCl to give the correct pH. In the case of a buffer like triethanolamine–HCl the base is a liquid, whereas triethanolamine hydrochloride is obtainable as a crystalline solid. If the latter is used, since it is easier to handle for weighing out, it then has to be adjusted to the correct pH using NaOH. This of course means that the final buffer solution contains NaCl. This procedure might be satisfactory provided the additional salt which contributed to the ionic strength did not affect the investigation under study. If it did so, then it would be best to adjust the triethanolamine base with HCl to the required pH.

**Table 3.** The ratios of acid/conjugate base (acid/salt) which give pH values in the range p*K* + 1 to p*K* − 1

| pH | Acid/Conjugate Base |
|---|---|
| p*K* −1.0 | 10.0 |
| −0.9 | 7.94 |
| −0.8 | 6.31 |
| −0.7 | 5.01 |
| −0.6 | 3.98 |
| −0.5 | 3.16 |
| −0.4 | 2.51 |
| −0.3 | 2.00 |
| −0.2 | 1.58 |
| −0.1 | 1.26 |
| p*K* | 1.00 |
| +0.1 | 0.79 |
| +0.2 | 0.63 |
| +0.3 | 0.50 |
| +0.4 | 0.40 |
| +0.5 | 0.32 |
| +0.6 | 0.25 |
| +0.7 | 0.20 |
| +0.8 | 0.16 |
| +0.9 | 0.13 |
| +1.0 | 0.10 |

Often buffers are required with additional components such as EDTA, 2-mercaptoethanol, or PMSF. In general the final pH should be adjusted after addition of the other components. However, in the case of PMSF, which is unstable in alkali it would not be appropriate to add it to a buffer on the alkaline side of its final pH and then adjust down since much of the PMSF might be hydrolysed during the adjustment.

Concentrated stock solutions can often be convenient since they require less storage space, and also it is possible to add bacteriocidal agents such as sodium azide, which may be sufficiently diluted in the working buffer not to affect the enzyme under investigation. It should be noted that buffers may change their pH significantly when undergoing large dilutions (Section 2.1).

When adjusting the pH of a buffer with acid or alkali as appropriate, it is often advantageous to use cations or anions which are unlikely to bind to enzymes. Many enzymes are affected by monovalent ions (e.g. pyruvate kinase is inhibited by $Na^+$ and activated by $K^+$, and salivary amylase is activated by $Cl^-$) and it is important to bear this in mind when choosing the counter ion. In this respect the larger cations from tetramethylammonium or tetraethylammonium hydroxide may be preferred to those from NaOH, and larger anions such as acetate prefered to chloride.

Allowance for temperature changes affecting the pH of a buffer should be made particularly for those buffers having a large temperature coefficient. If a buffer is to be used at a temperature differing from room temperature both the pH standard solution and the solution being adjusted should be cooled (or warmed) to the required temperature. The pH meter should then be standardized with the temperature adjustment control set to the temperature required. The solution can then be adjusted to the required pH at the correct temperature.

Occasionally there are other factors which influence the choice of buffer. Some buffers interfere with protein estimations (Section 3.2.1). If an enzyme solution is to be concentrated by freeze-drying, a volatile buffer may be required; a list of suitable buffers is given in (6). Enzymes are sometimes assayed in the presence of miscible organic solvents, i.e. in partial aqueous solution. The organic solvents affect the structure of water and also the ionization of the buffer components and functional groups on enzymes. These effects can be quite complex and are discussed further by Blanchard (4) and by Dempsey and Perrin (7). NMR is now being used more extensively to study individual functional groups on enzymes. These are most satisfactorily studied in $D_2O$ rather than in $H_2O$, since the resonances from the protons in water tend to swamp those from the enzyme. A list of deuterated buffers is given in (4).

## 3. Methods for protein determination

One of the theoretical difficulties with measuring protein concentration is the chemical basis of the method. All proteins have in common the peptide bond between each amino acid residue, and yet it is not easy to find a versatile and sensitive method which measures exclusively the peptide bond. Most methods give different results with different proteins depending on their amino acid compositions, and some reagents also react with certain amino acid side-chains.

Since the 1950s when Lowry *et al.* (8) modified the method originally devised by Wu (9) and Folin and Ciocalteu (10), the Lowry method has become the most widely used for protein estimation. However, during the 1980s dye-binding assays based on Coomassie Blue (11, 12) have become more popular because of their simplicity and because of certain shortcomings in the Lowry method. One of the oldest procedures, the Biuret test (13), is still used; it is quite satisfactory apart from the disadvantage of being relatively insensitive. Measurement of $A_{280}$ is very useful for a quick, approximate estimation of protein in non-turbid solutions, which is often all that is required. It has the added advantage of being non-destructive. Detailed aspects of the various methods are given below.

## 3.1 General considerations

Whatever method for protein determination is used, each depends to varying degrees on the amino acid composition of the protein concerned, and usually involves a comparison with a protein standard. Unless the protein being assayed is the same as that of the standard the accuracy of the result will be reduced. The only method of direct measurement of protein is to weigh the protein in question after complete removal of water and contaminants. This is very rarely possible or necessary. For most enzyme work it is important only to know the relative protein concentration and not the absolute concentration. However, it is well to bear in mind that some proteins do behave significantly different from the 'standard protein'; for example, the published $A_{280}$ for 1% protein solutions vary from 0 for some parvalbumins to 26.5 for lysozyme, though most fall in the range 4–15 (14). The standard normally used is bovine serum albumin which has $A_{280}^{1\%} = 6.6$ (15). The Lowry method for protein determination also depends to a significant extent on the aromatic amino acid content, and the colour yield for a protein like gelatin is about one quarter that of trypsin (16). Which method can be applied in a given situation may depend on the state of the protein-containing material. Highly purified proteins in aqueous solution can be estimated more easily than crude homogenates or membrane-bound proteins where many additional components in the mixture may interfere. For further reviews of protein estimation see (16–20).

## 3.2 Specific methods

### 3.2.1 Ultraviolet absorption

The ultraviolet absorption of proteins is due to a number of chromophoric groups; at 280 nm it is due almost entirely to the aromatic amino acids tryptophan and tyrosine. At around 260 nm phenylalanine absorbs, and histidine, methionine, cysteine, and cystine absorb between 225 and 240 nm. At wavelengths less than 225 nm the peptide bond itself is the principal absorbing group, but the above amino acids also contribute. Therefore, if the only consideration were to select a wavelength at which the absorption of a protein is independent of its amino acid composition, and at which it absorbs most strongly, then 192 nm would be the optimum. The absorption band for the peptide bond extends up to about 225 nm. There are a number of reasons why these shorter wavelengths are not the most frequently used. At 192 nm dissolved oxygen absorbs strongly, and thus measurements would have to be carried out in the absence of oxygen. 192 nm is below the working range of many spectrophotometers, since it requires a particularly good light source and stray radiation can be a problem. Wavelengths nearer 225 nm are used since, although the peptide bond absorbs less strongly (*Table 4*), the interference by oxygen is minimized.

**Table 4.** The absorption of protein solutions in the ultraviolet.

| **Wavelength (nm)** | **Absorbance of 0.1% solution** |
|---|---|
| 191–194 | $\simeq$68 |
| 200 | $\simeq$45 |
| 205 | $\simeq$32 |
| 206 | $\simeq$29 |
| 210 | $\simeq$21 |
| 215 | $\simeq$15 |
| 220 | $\simeq$11 |
| 224 | $\simeq$ 8 |
| 233 | $\simeq$ 4 |
| 280 | 0.3–1.8 |
| 320 | $\simeq$ 0.0 |

In a typical protein solution used in enzyme studies, the main interfering substances likely to affect absorbance measurements are nucleic acids, nucleotides such as ATP or NAD(P), haem-containing compounds such as cytochromes, reagents containing sulphydryl groups such as 2-mercapto-ethanol, dithiothreitol, glutathione or cysteine, and buffers. In general, the shorter the wavelength the larger the range of interfering substances. It is perhaps for this reason that although $A_{280}$ is less sensitive than some, and is dependent on the tyrosine and tryptophan content, it is the one most used. The absorption spectrum of tyrosine is pH dependent, but at 280 nm both the protonated and unprotonated forms have similar absorption coefficients and so measurements do not have to be made at a particular pH. Some spectrophotometers used to monitor column effluents use 206 nm, which gives much greater sensitivity provided there is no interference.

Before measuring the ultraviolet absorbance of a protein solution, it is important to check that the protein solution is not turbid, as this will cause light scattering. Clarification can be done by centrifugation. A clear protein solution (unless it contains prosthetic groups) will normally have zero $A_{320}$. A significant $A_{320}$ is usually due to light scattering and should be subtracted from $A_{280}$. After the absorbance of a suitable dilution of the protein has been made, and corrected if necessary, the approximate relationships for a protein at 1 mg/ml of $A_{280} \simeq 1.0$, or $A_{215} \simeq 15$, or $A_{206} \simeq 30$ can be used.

Various corrections have been devised to allow for interfering substances. These involve making measurements at two wavelengths, at the second of which there is a large difference in the absorbance coefficient between the interfering substance and the protein. The oldest of these is one used by Warburg and Christian to correct for nucleic acid interference, and is expressed as a formula by Layne (21):

$$\text{protein (mg/ml)} = 1.55\ A_{280} - 0.76\ A_{260}$$

A similar correction (22) but using the $A_{230}$ is:

$$\text{protein (mg/ml)} = 0.183\ A_{230} - 0.075\ A_{260}$$

If a particular interfering substance is known to be present then a suitable correction may be devised along the above lines.

Although the $A_{205}$ is due mainly to the peptide bond and the absorbance coefficients for most proteins lie in the range 28.5–33 there is some variation due to the tyrosine and tryptophan content. A correction for this has been devised (23).

$$A_{205}^{1\%} = 10\ [27 + (120 \times A_{280}/A_{205})]$$

The method of Waddell (24) is based on differential measurements at 215 nm and 225 nm. By using these wavelengths some of the $A_{215}$ due to buffers and other components is corrected for. The protein solution is diluted with sodium chloride solution (9 g/l) until the $A_{215}$ is less than 1.5. The relationship used is:

$$\text{protein (mg/ml)} = 0.144\ (A_{215} - A_{225})$$

Other less widely used corrections are given in (17 and 18). A comprehensive list of absorbance coefficients may be found in (15, 25–28).

### 3.2.2 The Biuret method

This method is simple to perform and is relatively unaffected by the amino acid composition of the protein. However, it is not often used in enzyme work because of its low sensitivity, which may mean using a large proportion of a valuable sample in order to estimate its protein concentration. The method depends on the formation of a purple copper complex with adjacent peptide bonds.

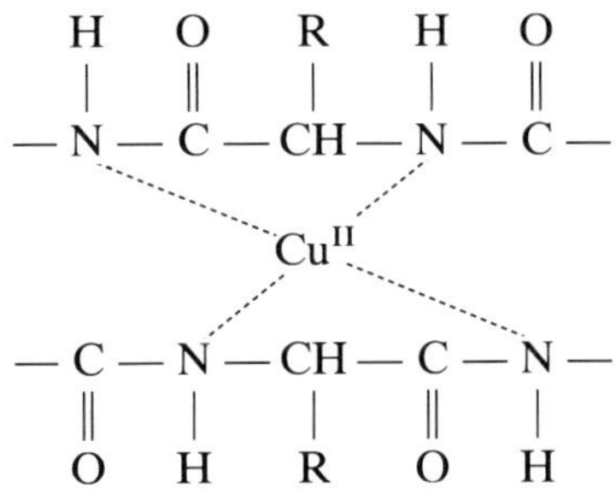

```
   H    O    R    H    O
   |    ||   |    |    ||
 — N — C — CH — N — C —
     .          .
        .    .
         Cu^II
        .    .
     .          .
 — C — N — CH — C — N —
   ||   |    |    ||   |
   O    H    R    O    H
```

Relatively few substances interfere with the biuret estimation; those which do include bile pigments, sucrose, Tris, glycerol, and ammonium ions. Sucrose, Tris, and glycerol can usually be corrected for by their inclusion in the blank and protein standard. The interference by ammonium ions means that it is unsatisfactory for precipitates obtained from ammonium sulphate fractionation.

The usual wavelength at which the colours are read is 540 nm; greater sensitivity can be achieved at 310 nm, but at the latter wavelength many more substances interfere (29). The Biuret reagent is easily prepared and is stable for months at room temperature if stored in a plastic bottle. Samples containing 1–10 mg protein can be measured; the method is outlined in *Protocol 1*.

---

**Protocol 1.** The biuret method for protein estimation

1. Dissolve 1.5 g $CuSO_4 . 5H_2O$, and 6.0 g sodium potassium tartrate in 500 ml water.
2. Add 300 ml 10%(w/v) NaOH, and make up to 1 l with water. If 1 g KI is also added, the reagent will keep indefinitely in a plastic container.
3. To 0.5 ml of unknown containing up to 3 mg of protein, add 2.5 ml reagent.
4. Allow to stand for 20–30 min before reading $A_{540}$.

---

### 3.2.3 The Lowry method

The active constituent of the Folin Ciocalteu reagent is phosphomolybdic-tungstic mixed acid. Proteins bring about a reduction of this reagent giving rise to a number of reduced species which have a characteristic blue colour having $\lambda_{max}$ 745–750 nm. Copper ions present in the first reagent chelate with the peptide bond and this facilitates the electron transfer between the phosphomolybdic-tungstic acid and the protein, and hence the formation of the chromagen. The amino acid residues with which it reacts are tryptophan, tyrosine, cysteine, cystine, histidine, and also the peptide bond itself. The colour yields of individual proteins show an approximately additive relationship to the number of reactive groups, but the sequence, chain length, and exposure of particular groups also have an effect. It is not possible to predict accurately the colour yield for a given protein based on its structure. For most proteins the colour yield only varies by a factor of 1.2, but in extremes, e.g. gelatin, which is low in the above amino acid residues compared to trypsin, a ratio of 2.5 is obtained. A detailed study of the chemical basis of the Lowry method has been made (30).

The main advantage of the Lowry method is its high sensitivity (16), and its main disadvantage is the number of interfering substances. The range over which the colour yield is linear is limited, but this can be overcome using suitable dilutions. A comprehensive list of substances which interfere with the Lowry method is given in (16); those most likely to be encountered in enzyme work include tyrosine, tryptophan, cysteine, cystine, Tris, Tricine, Bicine, Hepes, Hepps, Pipes, Bes, Mops, Mes, Caps, Ada, Aces, Taps, Tes, Ches, Triton X-100, dithioerythritol, reduced glutathione, 2-mercaptoethanol, potassium chloride, potassium phosphate, EDTA, and sucrose. The interfering substances may either increase the blank, decrease the colour yield, or cause precipitation. Various methods have been adopted to overcome interference, the simplest of which is to correct by including the interfering substance in the blank and the protein standards. This is only feasible if the approximate concentration of the interfering substance is known, and if the interference is not too pronounced. Many of the interfering substances can be removed by dialysis, but this is generally too cumbersome for routine use. Sulphydryl-containing compounds can be removed by oxidation with hydrogen peroxide (31). Lipids which interfere can be extracted with chloroform. An alternative is to precipitate the protein with trichloracetic acid (TCA) leaving the interfering substance in solution; however, if the protein concentration is low complete precipitation will not occur, e.g. ≃70% precipitation at 50 μg protein/ml and ≃50% at 20 μg protein/ml occurs using 5–7% TCA (16). A higher recovery of the protein after precipitation can be brought about by including deoxycholate (19) or yeast-soluble RNA (32) before the addition of TCA. The deoxycholate treatment is simplest but is not quantitative if detergents such as digitonin or SDS are present in significant amounts. The interference by sucrose can be overcome by increasing the tartrate concentration in the first reagent fourfold, and interference by lipids by including 1% sodium dodecyl sulphate in the first reagent. If sodium dodecyl sulphate is included, then it is important to use sodium tartrate and not sodium potassium tartrate, since potassium dodecyl sulphate is relatively insoluble and will precipitate. Sodium dodecyl sulphate is included when lipoproteins or membrane-bound proteins are estimated (33). The Lowry method is outlined in *Protocol 2*.

---

**Protocol 2.** The Lowry method for protein estimation

*Reagents*

- *Reagent A:* 0.1% (w/v) $CuSO_4$. $5H_2O$, 0.2% (w/v) Na or Na/K tartrate, 10% (w/v) $Na_2CO_3$. Both the $Na_2CO_3$ and the $CuSO_4$/tartrate are dissolved separately in approximately half times the final volume. The $Na_2CO_3$ solution is then slowly added to the copper tartrate with stirring and then made up to the final volume. At 10°C it can be stored indefinitely but near 0°C the carbonate tends to crystallize out.
- *Reagent B:* Mix 1 vol of Reagent A with 2 vol 5% (w/v) SDS and 1 vol 0.8 M NaOH. (Stable for 2–3 weeks at room temperature.)
- *Reagent C:* 1 vol Folin Ciocalteu reagent diluted with 5 vol $H_2O$.

*Procedure*

1. Mix 1 ml protein solution (5–100 μg) with 1 ml Reagent B, and stand at room temperature for 10 min.

**Protocol 2.** *Continued*

2. Add 0.5 ml Reagent C, mix immediately and stand for 30 min at room temperature.
3. Read $A_{750}$.
4. For calibration curve use 5–100 μg protein.

---

### 3.2.4 Dye-binding assays

Amido Black, bromosulphalein, and Coomassie Blue have all been used in protein estimation, but Coomassie Blue G250 is easily the most widely used for proteins in solution. Two methods were published, in 1976 (11) and 1977 (12), and both are based on the change in the absorption spectrum of Coomassie Blue on binding to proteins. The protonated form of Coomassie Blue is a pale orange-red colour, whereas the unprotonated form is blue. When proteins bind to Coomassie Blue in acid solution their positive charges suppress the protonation and a blue colour results. It has been found that about 1.5–3 dye molecules/charge bind (34), but that hydrophobic interactions between the dye and protein are also important.

The unprotonated form of Coomassie Blue G250:

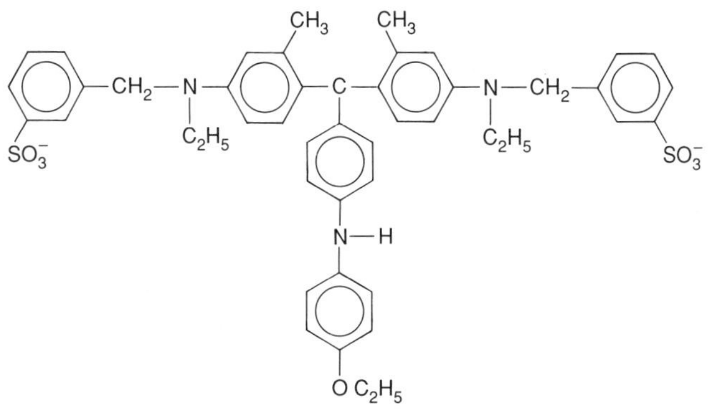

The method of Bradford (11) uses phosphoric acid, the absorbance is measured at a single wavelength (595 nm), and the response to increasing protein concentration is slightly non-linear. The method of Sedmak and Grossberg (12) uses perchloric acid and the absorbance is measured at two wavelengths (465 and 620 nm). The absorbance ratio $A_{620}/A_{465}$ shows a linear response for the range 0–50 μg protein. On the whole the Bradford method has proved the more popular of the two, and a modification of the original method (35) which uses an increased dye concentration gives greater sensitivity and less variability in the colour yield. The protocol is outlined in *Protocol 3*.

**Protocol 3.** Dye-binding protein assay (17)

In making up Coomassie Blue G250 solutions it is important to note the source, since Serva blue contains five times as much Coomassie Blue per gram as that of Eastman.

1. Dissolve 0.923 g of powder (Eastman) in 200 ml 88% phosphoric acid and 100 ml 95% ethanol. If Serva blue is used 1 g in 200 ml is used. These solutions are stable indefinitely at room temperature.
2. Dilute 150 ml of the dye solution (Eastman) to 600 ml with $H_2O$ and filter through Whatman No. 1 filter paper before use. Alternatively, mix 30 ml of dye solution (Serva) with 80 ml 88% phosphoric acid and 40 ml 95% ethanol, dilute to 600 ml with $H_2O$ and filter through Whatman No. 1 filter paper before use.
3. Mix 1.0 ml sample (1–25 μg protein) with 1.5 ml dye reagent, and stand for 5 min.
4. Read $A_{595}$.
5. For calibration curve use 1–20 μg protein.

*Note:* It is advisable to use glass or plastic cuvettes, since the dye binds more strongly to quartz. Cuvettes can be cleaned in either SDS solution or acid.

The dye-binding methods have the advantage over the Lowry method of being unaffected by sucrose, buffers, reducing agents, and chelating agents. The methods only require a single reagent and are quicker to perform. However, the colour is stable for a shorter period of time as the protein–dye complex-may precipitate. Detergents and alkali may interfere with the binding, and although guanidine hydrochloride does not affect the blank, it does affect the binding. The colour yield does show some variation with the protein and relates to some extent to the number of positive charges (34). Peptides having $M_r < 3000$ cannot be measured by this method. This can be an advantage over the Lowry method when protein is being measured in crude cell extracts, since the amino acids and small peptides present do not interfere.

## 3.3 Precautions

(a) It is always important to use clean test-tubes and cuvettes. The more sensitive the method the more critical this becomes. Tubes should be either acid- or detergent-washed before use. A fingerprint may contain up to 0.5 μg protein (36). Protein readily adsorbs on to glassware.

(b) Stock solutions of protein standards are generally made up at 10 mg/ml and diluted before use. Protein solutions which are less than 1 mg/ml lose a significant amount of protein by adsorption.

(c) Protein standards are generally made up by weighing the protein which, unless it has been dried over a desiccant and weighed in a non-humid atmosphere, will always contain some moisture. It is therefore best to check the protein standard by measuring $A_{280}$.

## 3.4 Solubilization

Proteins in homogenates, particulate fractions, membrane-bound proteins, and lipoproteins have to be solubilized before estimation. For the Lowry method the inclusion of SDS or deoxycholate and NaOH in the initial reagent, as in the recipe given above, is usually sufficient to solublize membrane-bound proteins or lipoproteins. Insoluble proteins or pellets can also be dissolved by adding 0.2 ml of 0.25% (w/v) sodium deoxycholate in 0.1 M NaOH and heating the mixture at 80°C for 5 min.

Anionic detergents, such as SDS or deoxycholate, and alkali affect the binding of Coomassie Blue to proteins, but Triton X-100 (0.2%) can be used to solublize proteins before carrying out this assay.

## 3.5 Other methods

There are several less widely used methods which have not been described here. A turbidometric method using tannin appears insensitive to variations in the protein side-chains and can be used in the range 10–100 μg (39). It has the added advantage that the reagents are cheap. A recent method is based on bicinchoninic acid (19 and 37). This involves the reduction of $Cu^{2+}$ to $Cu^{+}$ in alkali by protein, and the $Cu^{+}$ is detected by complex formation with bicinchoninic acid. This method is sensitive to reducing agents like the Lowry method, but is much less sensitive to detergents or neutral salts.

A comprehensive list is given in (18). Some of the most sensitive methods are those which entail fluorescence measurements. Of these the ophthalate method is the best (38). It can be used to measure 10 ng protein, and if the amino acid composition of the protein is known it can be used to calculate the absolute amount of protein since the fluorescence shows an additive relationship for the different fluorophores.

# References

1. Good, N. E., Winget, G. D., Winter, W., Connolly, T. N., Izawa, S., and Singh, R. M. M. (1966). *Biochemistry*, **5,** 467.
2. Good, N. E. and Izawa, S. (1972). *Methods Enzymol.*, **24,** 53.
3. Ferguson, W. J., Braunschweiger, K. I., Braunschweiger, W. R., Smith, J. R., McCormick, J. J., Wasmann, C. C., Jarvis, N. P., Bell, D. H., and Good, N. E. (1980). *Anal. Biochem.*, **104,** 300.
4. Blanchard, J. S. (1984). *Methods Enzymol.*, **104,** 104.
5. Ellis, K. J. and Morrison, J. F. (1982). *Methods Enzymol.*, **87,** 405.

6. Perrin, D. D. and Dempsey, B. (1974). *Buffers for pH and Metal Ion Control*, p. 50. Chapman and Hall, London.
7. Perrin, D. D. and Dempsey, B. (1974). *Buffers for pH and Metal Ion Control*, Chapter 6. Chapman and Hall, London.
8. Lowry, O. H., Rosebrough, N. J., Farr, A. L., and Randall, R. J. (1951). *J. Biol. Chem.*, **193,** 265.
9. Wu, H. (1922). *J. Biol. Chem.*, **51,** 33.
10. Folin, O. and Ciocalteu, V. (1927). *J. Biol. Chem.*, **73,** 627.
11. Bradford, M. M. (1976). *Anal. Biochem.*, **72,** 248.
12. Sedmak, J. J. and Grossberg, S. E. (1977). *Anal. Biochem.*, **79,** 544.
13. Gornall, A. C., Bardawill, C. J., and David, M. M. (1949). *J. Biol. Chem.*, **177,** 751.
14. Scopes, R. K. (1987). *Protein purification: principles and practice*, (2nd edn). Springer, New York.
15. Kirschenbaum, D. M. (1976). In: *Handbook of Biochemistry and Molecular Biology*, Vol. 2 (3rd edn), (ed. G. D. Fasman), p. 383. CRC Press, Cleveland.
16. Peterson, G. L. (1979). *Anal. Biochem.*, **100,** 201.
17. Peterson, G. L. (1983). *Methods Enzymol.*, **91,** 95.
18. Thorne, C. J. R. (1978). *Techniques in Protein and Enzyme Biochemistry*, B104, p. 1. Elsevier/North Holland Biomedical Press, Amsterdam.
19. Harris, D. A. (1987). In: *Spectrophotometry and Spectrofluorimetry: a practical approach*, (ed. D. A. Harris and C. L. Bashford), Chapter 3. IRL Press.
20. Leggett Bailey, J. (1967). *Techniques in Protein Chemistry*, (2nd edn), Chapter 11. Elsevier, Amsterdam.
21. Layne, E. (1957). *Methods Enzymol.*, **3,** 447.
22. Kalb, V. F. and Bernlohr, R. W. (1977). *Anal Biochem.*, **82,** 362.
23. Scopes, R. K. (1974). *Anal. Biochem.*, **59,** 277.
24. Waddell, W. J. (1956). *J. Lab. Clin. Med.*, **48,** 311.
25. Kirschenbaum, D. M. (1975). *Anal. Biochem.*, **68,** 465.
26. Kirschenbaum, D. M. (1977). *Anal. Biochem.*, **80,** 193.
27. Kirschenbaum, D. M. (1977). *Anal. Biochem.*, **81,** 220.
28. Kirschenbaum, D. M. (1977). *Anal. Biochem.*, **82,** 83.
29. Itzhaki, R. F. and Gill, D. M. (1964). *Anal. Biochem.*, **9,** 401.
30. Legler, G., Muller-Platz, C. M., Mentges-Hettkamp, M., Pflieger, G., and Julich, E. (1985). *Anal. Biochem.*, **150,** 278.
31. Geiger, P. J. and Bessman, S. P. (1972). *Anal. Biochem.*, **49,** 467.
32. Cabib, E. and Polacheck, I. (1984). *Methods Enzymol.*, **104,** 415.
33. Markwell, M. A., Haas, S. M., Tolbert, N. E., and Bieber, L. L. (1981). *Methods Enzymol.*, **72,** 296.
34. Tal, M., Silberstein, A., and Nusser, E. (1980). *J. Biol. Chem.*, **260,** 9976.
35. Read, S. M. and Northcote, D. H. (1981). *Anal. Biochem.*, **116,** 53.
36. McKnight, G. S. (1977). *Anal. Biochem.*, **78,** 86.
37. Smith, P. K., Krohn, R. I., Hermanson, C. T., Mallia, A. K., Gartner, F. H., Provenzano, M. D., Mujimoto, E. K., Goeke, N. H., Olson, B. J., and Klenk, D. G. (1985). *Anal. Biochem.*, **150,** 76.
38. Viets, J. W., Deen, W. M., Troy, J. L., and Brenner, B. M. (1978). *Anal. Biochem.*, **88,** 513.
39. Mejbaum-Katzenellenbogen, W. Dobryszka, W. R. (1959). *Clin. Chim. Acta*, **4,** 515.

# A

# Suppliers of specialist items

## Enzymes and chemicals

**Aldrich Chemical Company Limited,** The Old Brickyard, New Road, Gillingham, Dorset, SP8 4JL, UK.

**Amersham International Ltd.,** Amersham Place, Little Chalfont, Bucks, HP7 9NA, UK.

**B.D.H. Chemicals Limited,** Broom Road, Poole, Dorset, BH12 4AN, UK.

**Biozyme Laboratories Ltd.** Unit 6, Gilchrist-Thomas Estate, Blaenavon, Gwent, NP4 9RL, Wales, UK.

**Boehringer Mannheim House,** Bell Lane, Lewes, East Sussex, BN7 1LG, UK.

**Calbiochem Co.,** P.O. Box 12087, San Diego, C.A. 92112–4189, USA.

**Cambridge Bioscience,** 42 Devonshire Road, Cambridge, CB1 2BL, UK.

**DuPont (U.K.) Ltd.,** Biotechnology Systems Divisions, Wedgwood Way, Stevenage, Herts, SG1 4QN, UK.

**Eastman-Kodak Ltd.,** Acornfield Road, Knowsley Industrial Park North, Liverpool, L33 7UF, UK.

**Fluka Chemicals Ltd.,** Peakdale Road, Glossop, Derbys, SK13 9YB, UK.

**F.S.A. (Fisons) Laboratory Supplies,** Bishop Meadow Road, Loughborough, LE11 0RG, UK.

**Gibco B.R.L., Life Technologies Ltd.,** Trident House, P.O. Box 35, Renfrew Rd., Paisley, PA3 4EF, Renfrewshire, Scotland, UK.

**I.C.N. Biomedicals Ltd.,** Eagle House, Peregrine Business Park, Gomm Road, High Wycombe, Bucks, HP13 7DL, UK.

**Northumbria Biologicals Ltd.,** Nelson Industrial Estate, Cramlington, Northumberland, NE23 9BL, UK.

**Novabiochem (U.K.) Ltd.,** 3 Heathcoat Building, Highfields Science Park, University Boulevard, Nottingham, NG7 2QJ.

**Pharmacia – L.K.B.,** Davy Avenue, Milton Keynes, Bucks, MK5 8PH, UK.

**Pierce Chemicals, Life Science Laboratories Ltd.,** Sedgewick Road, Luton, LU4 9DT, UK.

**Promega Corporation,** 2800 Woods Hollow Road, Madison, W.I. 53711–5399, USA.

**Rhone-Poulenc Laboratory Products,** Liverpool Road, Eccles, Manchester, M30 7RT, UK.

**Serva** (see Cambridge Bioscience)

**Sigma Chemical Co.,** Fancy Road, Poole, Dorset, BH17 7NH, UK.

## Radiochemical assays

### (a) Radioanalytical scanning systems

**Ambis Systems,** 3939 Ruffing Road, San Diego, California 92123, USA.

**Ambis Systems,** Lablogic, St. Johns House, 131 Psalter Lane, Sheffield, S11 8UX, UK.

**Berthold Instruments,** Prof. Dr. Berthold, P.O. Box 160, D-7547 Wildbad, Germany.

**Berthold Instruments (UK),** 35 High Street, Sandridge, St. Albans, Herts, AL4 9DO, UK.

### (b) Cell harvesters

**Brandel, Biomedical Research and Development Laboratory Inc.,** 8561 Atlas Drive, Gaithersburg, M.D. 20877, USA.

**Brandel,** Semat, 1 Executive Park, Hatfield Road, St. Albans, Herts, AL1 4TA, UK.

**Dynatech, Dynatech Laboratories Ltd.,** Daux Road, Billingshurst, Sussex, RH14 9SJ, UK.

**Illacon, Illacon Ltd.,** Gilbert House, River Walk, Tonbridge, Kent, TN9 1DT, UK.

**Inotech, Inotech Biosystems International,** P.O. Box 21064, Lansing, M.I. 48909, USA.

**Inotech,** A.G., Postfach CH-5605, Dottikon, Switzerland.

**Inotech,** Berthold Instruments (U.K.), 35 High Street, Sandridge, St Albans, Herts, AL4 9DO, UK.

**Millipore Intertech,** P.O. Box 255, Bedford, M.A. 01730, USA.

**Millipore (U.K.) Ltd.,** The Boulevard, Blackmoor Lane, Watford, Herts, WD1 8YW, UK.

**Skatron A/S,** P.O. Box 8, N-3401, Lier, Norway.

**Skatron Ltd.,** P.O. Box 34, Studlands Park Avenue, Newmarket, Suffolk, CB8 7DB, UK.

### (c) Robotics and sample processors

**Alpha Laboratories,** 40 Parham Drive, Eastleigh, Hampshire, SO5 4NU, UK.

**Anachem,** Charles Street, Luton, Beds, LU2 0EB, UK.

**Beckman Instruments (U.K.) Ltd.,** Progress Road, Sands Industrial Estate, High Wycombe, Bucks, HP12 4JL, UK.

**Beckman Instruments Inc.,** Spinco Division, 1050 Page Mill Road, Palo Alto, California 94304, USA.

**Gilson Medical Electronics Inc.,** Box 27, 3000 West Beltine Highway, Middleton, W.I. 53562, USA.

**Hamilton Company,** P.O. Box 10030, Reno, Nevada 89520, USA.

**Hamilton, I.D.S., Boldon Business Park,** Boldon, Tyne and Wear, NE35 9PD, UK.

**Kemble Instrument Company Ltd.,** Marchants Way, Burgess Hill, West Sussex, RH15 8QY, UK.

**Quatro Biosystems,** Broadoak Business Centre, Ashburton Road West, Trafford Park, Manchester, M17 1RW, UK.

**Tecan A.G., Landhaus Holgass,** C.H.-8634 Hombrechtikon, Switzerland.

**Tecan U.K. Ltd.,** The Glebe House, Sand Pit Lane, Dunsden, Reading, RG4 9PG, UK.

**Tecan U.S. Ltd.,** P.O. Box 8101, Hillsborough, North Carolina 27278, USA.

**Welltech, Denley Instruments Ltd.,** Natts Lane, Billingshurst, West Sussex, RH14 9EY, UK.

**Zymark Corporation,** Zymark Center, Hopkinson, M.A. 01748, USA.

**Zymark Ltd.,** The Genesis Centre, Science Park South, Birchwood, Warrington, WA3 7BH, UK.

## HPLC assays

### (a) Stationary phases and columns

**Anachem Limited,** Charles Street, Luton, Beds, LU2 0EB, UK.

**Beckman Instruments Inc.,** Altex Division, 2350 Camino Ramon, P.O. Box 5101, San Ramon, California 94583–0701, USA.

**Beckman Ltd.,** Progress Road, High Wycombe, Buckinghamshire, HP12 4JL, UK.

**Bio-Rad Laboratories Ltd.,** Bio-Rad House, Mayland Avenue, Hemel Hempstead, Hertfordshire, HP2 7TD, UK.

**Bio-Rad Laboratories,** 1414 Harbour Way, Richmond, C.A. 94804, USA.

**DuPont (U.K.) Ltd.,** Wedgwood Way, Stevenage, Herts, SG1 4QN UK.

**Hichrom Limited,** 6 Chiltern Enterprise Centre, Station Road, Theale, Reading, Berkshire, RG7 4AA, UK.

**Pharmacia Inc.,** 800 Centennial Avenue, Piscataway, N.J. 08854, USA.

**Pharmacia-L.K.B.,** Davy Avenue, Milton Keynes, Bucks, MK5 8PH, UK.

## (b) Instrumentation

**Anachem Limited,** Charles Street, Luton, Beds, LU2 0EB, UK.

**Applied Chromatography Systems Limited,** Concorde Street, Luton, Beds, LU2 0JF, UK.

**Beckman-R11C Limited,** Progress Road, High Wycombe, Bucks, HP12 4JL, UK.

**Chemlab Instruments Limited,** Upminster Road, Hornchurch, Essex, RM11 3XJ, UK.

**Chrompak (U.K.),** Shrubbery Road, London SW16, UK.

**DuPont (U.K.) Limited,** Wedgwood Way, Stevenage, Herts, SG1 4QN, UK.

**Dyson Instruments Limited,** Hetton Lyons Industrial Estate, Hetton, Tyne and Wear, DH5 0RN, UK.

**Gilson Medical Electronics Inc.,** Box 27, 3000 West Bettline Hwy., Middleton, W.I. 53562, USA.

**Hewlett-Packard Company,** (Analytical Products), 3000 Hanover Street, MS20B3 Palo Alto, C.A. 94304, USA.

**Hewlett-Packard Limited,** Nine Mile Ride, Wokingham, Berkshire, RG11 3LL, UK.

**H.P.L.C. Technology Limited,** 10 Waterloo Street West, Macclesfield, Cheshire SK11 5PJ, UK.

**L.D.C.-Milton Roy,** Diamond Way, Stone Business Park, Stone, Staffs, ST15 0HH, UK.

**L.K.B. Instruments Limited,** 232 Addington Road, Selsdon, Surrey, CR2 8YD, UK.

**Perkin Elmer Corporation,** 761 Main Avenue, Norwalk, C.T. 06859–0012, USA.

**Perkin Elmer Limited,** Post Office Lane, Beaconsfield, Bucks, UK.

**Pye Unicam Limited,** Philips Analytical Division, York Street, Cambridge CB1 2PX, UK.

**Shimadzu Scientific Inc.,** 7102 Riverwood Drive, Columbia, M.A. 21046, USA.

**Varian Associates Limited,** Manor Road, Walton-on-Thames, Surrey, UK.

**Varian Instrument Group,** 220 Humboldt Court, Sunnyvale, C.A. 94089, USA.

**Waters Chromatography Division,** Millipore Corporation, 34 Maple Street, Milford, M.A. 01757, USA.

## Oxygen electrode and pH-stat assays

**Anachem Limited,** Charles Street, Luton, Beds, LU2 0EB, UK.

**Clandon Scientific Ltd.,** Aldershot, Hants, GU12 5QR, UK.

**Insted Labs Inc.,** Horsham, Penn. 19044, USA.

**Radiometer Limited,** Crawley, West Sussex, RH10 2PY, UK.

**Rank Bros. Ltd,** Bottisham, Cambs., UK.

**Transdyne General Corp.,** Ann Arbor, Mich. 48106, USA.

**Y.S.I. Inc.,** Yellow Springs, Ohio, USA.

## Electrophoresis equipment and chemicals

**Aldrich Chemical Co.,** P.O. Box 355, Milwaukee, W.I. 53201, USA.

**Ambis System Inc.,** 3939 Ruffin Road, San Diego, C.A. 92123, USA.

**American Biorganics Inc.,** 2236 Liberty Dr., Niagara Falls, N.Y. 14304, USA.

**Bio-Rad Labs.,** Chemical Division, 3300 Regatta Blvd., Richmond, C.A. 94804, USA.

**Biomed Instruments Inc.,** 1020 South Raymond Avenue, Suite B, Fullerton, C.A. 92521, USA.

**Buchler Instruments Inc.,** A Labconco Co., 8811 Prospect Avenue, Kansas City, M.O. 64132, USA.

**C.B.S. Science Company Inc.,** P.O. Box 856, Del Mar, C.A. 92014, USA.

**Caframo Ltd.,** P.O. Box 70, Airport Road, Wiarton, O.N. N0H 2T0, Canada.

**Calbiochem Corp.,** 10933 North Torrey Pines Road, La Jolla, C.A. 92037, USA.

**Carolina Biological Supply Co.,** 2700 York Road, Burlington, N.C. 27215, USA.

**Cole-Parmer Instrument Co.,** 7425 North Oak Park Avenue, Chicago, I.L. 60648, USA.

**E-C Apparatus,** 3831 Tyrone Blv. N St. St Petersburg, F.L. 33709, USA.

**Eastman Kodak,** Lab. & Res. Products Division, 343 State Street, Building 701, Rochester, N.Y. 14652, USA.

**EG&G Berthold,** 4 Tech Circle, Natick, M.A. 03063, USA.

**F.M.C. Corp.,** FMC Bio Products, 5 Maple Street, Rockland, M.E. 04841, USA.

**Hellma Cells Inc.,** P.O. Box 544, Borough Hall Station, Jamaica, N.Y. 11424, USA.

**Hoefer Scientific Instruments,**, P.O. Box 77387, 654 Minnesota Street, San Francisco, C.A. 94107, USA.

**Idea Scientific Co.,** P.O. Box 13210, Minneapolis, M.N. 55414, USA.

**Integrated Separation Sys (A Div. of Enprotech Corp),** 1 Westinghouse Plaza, Hyde Park, M.A. 02136, USA.

**Intel Biotechnologies Inc., (A Kodak Co.),** P.O. Box 9558, 25 Science Park, New Haven, C.T. 06535, USA.

**Isolab,** Drawer 4350, Akron, O.H. 44321, USA.

**Jordan Scientific Co.,** 4215 South State Road, 446 Bloomington, I.N. 47401, USA.

**Jule Biotech,** 25 Science Park, 695 New Haven, C.T. 06511, USA.

**Medtronic Inc.,** 7000 Central Ave. N.E., Minneapolis, M.N. 55432, USA.

**Midwest Scientific,** 228 Meremec Station Road, Valley Park, M.O. 63088, USA.

**Milligen/Biosearch** (A division of Millipore), 186 Middlesex Turnpike, Burlington, M.A. 01830, USA.

**MRA Corp.,** 1058 Cephas Road, Clearwater, F.L. 33515, USA.

**Novex,** Encinitas, C.A. 92024, USA.

**Owl Scientific Plastics Inc.,** Science Products Division, P.O. Box 566, Cambridge, M.A. 02139, USA.

**Perkin-Elmer Cetus Instruments,** 761 Main Avenue, Norwalk, C.T. 06859-0251, USA.

**Pharmacia L.K.B. Biotechnology,** 800 Centennial Avenue, Piscataway, N.J. 08854–9932, USA.

**Princeton Seps. Inc., Sci. & Diagnostic Divs.,** P.O. Box 300, Adelphia, N.J. 07710, USA.

**Quantimetrix Medical Instruments,** 11953 Prairie Avenue, Hawthorne, C.A. 90250, USA.

**Schleicher & Schuell Inc.,** 10 Optical Avenue, Keen, N.H. 03431, USA.

**Serva Biochemicals,** 50 A & S Drive, Paramus, N.J. 07652, USA.

**Stratagene Inc.,** 11099 N. Torrey Pines Road, La Jolla, C.A. 92037, USA.

**Tyler Research Instruments Corp.,** 8306 Davies Road, Edmonton, Alberta T6E 4Y5, Canada.

**United States Biochemical Corp.,** P.O. Box 22400, Cleveland, O.H. 44122, USA.

**West Coast Scientific Inc.,** Slab 2542 Barrington Ct. Hayward, C.A. 94545, USA.

# Enzyme index

This book is not intended to provide a compilation of assays for individual enzymes. However, as many assays, or references to them, are given in the text, the editors felt that inclusion of an Enzyme Index would be useful. Wherever possible, EC numbers are shown for individual enzymes. Many of the protocols described in the text are generally applicable to a group of enzymes, and these groups have been given separate listings. A subject index follows this listing.

# Subject index

Listings of individual enzymes or groups of enzymes are collated separately in the Enzyme index, which immediately precedes this index.